THE SOMATOSENSORY SYSTEM

Deciphering the Brain's Own Body Image

METHODS & NEW FRONTIERS IN NEUROSCIENCE

Series Editors
Sidney A. Simon, Ph.D.
Miguel A.L. Nicolelis, M.D., Ph.D.

Published Titles

THE SOMATOSENSORY SYSTEM

Deciphering the Brain's Own Body Image

Edited by Randall J. Nelson, Ph.D.

Professor of Anatomy and Neurobiology
University of Tennessee Health Science Center
Memphis, Tennessee

CRC PRESS

Boca Raton   London   New York   Washington, D.C.

BS

Library of Congress Cataloging-in-Publication Data

The somatosensory system : deciphering the brain's own body image / [edited by] Randall J. Nelson.
    p. ; cm. -- (Methods and new frontiers in neuroscience)
    Includes bibliographical references and index.
    ISBN 0-8493-2336-3 (alk. paper)
    1. Somesthesia. 2. Body schema.
    [DNLM: 1. Somatosensory Cortex--physiology. 2. Body Image. 3. Models, Neurological. 4. Perception. WL 307 S693 2001] I. Nelson, Randall J. II. Methods & new frontiers in neuroscience series
    QP448 .S65 2001
    612.8'8--dc21
                                         2001003827

Visit the CRC Press Web site at www.crcpress.com

© 2002 by CRC Press LLC

No claim to original U.S. Government works
International Standard Book Number 0-8493-2336-3
Library of Congress Card Number 2001003827
Printed in the United States of America 1 2 3 4 5 6 7 8 9 0
Printed on acid-free paper

10/29/03

Methods & New Frontiers in Neuroscience

Sidney A. Simon, Ph.D.
Miguel A.L. Nicolelis, M.D., Ph.D.
Series Editors

Our goal in creating the **Methods & New Frontiers in Neuroscience** series is to present the insights of experts on emerging experimental techniques and theoretical concepts that are, or will be, at the vanguard of neuroscience. Books in the series cover topics ranging from methods to investigate apoptosis, to modern techniques for neural ensemble recordings in behaving animals. The series also covers new and exciting multidisciplinary areas of brain research, such as computational neuroscience and neuroengineering, and describes breakthroughs in classical fields like behavioral neuroscience. We want these books to be the books every neuroscientist will use in order to get acquainted with new methodologies in brain research. These books can be given to graduate students and postdoctoral fellows when they are looking for guidance to start a new line of research.

Each book is edited by an expert and consists of chapters written by the leaders in a particular field. Books are richly illustrated and contain comprehensive bibliographies. Chapters provide substantial background material relevant to the particular subject. Hence, they are not only "methods books," but they also contain detailed "tricks of the trade" and information as to where these methods can be safely applied. In addition, they include information about where to buy equipment and about web sites helpful in solving both practical and theoretical problems

We hope that as the volumes become available, the effort put in by us, by the publisher, by the book editors, and by individual authors will contribute to the further development of brain research. The extent that we achieve this goal will be determined by the utility of these books.

Preface

Randall J. Nelson, Ph.D.

The somatosensory system is unique among the sensory systems in that it must convey information to the central nervous system (CNS) about both external and internal sensory environments. Objects coming in contact with skin surfaces (exteroceptive inputs) must be represented accurately in time and space so that appropriate behaviors can be planned and executed to either maintain that contact, as in tactile exploration, or disengage from that contact, as in avoidance. Moreover, the CNS must also maintain spatial and temporal representations of the body's own components, such as muscle tension and joint angle (proprioceptive inputs). Finally, to operate efficiently, it has been suggested that the CNS maintains internal representations of expected sensory feedback from movements and anticipated results of the actions toward which the movements themselves are directed. It is thought that all of these representations of inputs, intentions, and actions are continually updated and that the CNS has the ability to "multiplex," that is, select the information channel(s) at any given time that are most relevant for current behaviors.

Over the last several years, considerable effort has been directed toward demonstrating the location and characteristics of these representations, decoding the signals of neurons or ensembles of neurons within these areas, and relating the time-variant signatures of CNS activity to the actual behaviors being produced. In addition, advances in the understanding of the use of haptic information during behavior have paralleled those described above. Thus, recent technological and conceptual advances in the field have allowed great strides to be made in the description and understanding of how the CNS manages information about its own body image. This knowledge, apart from its obvious scientific merit, is quickly leading to clinical applications in the fields of neurorehabilitation after peripheral nerve injury and during recovery from stroke.

The purpose of this volume has been to gather, in one place, information about new avenues of pursuit in understanding how the brain deals with its own body image. The information compiled was intended to be of interest to a broad spectrum of the scientific and medical communities, including those interested in sensory processing, sensorimotor integration, ergometry, robotics, rehabilitative medicine, neurology, and neurosurgery, as well as the neuroscience community in general.

This volume begins by considering the organization of the central representation(s) of the somatosensory periphery, what might be considered to be the "body image." We begin with a chapter by Jon Kaas and his colleagues. They lead us through the diverse ways in which the sensory periphery is represented in the brain. In doing so, they emphasize both the similarities and differences across several species and provide us with a series of general tenets about somatosensory organization. Next, Harold Burton focuses on the homologies, revealed by imaging techniques, between the body representations in humans and monkeys. The strength of this approach can be readily appreciated when one thinks about being able to confirm observations in the former by direct neural recordings in the latter.

To understand what is being represented centrally, we must have some knowledge of the periphery and its contributions to the body image. Here, at the sensory interface with the external environment, events are not merely being transduced. Ken Johnson and Takashi Yoshioka do a masterful job of elucidating the multiple parallel channels by which information is directed centrally and of reminding us that a certain amount of "filtering" can and does occur at this very first stage of somatosensory processing. Susan Lederman and Roberta Klatzky usher us into the realm of perceptual psychology in their chapter. Here, they deal with manual exploration of the sensory environment and how subjects manipulate that environment when contact with it is "indirect." Their concentration on behavioral responses in different application domains reminds us that the somatosensory system is ultimately used for guidance, as well as sensation.

But what happens to guidance and sensation under unusual environmental conditions? What is the effect of orientation on the body image? James Lackner and Paul DiZio examine the exteroceptive and interoceptive contributions to orientation and localization. They also discuss self-calibration under conditions such as micro- and macro-gravity. Adaptive changes necessary to achieve appropriate reaching movements can be made with the help of input from other sensory systems. Arthur Prochazka and Sergiy Yakovenko then provide a framework for our understanding by examining the multiple levels of adaptive control that can modify locomotor patterns to achieve stability in the face of changes in terrain, initial conditions, and biomechanical parameters. They also consider predictive control in relation to behavioral goals.

Predictive control, by its nature, involves both learning of what is predictable and attention to current conditions to determine if things have suddenly become unpredictable. The next two chapters deal with higher-order concepts and their influence on sensory responsiveness. Mathew Diamond and his colleagues tackle the problem of "sensory learning" and explore whether the somatotopic organization of the cortical body image is the framework on which perceptual learning and memory are embellished. Here they argue that sensory maps can shape learning and sensory experience can reshape maps (see the chapter by Sherre Florence). Steven Hsiao and Francisco Vega-Bermudez focus their spotlight on attention and its ability to modify somatosensation under certain behavioral conditions. They remind us scientifically of things we know empirically. For example, we ignore most sensory inputs because attention focuses our mental efforts on sensations that are important to what we are doing now or what we expect to happen soon.

If we wish to study sensory inputs that are important for what we are doing now or for what we expect to happen in the near future, what tools do we use? Esther Gardner and her co-workers describe a new way of combining real-time video images with neural recording to correlate brain activity with somatosensory inputs linked to behavior. Through this approach, we can sample the rich banquet of signals that gives rise to perceptions that ultimately shape actions. But how do we determine how good the central representations of peripheral sensory signals are? Mikhail Lebedev, Yu Liu, and I examine ways of determining and demonstrating the "fidelity" with which peripheral somatosensory signals are represented centrally. We also show several behavioral conditions in which that fidelity changes on a trial-by-trial basis.

To predict how the somatosensory system deals with peripherally driven changes in its body image, models of the system are needed. These help to explain what we know about the system and predict what we don't know, so that we can test the predictions experimentally. Paul Cisek briefly describes the history of several models of movement guidance and presents, in more detail, one which is consistent with many of the phenomena associated with proprioceptively guided motor behavior. In a sense, it unifies many thoughts while leaving the door open for further modification as new data become available. Miguel Nicolelis and colleagues take a critical look at strictly feedforward models of somatosensation and perception and suggest, instead, that recent data support the concept of multiple ascending and descending systems that are linked as a distributed network for tactile processing. Closely linked with this concept are the influences of predictions and expectations, both of which sculpt central representations of the sensory periphery in time and space.

It is that central "sculpting" that is considered in detail by Sherre Florence. In her chapter, the spatial and temporal characteristics of body image "remodeling" are described with an emphasis on possible mechanisms and outcomes. From this work, it is clear that our understanding of how and why the central representations of the body reorganize may lead ultimately to new and more efficient strategies for rehabilitation following peripheral or "central" interruption of sensory inputs to processing areas. To complete the consideration of the functional implications of the brain's ability to reorganize its own body image, we turn full circle to the final chapter, once again by Jon Kaas. In it, we are reminded that training procedures improve perceptual and motor skills and that reorganization may be the underlying cause of the ameliorating properties of behavioral therapies.

No single volume can do justice to the rich experimental and conceptual heritage of work on the somatosensory system and its relationship to the brain's own body image. Herein, however, are chapters dealing with some of the current thoughts that have driven consideration of the link between the external somatosensory environment and the central representations thereof. I am privileged that each chapter has been overseen by not only a respected member of the field, but also by a friend. I wish to especially thank Miguel Nicolelis for initially encouraging me to take on this project and Barbara Norwitz, Life Sciences Publisher, and CRC Press for patiently dealing with the trials and tribulations experienced by me as a novice editor. Finally, I wish to thank my wife Heide and my children, Beverly and Christopher, for understanding when I brought the "big red notebook" home.

Author/Editor

Randall J. Nelson received his undergraduate training in psychology at Duke University with William C. Hall, receiving a B.S. from that institution in 1975. He did his graduate training in anatomy at Vanderbilt University with Jon H. Kaas and received a Ph.D. in 1980. Dr. Nelson did postdoctoral training with Michael M. Merzenich in the Department of Otolaryngology at the University of California at San Francisco. He subsequently worked at the National Institute of Mental Health under the direction first of Edward V. Evarts in the Laboratory of Neurophysiology and then under the direction of Mortimer Mishkin in the Laboratory of Neuropsychology. Dr. Nelson joined the faculty at the University of Tennessee, Health Science Center, in Memphis in 1984. He is currently Professor of Anatomy and Neurobiology.

Contributors

K. Srinivasa Babu, Ph.D
Department of Physiology
and Neuroscience
New York University School
of Medicine
New York, New York

Harold Burton, Ph.D.
Department of Anatomy
and Neurobiology
Washington University School
of Medicine
St. Louis, Missouri

Maria Chu
Department of Physiology
and Neuroscience
New York University School
of Medicine
New York, New York

Paul E. Cisek, Ph.D.
Department of Physiology
University of Montreal
Montreal, Quebec, Canada

Daniel J. Debowy
Department of Physiology
and Neuroscience
New York University School
of Medicine
New York, New York

Mathew E. Diamond, Ph.D.
Cognitive Neuroscience Sector
International School for Advanced
Studies
Trieste, Italy

Paul DiZio, Ph.D.
Ashton Graybiel Spatial
Orientation Laboratory
Brandeis Universtiy
Waltham, Massachusetts

Erika Fanselow
Department of Neurobiology
Duke University Medical Center
Durham, North Carolina

Sherre L. Florence, Ph.D.
Department of Psychology
Vanderbilt University
Nashville, Tennessee

Esther P. Gardner, Ph.D.
Department of Physiology
and Neuroscience
New York University School
of Medicine
New York, New York

Justin A. Harris, Ph.D.
Cognitive Neuroscience Sector
International School for Advanced
Studies
Trieste, Italy

Craig Henriquez, Ph.D.
Department of Biomedical Engineering
Duke University Medical Center
Durham, North Carolina

Steven S. Hsiao, Ph.D.
Krieger Mind/Brain Institute
Department of Neuroscience
The Johns Hopkins University
Baltimore, Maryland

Edward H. Hu
Department of Physiology
 and Neuroscience
New York University School
 of Medicine
New York, New York

Neeraj Jain, Ph.D.
Department of Psychology
Vanderbilt University
Nashville, Tennessee

Kenneth O. Johnson, Ph.D.
Krieger Mind/Brain Institute
Department of Neuroscience
The Johns Hopkins University
Baltimore, Maryland

Jon H. Kaas, Ph.D.
Department of Psychology
Vanderbilt University
Nashville, Tennessee

Roberta L. Klatzky, Ph.D.
Department of Psychology
Carnegie Mellon University
Pittsburgh, Pennsylvania

James R. Lackner, Ph.D.
Ashton Graybiel Spatial
 Orientation Laboratory
Brandeis Universtiy
Waltham, Massachusetts

Mikhail A. Lebedev, Ph.D.
Laboratory of Systems Neuroscience
National Institutes of Mental Health
National Institutes of Health
Bethesda, Maryland

Susan J. Lederman, Ph.D.
Department of Psychology
Queen's University
Kingston, Ontario

Yu Liu, M.D., Ph.D.
Department of Anatomy
 and Neurobiology
University of Tennessee, Memphis
Memphis, Tennessee

Michelle Natiello
Department of Physiology
 and Neuroscience
New York University School
 of Medicine
New York, New York

Randall J. Nelson, Ph.D.
Department of Anatomy
 and Neurobiology
University of Tennessee, Memphis
Memphis, Tennessee

Miguel A. L. Nicolelis, M.D., Ph.D.
Department of Neurobiology
Duke University Medical Center
Durham, North Carolina

Rasmus S. Petersen, Ph.D.
Cognitive Neuroscience Sector
International School for Advanced
 Studies
Trieste, Italy

Arthur Prochazka, Ph.D.
Centre for Neuroscience
University of Alberta
Edmonton, Alberta, Canada

Hui-Xin Qi, Ph.D.
Department of Psychology
Vanderbilt University
Nashville, Tennessee

Shari Reitzen
Department of Physiology
 and Neuroscience
New York University School
 of Medicine
New York, New York

Jill Sakai
Department of Physiology
  and Neuroscience
New York University School
  of Medicine
New York, New York

Francisco Vega-Bermudez, M.D., Ph.D.
Krieger Mind/Brain Institute
Department of Neuroscience
The Johns Hopkins University
Baltimore, Maryland

Sergiy Yakovenko
Centre for Neuroscience
University of Alberta
Edmonton, Alberta

Takashi Yoshioka, Ph.D.
Krieger Mind/Brain Institute
Department of Neuroscience
The Johns Hopkins University
Baltimore, Maryland

Table of Contents

1 The Organization of the Somatosensory System in Primates

Jon H. Kaas, Neeraj Jain, and Hui-Xin Qi

CONTENTS

1.1 INTRODUCTION

An understanding of how the somatosensory system processes and uses sensory information depends in large part on knowing how the system is organized. One might think that our understanding of the system would have been complete sometime ago, but actually our knowledge is far from complete, and considerable research is yet needed. There are several reasons for our incomplete understanding. Most importantly, evolution has created considerable variability in the organization of the mammalian somatosensory system. Part of the variability stems from adaptations that lead to the use of different parts of the body to gather important sensory information. Many mammals use the whiskers on the face, nose, and mouth for exploring the environment, but this is a bit risky for primates, which have emphasized binocular vision with forward directed eyes and a short snout. Explorations with the face risk damage to the eyes, and reaching and feeling with the hand became a suitable substitute. Some New World monkeys also use their highly sensitive prehensile tail. But primates have adapted in other ways as well, and have evolved into one of the most varied orders in terms of brain size and organization. Prosimian primates occupy a range of environmental niches, but have only moderate brain expansion. New World monkeys vary from small marmosets which have brains that are yet large relative to body size, to spider and capuchin monkeys with large brains that at least superficially resemble those of Old World macaque monkeys. Old World monkeys are less varied, and they have impressively large, complex brains. Chimpanzees have even larger brains, similar in size to those of our early hominoid ancestors, while our greatly expanded brains have allowed us to populate and dominate the planet. While basic aspects of somatosensory processing are likely to be similar, it seems unlikely that these greatly varied primate brains would all process somatosensory information in exactly the same way. Thus, we should try to determine how somatosensory systems of primates differ and what the common features are. In particular, do primates with larger, more complex brains have more extensive and more complex processing? Also, since the extant somatosensory systems evolved through gradual modifications of an ancestral system, what features have been commonly retained as parts of the basic framework? Here we begin to address these questions. We start by considering the basic features of the mammalian somatosensory system. To do this, we need to determine what features are commonly found in the somatosensory systems of non-primate mammals.

1.2 THE BASIC SOMATOSENSORY SYSTEM

The basic somatosensory system includes afferents from the peripheral receptors that terminate in the spinal cord or brain stem, a relay of second-order neurons from these structures to the thalamus, projections from the thalamus to different areas of neocortex, and a number of interconnected cortical somatosensory areas that project to motor cortex, other cortical areas, and subcortical structures. Afferents include those from several types of mechanoreceptors in the skin, muscle-spindle and joint receptors that signal movement and position, and afferents mediating temperature and pain. Here, our concern is with the parts of the system that are involved in tactile discriminations. Thus, we concentrate on the parts of the somatosensory system devoted to the inputs from mechanoreceptors in the skin for tactile information and to the muscle-spindle and joint receptors for information about position and movement, since tactile discriminations involve active exploration of surfaces and objects.

Afferents from the mechanoreceptors in the skin and muscle spindles terminate in a somatotopic pattern (for review, including humans and other primates, see Florence et al., 1989; Coq et al., 2000) in the dorsal column-trigeminal complex in the lower brain stem. Some of the muscle-spindle afferent terminations are segregated in separate subnuclei (the external cuneate "nucleus" for the forelimb). There is evidence from cats (Dykes et al., 1982) that the rapidly adapting (RA) and slowly adapting (SA) classes of afferents terminate in separate clusters of cells in the dorsal column nuclei. Glaborous skin has two types of slowly adapting afferents (SAI and SAII) associated with Merkel cell and Ruffini endings, respectively, and two types of rapidly adapting afferents (RAI and RAII), associated with Meissner corpuscles and Pacinian corpuscles, respectively. Hairy skin also has hair follicle receptors.

The mechanoreceptor and other afferents related to pain and temperature sensibilities also terminate in the dorsal horn of the spinal cord and the brain stem equivalent, where second-order neurons cross to form the ascending spinothalamic tract. The ascending spinothalamic information is clearly important, but after a thalamic relay, its role in the traditional areas of somatosensory cortex seems to be one of modulation rather than activation. After sectioning of the afferents in the dorsal columns of rats, the deprived areas of at least S1 are totally and persistently deactivated (Jain et al., 1995). Similarly, in monkeys, dorsal column section (Jain et al., 1997) abolishes all evoked activity in all four areas of the anterior parietal cortex (areas 3b or S1, 3a, 1, and 2). We conclude that only the dorsal column-medial lemniscus system is capable of independently activating primary and most-likely secondary areas of somatosensory cortex. While other afferents may have an important modulating role, and they are certainly critical in mediating pain and temperature sensations, dorsal column afferents are capable of mediating tactile discriminations by themselves.

The second-order neurons in the dorsal column-trigeminal complex cross and ascend to the contralateral ventroposterior nucleus (VP) in the thalamus. The RA and SA response classes are preserved in this relay, and these two types constitute the major inputs to VP, where they activate separate groups of neurons in cats (Dykes, 1982), monkeys (Dykes et al., 1981), and perhaps other mammals. Relay of the muscle-spindle, and possibly other, afferents terminate rostrodorsally just outside VP (Wiener et al., 1987) in a part of the thalamus that Dykes (1983) has distinguished as the ventroposterior oralis nucleus (VPO) while others included it in the posterior (PO) complex (see Gould et al., 1989). The thalamic nucleus for muscle spindles in primates has been called the ventroposterior superior nucleus (VPS). Possibly VPO, PO or part of it, and VPS are homologous nuclei.

In all investigated mammals, VP projects to the somatosensory koniocortex or primary somatosensory cortex (S1). Evidence from cats (e.g., Macchi et al., 1959) and a number of other species (see Garraghty et al., 1991) suggests that in the typical non-primate pattern some neurons in VP also project to the second somatosensory area, S2, and to a more recently described adjacent area, the parietal ventral area, PV (Krubitzer and Kaas, 1987). Some of the same neurons that project to S1 also project to S2 (see Spreafico et al., 1981). This means that S1 and S2 can be independently activated from the thalamus and that S2 does not depend on its inputs from S1 for activation. The capacity of S1 and S2 for parallel processing of VP outputs has been demonstrated in opossums (Coleman et al., 1999), rabbits (Murray et al., 1992), cats (Burton and Robinson, 1987), and tree shrews (Garraghty et al., 1991) by recording evoked responses in S2 after deactivating S1. In addition to VP, a medial posterior nucleus, Pom, also projects to both S1 and S2 (see Krubitzer and Kaas, 1987 for review).

The somatosensory cortex of many, and perhaps most, non-primate mammals appears to consist of five areas that are at least predominantly somatosensory in function. These areas are S1, S2, PV, and the cortical strips along the rostral and caudal borders of S1 (Figure 1.1). In opossums, we called these strips the caudal and rostral somatosensory areas, SC and SR (Beck et al., 1996). They have also been called the rostral (R) and posterior medial (PM) fields (Slutsky et al., 2000). In cats (Dykes, 1983; Felleman et al., 1983a), raccoons (Feldman and Johnson, 1988) and the flying fox (Krubitzer and Calford, 1992), a fruit bat, the rostral field has been called area 3a, because it resembles area 3a of primates in position, architecture, connections, and responsiveness to the stimulation of deep tissues (muscle-spindle receptors). In rats and squirrels, the rostrally bordering (and intruding into S1) area is known as dysgranular cortex (Chapin and Lin, 1984; Gould et al., 1989). The

Opossum

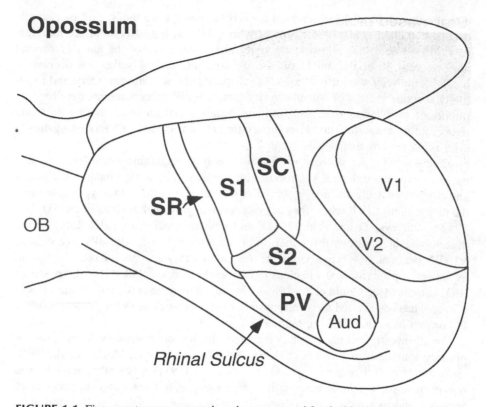

FIGURE 1.1 Five somatosensory areas have been proposed for the North American opossum (*Didelphis marsupialis*): a primary somatosensory area (S1), a secondary area (S2), a parietal ventral area (PV), and caudal (SC) and rostral (SR) somatosensory areas bordering S1. For reference, primary (V1) and secondary (V2) visual areas and auditory cortex (Aud.) are shown. The olfactory bulb (OB) is on the left. Based on Beck et al., 1996. In some opossums, such as *Monodelphis domestica*, PV may not be apparent (Huffman et al., 1999; Catania et al., 2000; Frost et al., 2000).

caudal area of the flying fox has been referred to as area 1/2 (Krubitzer and Calford, 1992) as it is in the same relative position as area 1 of primates and appears to be responsive to both cutaneous and deep receptors as area 2 of primates.

In all investigated mammals, S1 systematically represents the mechanoreceptors of the skin of the opposite side of the body (see Kaas, 1983). The body parts are usually represented from tail to tongue in a mediolateral sequence. The representations of different body parts often have a morphological counterpart in the cortex that can be visualized using appropriate histochemical techniques. These isomorphs of the body are best known for S1 of rats and mice, where an orderly arrangement of oval-like aggregates of neurons, one for each whisker on the side of the face, have long been described as the cortical barrels (Woolsey and Van Der Loos, 1970). Metabolic markers, cytochrome oxidase (CO), and succinic dehydrogenase have been used to reveal more of the isomorph including discrete cellular clusters for other whiskers and for the digits and pads of the forepaws and hindpaws (Dawson

Star-nosed mole

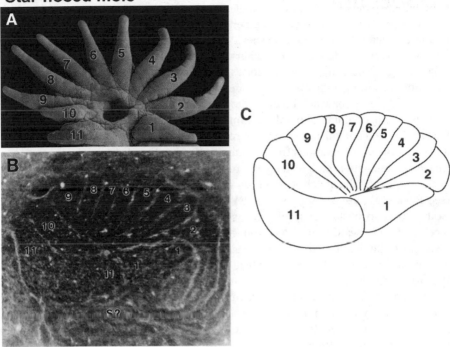

FIGURE 1.2 An example of isomorphs of body parts in primary somatosensory cortex S1. The star nose mole has 11 sensory rays on each half of its nose (A). These rays are represented in order in S1 and S2 of the contralateral cerebral hemisphere. In brain sections cut parallel to the cortical surface and processed for cytochrome oxidase (CO), the representation of each ray in S1 can be seen as a CO-dark stripe separated from neighbors by narrow CO-light septa, Numbered 1–11 in B corresponding to rays 1–11 in A. A second array of CO-dark stripes can be seen in S2 below S1 (unnumbered in A). The stripes are drawn in C. A and B are based on Catania and Kaas, 1995.

and Killackey, 1987; Li et al., 1990; Pearson et al., 1996). An isomorphic array of cortical bands, one for each of the 11 rays of the contralateral half of the nose, is especially prominent for the star-nosed mole (Figure 1.2; Catania and Kaas, 1995, 1996). Such isomorphs help identify S1, demonstrate existence of a single systematic representation in S1, and distinguish S1 from other representations. The star-nosed mole is unusual in that two other cortical representations, S2 and possibly PV, also have isomorphs of the rays of the nose. In addition, S2 also has a very prominent isomorph of the forepaw, while a comparable isomorph of the forepaw does not exist in S1 (Catania, 2000). This suggests a more pronounced or specialized role for S2 for forepaw afferents than for S1 in moles.

In a wide range of mammals, S1 projects directly to S2, PV, SR, and SC (see Beck et al., 1996). Thus, all of these areas are involved in further processing of information from S1, as well as processing inputs from the thalamus. S2 and PV receive VP inputs, while SR and SC receive most of their thalamic inputs from neurons just outside of VP (Pom and adjacent parts of the VP "shell"). Some of this thalamic relay may be of muscle-spindle information (see Gould et al., 1989), as it appears to be for area 3a of cats (see Felleman et al., 1983a). S2 has been described in a large range of mammalian species (see Nelson et al., 1979; Sur et al., 1981 for review) and it seems to be a universal or nearly universal (see Krubitzer et al., 1995) subdivision of the somatosensory cortex. S2 borders S1 laterally, with face representations adjoining. Other body parts are represented more distantly from the S1/S2 border, and the forepaw and hindpaw are represented along the rostral border of S2. Because of uncertainties about the organization of S2, some early investigators may have failed to distinguish S2 from the rostrally adjoining parietal ventral area, PV, and confounded observations from two areas. PV was first distinguished as a mirror image representation of S2 along the rostral border of S2 in squirrels (Krubitzer et al., 1986), and the area has now been identified in a range of mammalian species (see Beck et al., 1996). Both S2 and PV respond throughout to cutaneous stimuli. Both appear to be higher level processors of information from S1, but PV also receives feed forward projections from S2. Thus, PV is characterized by a convergence of thalamic VP and cortical S1 and S2 inputs. Projections from PV include an even more ventral and rostral cortical region that we have called parietal rostral, PR (Krubitzer et al., 1986). Thus, another area may be part of the basic array of somatosensory areas in mammals. PR may provide the major inputs to peripheral cortex, the relay to the hippocampus as way of creating somatosensory memories (Mishkin, 1979). S2 and PV also provide substantial inputs to primary motor cortex, M1 (Krubitzer et al., 1986).

1.3 THE SOMATOSENSORY SYSTEM IN PROSIMIAN PRIMATES

Prosimian primates are varied in appearance and lifestyle (Wolfheim, 1983; Fleagle, 1999). They include lemurs, lorises, and galagos. Tarsiers have also been classified with prosimians, but form a distinctly separate line of primate evolution, one of a specialized, visual, nocturnal predator. Of the prosimians, the somatosensory system has been significantly studied only in galagos. There have been limited studies of S1 organization in other prosimians (Krishnamurti et al., 1976; Carlson and Fitzpatrick, 1982; Fitzpatrick et al., 1982), and thus we know that this S1 has the typical mammalian organization. There have been no experimental studies on the somatosensory system of tarsiers, now considered an endangered primate. Thus, we are uncertain about the generality of conclusions about galagos to other tarsiers and prosimians, but all prosimians appear to have only a moderate level of neocortical expansion. From this, we would not expect large, complex, cortical somatosensory systems.

The somatosensory system of galagos seems very similar to the basic, generalized somatosensory system of other mammals, but not nearly as expanded or specialized as in other primates. Yet, the dorsal column relay nuclei in the brain stem seem somatotopically organized, much as in simian primates (Coq et al., 2000), and distinct simian-like ventroposterior (VP), ventroposterior inferior (VPI), and ventroposterior superior (VPS) nuclei are apparent in the thalamus (Kaas, 1982; Wu et al.,

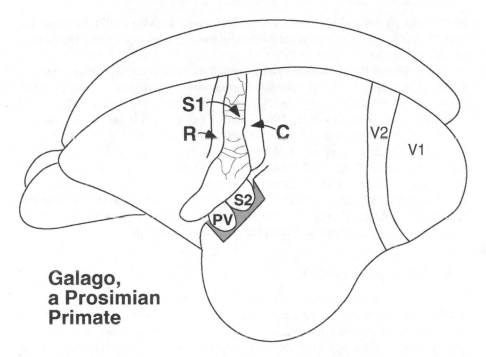

Galago, a Prosimian Primate

FIGURE 1.3 Somatosensory cortex of prosimian galagos. Five areas have been defined including S1 (area 3b), a rostral bordering area (R) or area 3a, a caudal bordering area (C), which may correspond to area 1, a secondary somatosensory area, S2, and the parietal ventral area, PV. Part of the temporal lobe has been cut away to show S2 and PV, which are located on the upper bank of the lateral sulcus. The primary (V1) and secondary (V2) visual areas are shown as reference. Connection patterns of S2 and PV suggest that several additional somatosensory areas exist in parietal cortex of galagos.

1996). Somatosensory cortex includes a primary area, S1, rostral and caudal bordering areas, R and C, and S2 and PV (Figure 1.3; Sur et al., 1980; Wu et al., 1995, 1996, 1997). In galagos, S1 occupies a strip of koniocortex that is similar to area 3b of monkeys, but not as well differentiated. The representation of the body surface in S1 resembles that of area 3b of monkeys, and thalamic inputs are from SA and RA neurons of the ventroposterior nucleus. All these observations indicate that this single S1 representation is area 3b, and not a configuration of fields or area 1 as proposed by some investigators. The rostral area, R, is almost certainly homologous with area 3a of other primates. The rostral area of galagos appears to be activated by muscle spindles, and it does have a motor component as does area 3a of monkeys. In addition, the rostral area receives thalamic input from the ventroposterior superior nucleus (VPS) just dorsal to VP. The caudal area is in the same relative position as area 1, and it has some of the architectonic features of area 1. In addition, it receives projections from area 3b, as does area 1. Nevertheless, the caudal area is not as highly responsive to cutaneous stimuli as area 1 in most monkeys. In anesthetized galagos, neurons are either unresponsive to cutaneous stimuli or require intense stimulation, much as the caudal area does in some non-primates (e.g., Slutsky et al.,

2000). Because this more intense stimulation would activate both cutaneous and deep receptors, muscle spindles may contribute to the activation of the caudal area, but this is uncertain.

Areas S2 and PV form additional representations of the body surface, as they do in other mammals. As in non-primate mammals, S2 receives major thalamic inputs from VP (Wu et al., 1996) and S2 remains responsive to cutaneous stimulation after S1 lesions (Garraghty et al., 1991). Thus, S2 of galagos is activated in parallel with S1 by the VP neurons. However, S1 projects to both S2 and PV, and S2 and PV both project to primary motor cortex conforming to the generalized mammalian pattern. Injections of neuronal tracers in S2 also label regions rostroventral to PV and caudal to the caudal area 'C,' suggesting the locations of additional somatosensory areas. Posterior parietal cortex is limited in extent compared to simian primates, and its organization is largely unknown. Overall, somatosensory cortex in prosimians seems only moderately changed from the basic nonprimate pattern.

1.4 THE SIMIAN ELABORATION

The somatosensory system is more complex in monkeys than in prosimian primates at the cortical level and perhaps at the thalamic level (Kaas and Pons, 1988; Kaas, 1993). At the level of the thalamus, we distinguish a ventroposterior nucleus (VP), a ventroposterior inferior nucleus (VPI), and a ventroposterior superior nucleus (VPS) as the main relay nuclei of the somatosensory information (Figure 1.4; Dykes et al., 1981; Nelson et al., 1981; Kaas et al., 1984; Cusick et al., 1985; Krubitzer and Kaas, 1992). The VP as defined by us occupies most of the traditional ventroposterior nucleus of primates. VP stands out in Nissl preparations as a structure

---

FIGURE 1.4 (opposite) The somatosensory system of monkeys. The diagram shows how afferents enter the spinal cord to rise in the dorsal column nuclei in the lower brain stem (not to scale), or activate second order neurons in the spinal cord that contribute to the ascending spinothalamic system. Slowly adapting and rapidly adapting mechanoreceptor inputs to gracile, cuneate, and trigeminal nuclei (N) are relayed to the ventroposterior nucleus (VP) of the thalamus. VP has two large subnuclei; a medial subnucleus (VPM) represents the face while a lateral subnucleus (VPL) represents the rest of the body. VPL contains hand and foot subnuclei, as well as a lateralmost tail subnucleus (not shown). Muscle-spindle receptor inputs to the dorsal column nuclei are relayed to the ventroposterior superior nucleus (VPS), while spinothalamic terminals relay to the ventroposterior inferior nucleus (VPI). VPI includes septa that extend dorsally into VP. A more caudal region of the somatosensory thalamus contains a nucleus specific for spinothalamic inputs responsive to pain and temperature (not shown). This nucleus has been called the ventral medial posterior nucleus (VMpo). VP projects densely to area 3b, less strongly to area 1, and sparsely to area 2 of anterior parietal cortex. VPS projects to area 3a and area 2. VPI projects widely to many somatosensory areas, but most densely to areas S2 and PV. Connections and recordings suggest the existence of two additional areas, the parietal rostral area, PR, and ventral somatosensory area, VS. Posterior parietal cortex, traditionally subdivided into "areas" 5 and 7, is complexly subdivided into areas with somatosensory, motor, and visual functions. In primates with a central sulcus, some or most of area 3b is in the central sulcus (CS).

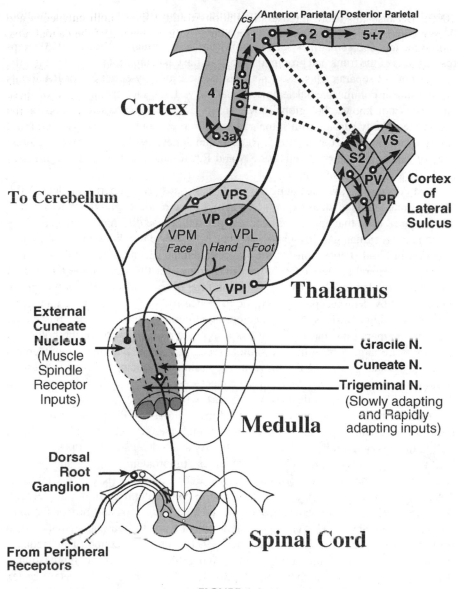

FIGURE 1.4

packed with densely stained neurons. These neurons are highly active metabolically and thus VP is also apparent as the nucleus with the most expression of cytochrome oxidase (CO) in the somatosensory thalamus. The medial subnucleus, ventroposterior medial (VPM), representing the face, is separated from the lateral portions, ventroposterior lateral (VPL), representing the rest of the body, by a cell-sparse fiber septum, the arcuate lamina and many early investigators considered them as separate nuclei. A second, less obvious band separates the VPL into a more medial subnucleus, representing the hand, from a more lateral subnucleus representing the foot, and a third more lateral fiber band, separating the foot from the tail representation, is sometimes apparent. Such fiber bands separating face, hand, foot, and tail representations are also sometimes seen in non-primate mammals (Welker, 1973). Representations of the trunk and proximal limbs cap the hand and foot subnuclei dorsally.

Within VP, separate clusters of neurons are activated by either SA or RA inputs via the dorsal column-medial lemniscus pathway. The exact arrangement of these clusters is not known, but globally VP forms a systematic representation of the contralateral body surface with some additional representation of the ipsilateral oral cavity (see Bombardiei et al., 1975; Rausell and Jones, 1991a). However, the presence of separate clusters of cells for SA and RA inputs indicates that two maps of the body are interdigitated in VP.

Just dorsal to VP, we distinguish a small ventroposterior superior nucleus, VPS. The neurons in this nucleus express less cytochrome oxidase, and it has less darkly Nissl-stained cells that appear less densely packed. These differences are especially obvious in prosimian galagos and the smaller New World monkeys, but they are less apparent in Nissl preparations of the larger thalamus of macaque monkeys where neurons in general are less densely packed. Perhaps for this reason, the territory of VPS was included in VP in early studies, and the dorsal part of VP was thought to be activated by deep receptors. Now, the evidence for VPS as a separate nucleus is more compelling. Besides the architectonic distinctions, which are obvious in CO and other preparations, the VPS contains a separate representation of the body, and the inputs are largely from muscle-spindle receptors, rather than mechanoreceptors of the skin. The inputs to VPS are from the brain-stem nuclei, devoted to muscle-spindle receptors, such as the external cuneate nucleus. Also, VPS projects in a distinctly different pattern than VP to the cortex (see below).

Just ventral to VP, a ventroposterior inferior nucleus (VPI) has long been recognized in primates, but often not in other mammals (see Krubitzer and Kaas, 1987). VPI is thin dorsoventrally, and has small, lightly stained cells in Nissl preparations. The nucleus expresses little CO. VPI receives spinothalamic inputs and projects widely to the somatosensory cortex, especially areas S2 and PV (Krubitzer and Kaas, 1992). VPI appears to form a crude map of the contralateral body in parallel with the map in VP, and neurons are responsive to both cutaneous and noxious stimuli (Apkarian and Shi, 1994). The projections of VPI are to superficial, rather than middle, cortical layers (Rausell and Jones, 1991b), and transection of afferents in the dorsal columns of the spinal cord, leaving the spinothalamic inputs intact, deactivates neurons throughout areas 3a, 3b, 1, and 2 of the somatosensory cortex (Jain et al., 1997, 1998, 2001a). Thus, the thalamic relay of spinothalamic information to anterior parietal cortex does not seem to be capable of independently activating cortical neurons, and VPI appears to have a role in modulating, rather than evoking, cortical activity. However, spinothalamic terminations are also found in a recently defined nucleus, the posterior ventral medial nucleus (VMpo), which may be specific for pain and temperature sensibilities; VMpo projects to anterior cortex in the lateral sulcus (Craig et al., 1994). Thus, the spinothalamic terminations in VPI probably have functional roles other than mediating the sensations of pain and temperature.

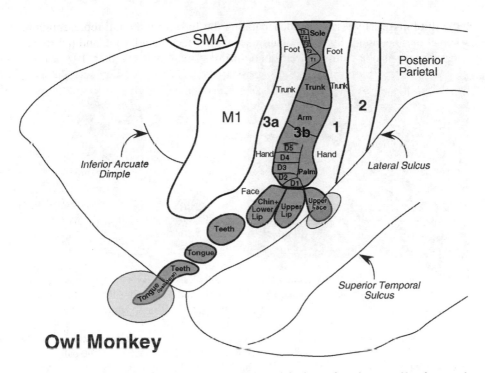

Owl Monkey

FIGURE 1.5 The representation of the contralateral body surface in area 3b of an owl monkey. This New World monkey has only a shallow central dimple rather than a central fissure, and thus most of area 3b is exposed on the dorsolateral surface of the parietal cortex. A large sector of area 3b represents the face and oral cavity, including both the contralateral and ipsilateral tongue and teeth. This part of the representation curves rostrally to form a rostrolateral extension of parietal cortex. A mediolateral strip of area 3b represents the body from tail to digits of the hand. Areas 3a and 1 form additional representations of the body along the rostral and caudal borders of area 3b, and area 2 forms a fourth representation just caudal to area 1. Portions of parietal cortex contain several functional subdivisions of somatosensory cortex, but they have not been well defined. Lateral somatosensory cortex in the lateral sulcus includes the secondary somatosensory area, S2, the parietal ventral area, PV, and the rostral parietal area, PR (all not shown). Primary motor cortex, M1, or area 4, receives inputs from several somatosensory areas, as does the supplementary motor area, SMA. See Merzenich et al., 1978; Jain et al., 1997, and Jain et al., 2001b for the representation in area 3b. Areas are shown on a dorsolateral view of the brain. The encircled portions of the tongue and upper face representations curve around on the orbital surface and the upper bank of lateral sulcus, respectively.

The anterior parietal cortex of monkeys contains four strip-like areas that extend mediolaterally (Figure 1.5). All four areas were originally thought to be within a single cortical representation, S1 (Marshall et al., 1937). But, it is now clear that each of these architectonic fields contains a separate representation of the body and is a distinct area of the somatosensory cortex (for early evidence, see Merzenich et al., 1978; Kaas et al., 1979). In addition, only the representation in area 3b is the homologue of the single representation that is commonly described as S1 in other

Owl Monkey 3b

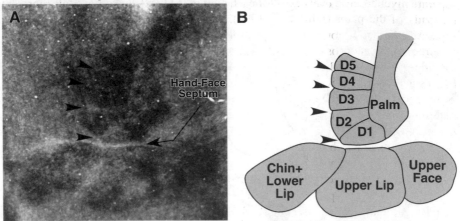

FIGURE 1.6 Isomorphs of body parts in area 3b of monkeys. By cutting somatosensory cortex into sections parallel to the surface of the brain and processing for myelin, area 3b can be recognized as a myelin-dense field (A). In addition, it is apparent that the field is subdivided by narrow myelin-light septa, creating a sequence of myelin-dense ovals. Recordings with microelectrode indicate that these ovals correspond to the representations of body parts. In A, the ovals for a owl monkey correspond to digits 1–5 of the hand, the palm, upper face, upper lip, and chin plus lower lip. The hand-face septum and interdigital septa (arrow heads) are marked. These territories are drawn in B for comparison with A. See Jain et al. (1998, 2001b) for details.

mammals (Kaas, 1983). This conclusion is supported by a multitude of observations, including basic similarities of the organization of representations in 3b and S1, their koniocellular nature, similar patterns of intracortical (Jones et al., 1978; Pons and Kaas, 1985; Krubitzer and Kaas, 1990; Huffman and Krubitzer, 2001) and callosal (Killackey et al., 1983) connections, and the dominant projection of VP to 3b and S1 (Jones, 1975). In addition, as for S1, only area 3b responds well to cutaneous stimuli under typical recording conditions in all primates. In the anterior parietal cortex of galagos, only area 3b responds well to cutaneous stimuli (see above). Even in marmoset monkeys, only area 3b consistently responds well to cutaneous stimuli (Carlson et al., 1986; Krubitzer and Kaas, 1990). Thus, there is only one representation that is S1-like in these primates, which is located in area 3b. As expected for S1, the receptive fields of neurons in area 3b are smaller than for any other somatosensory area.

As with other primary sensory areas, only area 3b is densely myelinated and only area 3b expresses a high level of cytochrome oxidase. In such preparations in New World monkeys, area 3b stands out as a narrow mediolateral strip of cortex that curves rostrally near the lateral fissure to extend far into the frontal lobe (Jain et al., 2001b). Furthermore, if cortex is cut parallel to the brain surface, and the sections processed for myelin or CO, area 3b can be seen to be composed of isomorphs of the parts of the contralateral body surface, with narrow septa separating the representations of various body parts (Figure 1.6), much as in the barrel field of mice and rats or the nose field of star-nosed moles. In the hand region of area 3b, separate myelin-dense ovals exist for each digit of the hand, and other ovals represent the pads of the palm (Jain et al., 1998). A major septum, the hand-face septum, separates the hand representation from the face representation. In the more lateral face region, a caudorostral row of three myelin-dense ovals corresponds in turn to the upper face, upper lip, and lower lip plus chin. A sequence of four more rostral ovals successively represents the contralateral teeth, contralateral tongue, ipsilateral teeth, and ipsilateral tongue. These isomorphic ovals make it possible to precisely determine parts of the representation from architecture alone. This can be very meaningful in studies of plasticity where such ovals might become responsive to different parts of the body (see Jain et al., 1998; Kaas and Florence, 2000) and in studies of neuronal connections (Fang et al., 2000). The presence of such isomorphic ovals is harder to detect in the cortex of larger brains with disrupting fissures. Yet, we have seen myelin-poor septa separating the representations of digits in area 3b of macaque monkeys (Jain et al., 1998), and such septa may be detectable in area 3b of humans.

Although modular or columnar organization is probably a common feature of cortical areas, in the somatosensory cortex of primates, modularity, in terms of segregated classes of cells, has only been demonstrated in area 3b. In the hand representation of area 3b of monkeys, separate patches of neurons receive either SA or RA thalamic inputs (Sur et al., 1981; 1984). These inputs may be segregated in other parts of area 3b as well.

Area 3a forms a strip of cortex just rostral to area 3b. Area 3a receives muscle-spindle receptor information (Phillips et al., 1971; Schwarz et al., 1973; Wiesendanger and Miles, 1982) from VPS in the thalamus (Cusick et al., 1985) and cutaneous receptor information from area 3b (i.e., Krubitzer and Kaas, 1990). A head and neck portion of area 3a also receives vestibular information from the thalamus (Akbarian et al., 1992). In microelectrode recording experiments in anesthetized monkeys, area 3a neurons are generally activated by lightly tapping body parts and other moderate levels of stimulation that would activate deep as well as cutaneous receptors (see Huffman and Krubitzer, 2001). Under light anesthesia, neurons responsive to light tactile stimulation are sometimes seen as well. Area 3a represents deep receptors in a somatotopic pattern that parallels that of cutaneous receptors in area 3b (Huffman and Krubitzer, 2001). The receptive fields in area 3a are large and can span multiple digits in the hand area.

Area 1 forms a second representation of the body surface in a strip of cortex just caudal to area 3b (Merzenick et al., 1978; Kaas et al., 1979; Nelson et al., 1980; Sur et al., 1982; Felleman et al., 1983b). Area 1 receives cutaneous information directly from VP and area 3b. Possibly as many as 20% of the relay neurons in VP project to both area 3b and area 1 (Lin et al., 1979; Cusick et al., 1985). Yet, the projections to area 1 from area 3b terminate in layer 4, while those from VP terminate largely above layer 4 (Jones et al., 1975), suggesting that the 3b inputs have an activating role and the VP inputs have a modulatory role. In support of this conclusion, lesions of area 3b abolish evoked responses in area 1 of squirrel monkeys and owl monkeys (Garraghty et al., 1990b). Area 1 contains a systematic map of cutaneous receptors that is essentially a mirror reversal of the map in area 3b (Merzenich et al., 1978; Nelson et al., 1980). Receptive fields for neurons in area 1 are quite small, although they tend to be larger than in area 3b.

Area 2 forms another strip-like representation of the body just caudal to area 1. While the caudal border of area 2 has been variously defined, correlations of architecture with microelectrode maps have provided a good estimate of the location of this border (see Pons et al., 1985; Lewis et al., 1999). Area 2 receives sparse cutaneous inputs from area 3b, dense cutaneous inputs from area 1 (Pons and Kaas, 1986), and muscle-spindle receptor inputs (Schwarz et al., 1973) from VPS in the thalamus (Pons and Kaas, 1985). Possibly as many as 40% of the VPS neurons that project to area 3a also project to area 2 (Cusick et al., 1985).

Neurons in area 2 generally respond to light touch (Pons et al., 1985), but they are also responsive to manipulations of the muscles and body that would activate deep receptors. Under conditions of deep anesthesia, area 2 neurons may respond poorly or not at all to somatosensory stimuli, or only to more intense stimuli that would activate deep receptors (Merzenich et al., 1978). The receptive fields for neurons in area 2 vary in size, but the receptive fields are larger than for neurons in area 3b or area 1. In the hand region of area 2, neurons with multidigit receptive fields predominate, unlike in areas 3b or 1. Area 2 has been characterized as having a systematic map of the body that is a mirror reversal of the one in area 1, but this is an oversimplification. Reversals of somatotopic organization do occur from area 1 to area 2, but area 2 also has more than one representation of digits of the hand and perhaps other body parts in at least macaque monkeys (Pons et al., 1985). The significance of this more complex somatotopic organization is not clear. Area 2 is interconnected with area 3a and more caudal somatosensory cortex, areas 5 and 7 (Pons and Kaas, 1985). The organization of this more caudal somatosensory cortex is not yet well understood.

In summary, area 3b in monkeys is clearly the homologue of S1 as defined in most mammals (Kaas, 1983). Area 3b activates area 1, which in turn activates area 2. Thus, these three areas are largely serial steps in a processing hierarchy, although area 2 directly receives important proprioceptive information from VPS. Area 3a is part of the group, but, significant area 3a outputs are directed to primary and premotor areas of motor cortex (Huerta and Pons, 1990; Darian-Smith et al., 1993; Huffman and Krubitzer, 2001). All four areas of anterior parietal cortex project to areas PV and S2 in the lateral sulcus (Krubitzer and Kaas, 1990; Qi et al., 2000).

Monkeys have at least three representations of the body surface in lateral parietal cortex of the lateral sulcus, areas S2, PV, and VS. S2 and PV either immediately border the lateral part of 3b or are separated from 3b by a narrow strip of area 1. This issue has been difficult to resolve with microelectrode mapping experiments, since lateral 3b represents the face and the immediately adjoining more lateral cortex also represents the face. This adjoining cortex could be parts of the face representations in S2 and PV, or include part of an area 1 face representation, as well as S2 and PV representations. More distant from the area 3b border and further into the lateral sulcus, S2 and PV represent the limbs and trunk, body parts that would not be represented in adjacent areas 1 or 3b. Thus, this more distant cortex is clearly not area 1. S2 and PV can be distinguished from each other because they form mirror image reversals of each other in somatotopic organization (Krubitzer and Kaas, 1990;

Krubitzer et al., 1995; Disbrow et al., 2000; Coq et al., 1999; Qi et al., 2000). The two areas adjoin along representations of the hand and face, while representations of the trunk and hindlimb are distant from each other.

In monkeys, S2 and PV differ from galagos and non-primate mammals in that they depend on inputs from the areas of anterior parietal cortex for activation (Pons et al., 1987; Garraghty et al., 1990a; Burton et al., 1990; however, see Zhang et al., 1996). Lesions of these areas, or possibly just areas 3a and 3b, abolish evoked activity in S2 and PV. This may be because the large relay neurons in the ventro-posterior nucleus project in parallel to both S1 (area 3b) and to S2 and PV in most mammals but do not to project to S2 and PV in monkeys. Instead, S2 and PV receive their major thalamic inputs from VPI (Krubitzer and Kaas, 1992), and this input appears to modulate, rather than activate, S2 and PV neurons. The receptive fields in both areas S2 and PV are quite large and often extend over large parts of the body, although occasionally small receptive fields that are restricted to a single digit of the hand have been noted. Although S2 neurons typically have large receptive fields, information about stimulus location may be retained in population code (Nicolelis et al., 1998).

The ventral somatosensory area, VS, is a proposed subdivision of the soma-tosensory cortex that lies deeper to S2 in the lateral sulcus. In New World monkeys, much of VS is in the fundus of the sulcus and on the lower bank of the sulcus adjacent to the auditory cortex (Cusick et al., 1989; Krubitzer et al., 1995; Qi et al., 2000). Some of the neurons in VS respond to auditory stimuli, but the majority of neurons respond to light tactile stimulation on the contralateral side of the body. The receptive fields in area VS are comparable in size to those in area S2. Overall, VS appears to form a crude mirror reversal of S2, with foot and hand representations bordering S2, and face and trunk representations more distant from their shared border. The connections of VS are not well understood, but S2 and PV have major interconnections (Krubitzer and Kaas, 1990; Qi et al., 2000). S2 and PV also project to a parietal rostral (PR) area just rostral to PV (Krubitzer and Kaas, 1990). Little is known about PR in terms of responsiveness to tactile stimuli and other connections, but it seems likely to be a higher order somatosensory area, possibly relaying to entorhinal cortex and then to the hippocampus (Mishkin, 1979; Friedman et al., 1986) as part of the corticolimbic pathway for touch.

The posterior parietal region of monkeys contains a complex of areas that appear to use somatosensory and visual information for encoding early stages of motor control (Andersen et al., 1990, 1997). The extent of the region varies greatly across simian species, with some New World monkeys having relatively little posterior parietal cortex. Unfortunately, little is known about species differences, and even in the most studied macaque monkeys, only parts of posterior parietal cortex are well understood. Traditionally, posterior parietal cortex has been divided into areas 5 and 7 of Brodmann (1909), and these regions have been further divided into 5a, 5b, 7a, and 7b, as well as even more subdivisions, but these designations have no precise significance. Often more recently proposed subdivisions are based on physiological and anatomical distinctions (see Lewis and Van Essen, 2000), and they are more likely to constitute functionally valid subdivisions of cortex. Most notably, medial (MIP), lateral (LIP), and ventral (VIP) areas of the intraparietal sulcus of macaques have been delimited as distinct visuomotor areas that utilize eye position, proprioceptive, vestibular, and visual information to help guide eye, hand, and body movements. More anterior regions of posterior parietal cortex appear to be more directly involved in somatosensory processing, but even neurons in area 5 combine visual and somatosensory signals (Graziano et al., 2000). Comparative studies of posterior parietal cortex organization are clearly needed.

1.5 SOMATOSENSORY CORTEX IN CHIMPANZEES AND HUMANS

Chimpanzees are our closest relatives, so we would expect major similarities in our somatosensory system. Nevertheless, chimpanzees have much smaller brains, so we might expect some differences. In general, larger brains have larger cortical areas, but they also have more cortical areas (Kaas, 2000). Humans clearly have larger subdivisions of anterior parietal cortex, but too little is known about the rest of the somatosensory systems in hominoid primates to determine if there are any differences in the number of areas. However, we do know from architectonic studies that both chimps and humans have four strip-like somatosensory areas in anterior parietal cortex similar to monkeys. As in monkeys, areas 3a, 3b, 1, and 2 have been identified by differences in cellular structure as well as by a number of related histological and histochemical criteria (Brodmann, 1909; Zilles et al., 1995; White et al., 1997; Qi et al., 1997, 1998; Geyer et al., 1997; Eskenasy and Clarke, 2000). Area 3b is relatively easy to delimit, but the proposed borders of areas 1 and 2 vary somewhat in different reports. A careful comparison, using several types of histochemical preparations of material from monkeys, chimps, and humans (Qi et al., 1998) may provide the most accurate placements of borders. Since microelectrode maps and connection patterns have been used only in monkeys to establish the significance of proposed borders, criteria validated from anatomical and physiological studies on macaque monkeys most usefully define areas in chimps and humans.

As in monkeys (e.g., Marshall et al., 1937), the anterior parietal cortex of humans was thought to contain a single representation, S1, in early investigations that used surface electrodes to produce electrically evoked sensations or record slow waves elicited by taps to the body (e.g., Penfield and Boldrey, 1937; Penfield and Rasmussen, 1950; Woolsey et al., 1979). Because of the inaccessibility of most of areas 3a and 3b in the central sulcus, and most of area 2 in the postcentral sulcus, most of the results from human cortex have been from area 1. These recordings revealed a general mediolateral representation of body parts from foot to tongue, but not the multiple representations of anterior parietal cortex. More recently, the mediolateral somatotopy of anterior parietal cortex has been further revealed by magnetoencephalography (Nakamura et al., 1998), largely confirming these earlier conclusions.

Recent studies have also revealed several new details about the somatotopic organization of anterior parietal cortex of humans. Early studies of sensations evoked by electrical stimulation of the brain with surface electrodes suggested that the genitals are represented on the medial wall of the cerebral hemisphere, just ventral to the toes in the anterior parietal cortex (3b and 1). A more recent study using electrical stimulation of the dorsal nerve of the penis and recordings of evoked potentials with surface electrodes placed the representation of the penis more laterally with hip and upper leg (Bradley et al., 1998). This seems to be different from monkeys where the posterior leg, genitals, and tail appear to be located in the medial wall, just ventral to the foot in areas 3b and 1 (Nelson et al., 1980). However, in a recent, yet unpublished, reinvestigation in macaques, we found the genitals to be represented just lateral to the foot and leg in area 3b. The use of electrical stimulation of the nerve in humans resulted in an overestimation of the proportional size of the representation of the penis compared to other body structures (Bradley et al., 1998).

Most importantly, the greater detail of recent results based on functional magnetic resonance imaging (fMRI) has provided good evidence for separate body surface representations in areas 3b and 1, as well as evidence for areas 3a and 2 as distinct fields. Thus, separate representations of the fingers have been demonstrated in areas 3b and 1, and fingers have been shown to be represented from D5 to D1 in a mediolateral sequence (Gelnar et al., 1998; Kurth et al., 2000). The hand portion of area 3b appears to correspond to a bulge in the central sulcus (Sastre-Janer et al., 1998). Evidence has also been obtained for a third representation of fingers in area 2 (Lin et al., 1996; Kurth et al., 1998, 2000; Moore et al., 2000). There is also evidence from cortical responses to electrical stimulation of muscle afferents (Gandevia et al., 1984), electrical stimulation of mixed cutaneous and deep nerves (Kaukoranta et al., 1986), and passive flexion of fingers (Mima et al., 1996, 1997) for muscle spindle receptor inputs to area 3a and area 2. Thus, the basic organization of anterior parietal cortex in humans appears to consist of four strip-like body representations as in monkeys.

Recent studies have started to demonstrate other areas of somatosensory cortex in humans as well. Disbrow et al. (2000) used fMRI to investigate the somatotopic organization of cortex in the lateral sulcus of humans, while using the same methods for gathering comparable data from monkeys. As for monkeys, they found evidence for two somatotopically organized fields in the cortex of the upper bank of the lateral sulcus, the second somatosensory area, S2, and the parietal ventral area, PV. In both fields, the face, hand, and foot were represented in order from near the lip of the sulcus to the depths of the sulcus (also see Burton et al., 1993; Maldjian et al., 1999 for S2 organization). The two representations form mirror images of one another, joining along the hand, foot, and face representations, with the shoulder and hip separated. Both areas represented the contralateral body surface; however, S2 also responds to the stimulation of the ipsilateral hand. The study also revealed adjoining regions of evoked activity, including a caudal region that may correspond to part of "area" 7b and the rostral parietal area, PR, of monkeys. Nearby portions of insular cortex appear to be involved in the processing of tactual memories (Bonda et al., 1996a).

Of course, we expect regions of posterior parietal cortex of humans to be involved in somatosensory functions as well (see Kaas, 1990); however, the organization of this cortex is not well understood. Patients with posterior parietal lobe injury tend to neglect visual and tactile stimuli presented contralaterally (see Moscovitch and Behrmann, 1994). Functions may include coding postures of the body. Blood flow increases in the superior parietal cortex and intraparietal sulcus of humans when they perform tasks requiring mental rotations of the hand (Bonda et al., 1995,1996b;

also see Seitz et al., 1997). As in monkeys (see Andersen et al., 1990; Lewis and Van Essen, 2000), posterior parietal cortex of humans is likely to contain a number of specialized areas with different patterns of connections and functional roles.

1.6 CONCLUSIONS

1. At the cortical level, the basic somatosensory system of mammals includes a primary field (S1) that is bordered rostrally and caudally by band-like somatosensory areas. This S1 is homologous to the area 3b representation of primates, the rostral area appears to correspond to area 3a, while the relationship of the caudal strip to areas 1 and 2 is uncertain. Somatosensory cortex also includes S2 and, in most mammals, PV. Usually S2 is independently activated by direct projections from VP of the thalamus.
2. Primates have clearly differentiated VP, VPI, and VPS nuclei in the somatosensory thalamus. Prosimian galagos have somatosensory areas 3a and 3b and a caudal somatosensory strip that may be area 1 or a combination of areas 1 and 2. S2 appears to be independently activated by direct VP projections as in many other mammals. Posterior parietal cortex of galagos is not as expansive as in simian primates. Thus, in general, somatosensory cortex in prosimian primates has not greatly changed from the basic mammalian plan.
3. Most monkeys have clearly defined areas 3a, 3b, 1, and 2, each with its own representation of body receptors. Area 2 appears to be more developed in Old World monkeys than in a number of New World monkeys. Processing is highly serial from area 3b to area 1 to area 2, while area 3a relates more to the motor cortex. Areas S2 and PV appear to depend on inputs from anterior parietal cortex for activation, and thus processing has become more serial in monkeys. Posterior parietal cortex has become more elaborate, where a number of visuomotor and somatosensory areas have been defined.
4. Chimpanzees and humans clearly have the four areas of anterior parietal cortex and both areas S2 and PV have been defined in humans. Posterior parietal cortex is expanded, and fMRI studies are starting to delimit functionally distinct regions.

REFERENCES

Akbarian, S., Grusser, O. J., and Guldin, W. O. Thalamic connections of the vestibular cortical fields in the squirrel monkey (*Saimiri sciureus*), *J. Comp. Neurol.*, 326, 423, 1992.

Andersen, R. A., Asanuma, C., Essick, G., and Siegel, R. M. Corticocortical connections of anatomically and physiologically defined subdivisions within the inferior parietal lobule, *J. Comp. Neurol.*, 296, 65, 1990.

Andersen, R. A., Snyder, L. H., Bradley, D. C., and Xing, J. Multimodal representation of space in the posterior parietal cortex and its use in planning movements, *Annu. Rev. Neurosci.*, 20, 303, 1997.

Apkarian, A. V. and Shi, T. Squirrel monkey lateral thalamus. I. Somatic nociresponsive neurons and their relation to spinothalamic terminals, *J. Neurosci.*, 14, 6779, 1994.

Beck, P. D., Pospichal, M. W., and Kaas, J. H. Topography, architecture, and connections of somatosensory cortex in opossums: evidence for five somatosensory areas, *J. Comp. Neurol.*, 366, 109, 1996.

Bombardiei, R. A. J., Johnson, J. I. J., and Campos, G. B. Species differences in mechanosensory projections from the mouth to the ventrobasal thalamus, *J. Comp. Neurol.*, 163, 41, 1975.

Bonda, E., Petrides, M., Frey, S., and Evans, A. Neural correlates of mental transformations of the body-in-space, *Proc. Natl. Acad. Sci. (U.S.A.)*, 92, 11180, 1995.

Bonda, E., Petrides, M., and Evans, A. Neural systems for tactual memories, *J. Neurophysiol.*, 75, 1730, 1996a.

Bonda, E., Frey, S., and Petrides, M. Evidence for a dorso-medial parietal system involved in mental transformations of the body, *J. Neurophysiol.*, 76, 2042, 1996b.

Bradley, W. E., Farrell, D. F., and Ojemann, G. A. Human cerebrocortical potentials evoked by stimulation of the dorsal nerve of the penis, *Somatosens. Mot. Res.*, 15, 118, 1998.

Brodmann, K. *Vergleichende Lokalisation der Grosshirnrinde in ihren Prizipien dargestellt auf Grund des Zellenbaues*, Brodmann, K., Ed., Barth, J.A., Leipzig, 1909.

Burton, H. and Robinson, C. J. Responses in the first or second somatosensory cortical area in cats during transient inactivation of the other ipsilateral area with lidocaine hydrochloride, *Somatosens Res.*, 4, 215, 1987.

Burton, H., Sathian, K., and Shao, D. H. Altered responses to cutaneous stimuli in the second somatosensory cortex following lesions of the postcentral gyrus in infant and juvenile macaques, *J. Comp. Neurol.*, 291, 395, 1990.

Burton, H., Videen, T. O., and Raichle, M. E. Tactile-vibration-activated foci in insular and parietal-opercular cortex studied with positron emission tomography: mapping the second somatosensory area in humans, *Somatosens. Mot. Res.*, 10, 297, 1993.

Carlson, M. and Fitzpatrick, K. A. Organization of the hand area in the primary somatic sensory cortex (SmI) of the prosimian primate, *Nycticebus coucang*, *J. Comp. Neurol.*, 204, 280, 1982.

Carlson, M., Huerta, M. F., Cusick, C. G., and Kaas, J. H. Studies on the evolution of multiple somatosensory representations in primates: the organization of anterior parietal cortex in the New World Callitrichid, Saguinus, *J. Comp. Neurol.*, 246, 409, 1986.

Catania, K. C. and Kaas, J. H. Organization of the somatosensory cortex of the star-nosed mole, *J. Comp. Neurol.*, 351, 549, 1995.

Catania, K. C. and Kaas, J. H. The unusual nose and brain of the star-nosed mole, *BioSci.*, 46, 578, 1996.

Catania, K. C., Collins, C. E., and Kaas, J. H. Organization of sensory cortex in the East African hedgehog (*Atelerix albiventris*), *J. Comp. Neurol.*, 421, 256, 2000.

Catania, K. C. Cortical organization in moles: evidence of new areas and a specialized S2, *Somatosens. Mot. Res.*, 17, 335, 2000.

Chapin, J. K. and Lin, C. S. Mapping the body representation in the SI cortex of anesthetized and awake rats, *J. Comp. Neurol.*, 229, 199, 1984.

Coleman, G. T., Zhang, H. Q., Murray, G. M., Zachariah, M. K., and Rowe, M. J. Organization of somatosensory areas I and II in marsupial cerebral cortex: parallel processing in the possum sensory cortex, *J. Neurophysiol.*, 81, 2316, 1999.

Coq, J.-O., Qi, H.-X., Catania, K. C., Collins, C. E., Jain, N., and Kaas, J. H. Organization of somatosensory cortex in New World Titi monkey, *Soc. Neurosci. Abstrs.*, 25, 1683, 1999.

Coq, J.-O., Strata, F., Caroni, P., and Kaas, J. H. Somatotopic organization of cuneatus and gracilis nuclei in the brainstem of primates, *Soc. Neurosci. Abstrs.*, 26, 148, 2000.

Craig, A. D., Bushnell, M. C., Zhang, E. T., and Blomqvist, A. A thalamic nucleus specific for pain and temperature sensation, *Nature*, 372, 770, 1994.

Cusick, C. G., Steindler, D. A., and Kaas, J. H. Corticocortical and collateral thalamocortical connections of postcentral somatosensory cortical areas in squirrel monkeys: a double-labeling study with radiolabeled wheatgerm agglutinin and wheatgerm agglutinin conjugated to horseradish peroxidase, *Somatosens. Res.*, 3, 1, 1985.

Cusick, C. G., Wall, J. T., Felleman, D. J., and Kaas, J. H. Somatotopic organization of the lateral sulcus of owl monkeys: area 3b, SII, and a ventral somatosensory area, *J. Comp. Neurol.*, 282, 169, 1989.

Darian-Smith, C., Darian-Smith, I., Burman, K., and Ratcliffe, N. Ipsilateral cortical projections to areas 3a, 3b, and 4 in the macaque monkey, *J. Comp. Neurol.*, 335, 200, 1993.

Dawson, D. R. and Killackey, H. P. The organization and mutability of the forepaw and hindpaw representations in the somatosensory cortex of the neonatal rat, *J. Comp. Neurol.*, 256, 246, 1987.

Disbrow, E., Roberts, T., and Krubitzer, L. Somatotopic organization of cortical fields in the lateral sulcus of *Homo sapiens*: evidence for SII and PV, *J. Comp. Neurol.*, 418, 1, 2000.

Dykes, R. W., Sur, M., Merzenich, M. M., Kaas, J. H., and Nelson, R. J. Regional segregation of neurons responding to quickly adapting, slowly adapting, deep, and Pacinian receptors within thalamic ventroposterior lateral and ventroposterior inferior nuclei in the squirrel monkey (*Saimiri sciureus*), *Neuroscience.*, 6, 1687, 1981.

Dykes, R. W., Rasmusson, D. D., Sretavan, D., and Rehman, N. B. Submodality segregation and receptive-field sequences in cuneate, gracile, and external cuneate nuclei of the cat, *J. Neurophysiol.*, 47, 389, 1982.

Dykes, R. W. Parallel processing of somatosensory information: a theory, *Brain Res. Rev.*, 6, 47, 1983.

Eskenasy, A. C. C. and Clarke, S. Hierarchy within human S1: Supporting data from cytochrome oxidase, acetylcholinesterase, and NADPH-diaphorase staining patterns, *Somatosens. Mot. Res.*, 17, 123, 2000.

Fang, P.-C., Jain, N., and Kaas, J. H. Intracortical connections in the region of the hand-face border in area 3b of New World monkeys, *Soc. Neurosci. Abstrs.*, 26, 1462, 2000.

Feldman, S. H. and Johnson, J. I. J. Kinesthetic cortical area anterior to primary somatic sensory cortex in the raccoon (*Procyon lotor*), *J. Comp. Neurol.*, 277, 80, 1988.

Felleman, D. J., Wall, J. T., Cusick, C. G., and Kaas, J. H. The representation of the body surface in SI of cats, *J. Neurosci.*, 3, 1648, 1983a.

Felleman, D. J., Nelson, R. J., Sur, M., and Kaas, J. H. Representations of the body surface in areas 3b and 1 of postcentral parietal cortex of Cebus monkeys, *Brain Res.*, 268, 15, 1983b.

Fitzpatrick, K. A., Carlson, M., and Charlton, J. Topography, cytoarchitecture, and sulcal patterns in primary somatic sensory cortex (SmI) prosimian primate, *Perodicticus potto, J. Comp. Neurol.*, 204, 296, 1982.

Fleagle, J. G. Ed., *Primate Adaptation and Evolution*, Academic Press, San Diego, 1999.

Florence, S. L., Wall, J. T., and Kaas, J. H. Somatotopic organization of inputs from the hand to the spinal gray and cuneate nucleus of monkeys with observations on the cuneate nucleus of humans, *J. Comp. Neurol.*, 286, 48, 1989.

Friedman, D. P., Murray, E. A., O'Neill, J. B., and Mishkin, M. Cortical connections of the somatosensory fields of the lateral sulcus of macaques: evidence for a corticolimbic pathway for touch, *J. Comp. Neurol.*, 252, 323, 1986.

Frost, S. B., Milliken, G. W., Plautz, E. J., Masterton, R. B., and Nudo, R. J. Somatosensory and motor representations in cerebral cortex of a primitive mammal (*Monodelphis domestica*): a window into the early evolution of sensorimotor cortex, *J. Comp. Neurol.*, 421, 29, 2000.

Gandevia, S. C., Burke, D., and McKeon, B. The projection of muscle afferents from the hand to cerebral cortex in man, *Brain.*, 107 (Pt 1), 1, 1984.

Garraghty, P. E., Pons, T. P., and Kaas, J. H. Ablations of areas 3b (SI proper) and 3a of somatosensory cortex in marmosets deactivate the second and parietal ventral somatosensory areas, *Somatosens. Mot. Res.*, 7, 125, 1990a.

Garraghty, P. E., Florence, S. L., and Kaas, J. H. Ablations of areas 3a and 3b of monkey somatosensory cortex abolish cutaneous responsivity in area 1, *Brain Res.*, 528, 165, 1990b.

Garraghty, P. E., Florence, S. L., Tenhula, W. N., and Kaas, J. H. Parallel thalamic activation of the first and second somatosensory areas in prosimian primates and tree shrews, *J. Comp. Neurol.*, 311, 289, 1991.

Gelnar, P. A., Krauss, B. R., Szeverenyi, N. M., and Apkarian, A. V. Fingertip representation in the human somatosensory cortex: an fMRI study, *Neuroimage*, 7, 261, 1998.

Geyer, S., Schleicher, A., and Zilles, K. The somatosensory cortex of human: cytoarchitecture and regional distributions of receptor-binding sites, *Neuroimage*, 6, 27, 1997.

Gould, H. J. D., Whitworth, R. H. J., and LeDoux, M. S. Thalamic and extrathalamic connections of the dysgranular unresponsive zone in the grey squirrel (*Sciurus carolinensis*), *J. Comp. Neurol.*, 287, 38, 1989.

Graziano, M. S. A., Cooke, D. F., and Taylor, C. S. R. Coding the location of the arm by sight, *Science*, 290, 1782, 2000.

Huerta, M. F. and Pons, T. P. Primary motor cortex receives input from area 3a in macaques, *Brain Res.*, 537, 367, 1990.

Huffman, K. J., Nelson, J., Clarey, J., and Krubitzer, L. Organization of somatosensory cortex in three species of marsupials, *Dasyurus hallucatus, Dactylopsila trivirgata*, and *Monodelphis domestica*: neural correlates of morphological specializations, *J. Comp. Neurol.*, 403, 5, 1999.

Huffman, K. J. and Krubitzer, L. Area 3a: Topographic Organization and cortical connections, *Cereb. Cortex*, in press, 2001.

Jain, N., Florence, S. L., and Kaas, J. H. Limits on plasticity in somatosensory cortex of adult rats: hindlimb cortex is not reactivated after dorsal column section, *J. Neurophysiol.*, 73, 1537, 1995.

Jain, N., Catania, K. C., and Kaas, J. H. Deactivation and reactivation of somatosensory cortex after dorsal spinal cord injury, *Nature*, 386, 495, 1997.

Jain, N., Catania, K. C., and Kaas, J. H. A histologically visible representation of the fingers and palm in primate area 3b and its immutability following long-term deafferentations, *Cereb. Cortex*, 8, 227, 1998.

Jain, N., Qi, H.-X., and Kaas, J. H. Long-term chronic multichannel recordings from sensorimotor cortex and thalamus of primates, *Prog. Brain Res.*, in press, 2001a.

Jain, N., Qi, H. X., Catania, K. C., and Kaas, J. H. Anatomical correlates of the face and oral cavity representations in somatosensory area 3b of monkeys, *J. Comp. Neurol.*, 429, 455, 2001b.

Jones, E. G. Lamination and differential distribution of thalamic afferents within the sensorimotor cortex of the squirrel monkey, *J. Comp. Neurol.*, 160, 167, 1975.

Jones, E. G., Coulter, J. D., and Hendry, S. H. Intracortical connectivity of architectonic fields in the somatic sensory, motor, and parietal cortex of monkeys, *J. Comp. Neurol.*, 181, 291, 1978.

Kaas, J. H., Nelson, R. J., Sur, M., Lin, C. S., and Merzenich, M. M. Multiple representations of the body within the primary somatosensory cortex of primates, *Science*, 204, 521, 1979.

Kaas, J. H. *The Somatosensory Cortex and Thalamus in Galago*, Haines, E. E., Ed., CRC Press, Inc., Boca Raton, FL., 1982, 169.

Kaas, J. H. What, if anything, is SI? Organization of first somatosensory area of cortex, *Physiol. Rev.*, 63, 206, 1983.

Kaas, J. H., Nelson, R. J., Sur, M., Dykes, R. W., and Merzenich, M. M. The somatotopic organization of the ventroposterior thalamus of the squirrel monkey, Saimiri sciureus, *J. Comp. Neurol.*, 226, 111, 1984.

Kaas, J. H. and Pons, T. P. The somatosensory system of primates, in *Comparative Primate Biology*, Steklis, H. P., Ed., Alan R. Liss, Inc., New York, 1988, 421.

Kaas, J. H. The somatosensory system, in *The Human Nervous System*, Paxinos, G., Ed., Academic Press, New York, 1990, 813.

Kaas, J. H. The functional organization of somatosensory cortex in primates, *Anat. Anz.*, 175, 509, 1993.

Kaas, J. H. and Florence, S. L. Reorganization of sensory and motor systems in adult mammals after injury, in *The Mutable Brain*, Kaas, J. H., Ed., Gorden and Breach Science Publishers, London, England, 2000, 165.

Kaas, J. H. Why brain size is so important: Design problems and solutions as neocortex gets bigger or smaller, *Brain and Mind*, 1, 7, 2000.

Kaukoranta, E., Hamalainen, M., Sarvas, J., and Hari, R. Mixed and sensory nerve stimulations activate different cytoarchitectonic areas in the human primary somatosensory cortex SI. Neuromagnetic recordings and statistical considerations, *Exp. Brain Res.*, 63, 60, 1986.

Killackey, H. P., Gould, H. J. D., Cusick, C. G., Pons, T. P., and Kaas, J. H. The relation of corpus callosum connections to architectonic fields and body surface maps in sensorimotor cortex of New and Old World monkeys, *J. Comp. Neurol.*, 219, 384, 1983.

Krishnamurti, A., Sanides, F., and Welker, W. I. Microelectrode mapping of modality-specific somatic sensory cerebral neocortex in slow loris, *Brain Behav. Evol.*, 13, 267, 1976.

Krubitzer, L. A., Sesma, M. A., and Kaas, J. H. Microelectrode maps, myeloarchitecture, and cortical connections of three somatotopically organized representations of the body surface in the parietal cortex of squirrels, *J. Comp. Neurol.*, 250, 403, 1986.

Krubitzer, L. A. and Kaas, J. H. Thalamic connections of three representations of the body surface in somatosensory cortex of gray squirrels, *J. Comp. Neurol.*, 265, 549, 1987.

Krubitzer, L. A. and Kaas, J. H. The organization and connections of somatosensory cortex in marmosets, *J. Neurosci.*, 10, 952, 1990.

Krubitzer, L. A. and Calford, M. B. Five topographically organized fields in the somatosensory cortex of the flying fox: microelectrode maps, myeloarchitecture, and cortical modules, *J. Comp. Neurol.*, 317, 1, 1992.

Krubitzer, L. A. and Kaas, J. H. The somatosensory thalamus of monkeys: cortical connections and a redefinition of nuclei in marmosets, *J. Comp. Neurol.*, 319, 123, 1992.

Krubitzer, L., Manger, P., Pettigrew, J., and Calford, M. Organization of somatosensory cortex in monotremes: in search of the prototypical plan, *J. Comp. Neurol.*, 351, 261, 1995.

Kurth, R., Villringer, K., Mackert, B. M., Schwiemann, J., Braun, J., Curio, G., Villringer, A., and Wolf, K. J. fMRI assessment of somatotopy in human Brodmann area 3b by electrical finger stimulation, *Neuroreport*, 9, 207, 1998.

Kurth, R., Villringer, K., Curio, G., Wolf, K. J., Krause, T., Repenthin, J., Schwiemann, J., Deuchert, M., and Villringer, A. fMRI shows multiple somatotopic digit representations in human primary somatosensory cortex, *Neuroreport*, 11, 1487, 2000.

Lewis, J. W., Burton, H., and Van Essen, D. C. Anatomical evidence for the posterior boundary of area 2 in the macaque monkey, *Somatosens. Mot. Res.*, 16, 382, 1999.

Lewis, J. W. and Van Essen, D. C. Mapping of architectonic subdivisions in the macaque monkeys with emphasis on parieto-occipital cortex, *J. Comp. Neurol.*, 428, 79, 2000.

Li, X. G., Florence, S. L., and Kaas, J. H. Areal distributions of cortical neurons projecting to different levels of the caudal brain stem and spinal cord in rats, *Somatosens. Mot. Res.*, 7, 315, 1990.

Lin, C. S., Merzenich, M. M., Sur, M., and Kaas, J. H. Connections of areas 3b and 1 of the parietal somatosensory strip with the ventroposterior nucleus in the owl monkey (*Aotus trivirgatus*), *J. Comp. Neurol.*, 185, 355, 1979.

Lin, W., Kuppusamy, K., Haacke, E. M., and Burton, H. Functional MRI in human soma tosensory cortex activated by touching textured surfaces, *J. Magn. Reson. Imaging*, 6, 565, 1996.

Macchi, G. F., Angeleri, and Guazzi. Thalamo-cortical connections of the first and second somatosensory areas in the cat, *J. Comp. Neurol.*, 111, 387, 1959.

Maldjian, J. A., Gottschalk, A., Patel, R. S., Pincus, D., Detre, J. A., and Alsop, D. C. Mapping of secondary somatosensory cortex activation induced by vibrational stimulation: an fMRI study, *Brain Res.*, 824, 291, 1999.

Marshall, W. H., Woolsey, C. N., and Bard, P. Cortical representation of tactile sensibility as indicated by cortical potentials, *Science*, 85, 388, 1937.

Merzenich, M. M., Kaas, J. H., Sur, M., and Lin, C. S. Double representation of the body surface within cytoarchitectonic areas 3b and 1 in SI in the owl monkey (*Aotus trivirgatus*), *J. Comp. Neurol.*, 181, 41, 1978.

Mima, T., Terada, K., Maekawa, M., Nagamine, T., Ikeda, A., and Shibasaki, H. Somatosensory evoked potentials following proprioceptive stimulation of finger in man, *Exp. Brain Res.*, 111, 233, 1996.

Mima, T., Ikeda, A., Terada, K., Yazawa, S., Mikuni, N., Kunieda, T., Taki, W., Kimura, J., and Shibasaki, H. Modality-specific organization for cutaneous and proprioceptive sense in human primary sensory cortex studied by chronic epicortical recording, *Electroencephalogr. Clin. Neurophysiol.*, 104, 103, 1997.

Mishkin, M. Analogous neural models for tactual and visual learning, *Neuropsychologia*, 17, 139, 1979.

Moore, C. I., Stern, C. E., Corkin, S., Fischl, B., Gray, A. C., Rosen, B. R., and Dale, A. M. Segregation of somatosensory activation in the human rolandic cortex using fMRI, *J. Neurophysiol.*, 84, 558, 2000.

Moscovitch, M. and Behrmann, M. Coding of spatial information in the somatosensory system: evidence from patients with neglect following parietal lobe damage, *J. Cog. Neurosci.*, 66, 151, 1994.

Murray, G. M., Zhang, H. Q., Kaye, A. N., Sinnadurai, T., Campbell, D. H., and Rowe, M. J. Parallel processing in rabbit first (SI) and second (SII) somatosensory cortical areas: effects of reversible inactivation by cooling of SI on responses in SII, *J. Neurophysiol.*, 68, 703, 1992.

Nakamura, A., Yamada, T., Goto, A., Kato, T., Ito, K., Abe, Y., Kachi, T., and Kakigi, R. Somatosensory homunculus as drawn by MEG, *Neuroimage*, 7, 377, 1998.

Nelson, R. J., Sur, M., and Kaas, J. H. The organization of the second somatosensory area (SmII) of the grey squirrel, *J. Comp. Neurol.*, 184, 473, 1979.

Nelson, R. J., Sur, M., Felleman, D. J., and Kaas, J. H. Representations of the body surface in postcentral parietal cortex of *Macaca fascicularis*, *J. Comp. Neurol.*, 192, 611, 1980.

Nelson, R. J. and Kaas, J. H. Connections of the ventroposterior nucleus of the thalamus with the body surface representations in cortical areas 3b and 1 of the cynomolgus macaque (*Macaca fascicularis*), *J. Comp. Neurol.*, 199, 29, 1981.

Nicolelis, M. A., Ghazanfar, A. A., Stambaugh, C. R., Oliveira, L. M., Laubach, M., Chapin, J. K., Nelson, R. J., and Kaas, J. H. Simultaneous encoding of tactile information by three primate cortical areas, *Nat. Neurosci.*, 1, 621, 1998.

Pearson, P. P., Oladehin, A., Li, C. X., Johnson, E. F., Weeden, A. M., Daniel, C. H., and Waters, R. S. Relationship between representation of hindpaw and hindpaw barrel subfield (HBS) in layer IV of rat somatosensory cortex, *Neuroreport*, 7, 2317, 1996.

Penfield, W. and Boldrey, E. Somatic motor and sensory representation in the cerebral cortex of man as studied by electrical stimulation, *Brain*, 60, 389, 1937.

Penfield, W. and Rasmussen, T. *The Cerebral Cortex of Man*, Macmillan, New York, 1950.

Phillips, C. G., Powell, T. P., and Wiesendanger, M. Projection from low-threshold muscle afferents of hand and forearm to area 3a of baboon's cortex, *J. Physiol. (Lond)*, 217, 419, 1971.

Pons, T. P. and Kaas, J. H. Connections of area 2 of somatosensory cortex with the anterior pulvinar and subdivisions of the ventroposterior complex in macaque monkeys, *J. Comp. Neurol.*, 240, 16, 1985.

Pons, T. P., Garraghty, P. E., Cusick, C. G., and Kaas, J. H. The somatotopic organization of area 2 in macaque monkeys, *J. Comp. Neurol.*, 241, 445, 1985.

Pons, T. P. and Kaas, J. H. Corticocortical connections of area 2 of somatosensory cortex in macaque monkeys: a correlative anatomical and electrophysiological study, *J. Comp. Neurol.*, 248, 313, 1986.

Pons, T. P., Wall, J. T., Garraghty, P. E., Cusick, C. G., and Kaas, J. H. Consistent features of the representation of the hand in area 3b of macaque monkeys, *Somatosens. Res.*, 4, 309, 1987.

Qi, H.-X., Jain, N., and Kaas, J. H. Histochemical organization of somatosensory area 3b and surrounding cortex in chimpanzees, *Soc. Neurosci. Abstrs.*, 23, 1007, 1997.

Qi, H.-X., Jain, N., Preuss, T. M., and Kaas, J. H. Comparative architecture of areas 3a, 3b, and 1 of somatosensory cortex in chimpanzees, humans, and macaques, *Soc. Neurosci. Abstrs.*, 24, 1125, 1998.

Qi, H. X., Lyon, D., and Kaas, J. H. Topography, connections, and architecture of parietal ventral somatosensory area in marmosets, *Soc. Neurosci. Abstrs.*, 26, 1685, 2000.

Rausell, E. and Jones, E. G. Histochemical and immunocytochemical compartments of the thalamic VPM nucleus in monkeys and their relationship to the representational map, *J. Neurosci.*, 11, 210, 1991a.

Rausell, E. and Jones, E. G. Chemically distinct compartments of the thalamic VPM nucleus in monkeys relay principal and spinal trigeminal pathways to different layers of the somatosensory cortex, *J. Neurosci.*, 11, 226, 1991b.

Sastre-Janer, F. A., Regis, J., Belin, P., Mangin, J. F., Dormont, D., Masure, M. C., Remy, P., Frouin, V., and Samson, Y. Three-dimensional reconstruction of the human central sulcus reveals a morphological correlate of the hand area, *Cereb. Cortex*, 8, 641, 1998.

Schwarz, D. W., Deecke, L., and Fredrickson, J. M. Cortical projection of group I muscle afferents to areas 2, 3a, and the vestibular field in the rhesus monkey, *Exp. Brain Res.*, 17, 516, 1973.

Seitz, R. J., Canavan, A. G., Yaguez, L., Herzog, H., Tellmann, L., Knorr, U., Huang, Y., and Homberg, V. Representations of graphomotor trajectories in the human parietal cortex: evidence for controlled processing and automatic performance, *Eur. J. Neurosci.*, 9, 378, 1997.

Slutsky, D. A., Manger, P. R., and Krubitzer, L. Multiple somatosensory areas in the anterior parietal cortex of the California ground squirrel (*Spermophilus beecheyii*), *J. Comp. Neurol.*, 416, 521, 2000.

Spreafico, R., Hayes, N. L., and Rustioni, A. Thalamic projections to the primary and secondary somatosensory cortices in cat: single and double retrograde tracer studies, *J. Comp. Neurol.*, 203, 67, 1981.

Sur, M., Nelson, R. J., and Kaas, J. H. Representation of the body surface in somatic koniocortex in the prosimian Galago, *J. Comp. Neurol.*, 189, 381, 1980.

Sur, M., Weller, R. E., and Kaas, J. H. Physiological and anatomical evidence for a discontinuous representation of the trunk in SI of tree shrews, *J. Comp. Neurol.*, 201, 135, 1981.

Sur, M., Nelson, R. J., and Kaas, J. H. Representations of the body surface in cortical areas 3b and 1 of squirrel monkeys: comparisons with other primates, *J. Comp. Neurol.*, 211, 177, 1982.

Sur, M., Wall, J. T., and Kaas, J. H. Modular distribution of neurons with slowly adapting and rapidly adapting responses in area 3b of somatosensory cortex in monkeys, *J. Neurophysiol.*, 51, 724, 1984.

Welker, W. I. Principles of organization of the ventrobasal complex in mammals, *Brain Behav. Evol.*, 7, 253, 1973.

White, L. E., Andrews, T. J., Hulette, C., Richards, A., Groelle, M., Paydarfar, J., and Purves, D. Structure of the human sensorimotor system. I: Morphology and cytoarchitecture of the central sulcus, *Cereb. Cortex*, 7, 18, 1997.

Wiener, S. I., Johnson, J. I., and Ostapoff, E. M. Organization of postcranial kinesthetic projections to the ventrobasal thalamus in raccoons, *J. Comp. Neurol.*, 258, 496, 1987.

Wiesendanger, M. and Miles, T. S. Ascending pathway of low-threshold muscle afferents to the cerebral cortex and its possible role in motor control, *Physiol. Rev.*, 62, 1234, 1982.

Wolfheim, J. H. Ed., *Primates of the World*, University of Washington Press, Seattle, 1983.

Woolsey, T. A. and Van der Loos, H. The structural organization of layer IV in the somatosensory region (SI) of mouse cerebral cortex. The description of a cortical field composed of discrete cytoarchitectonic units, *Brain Res.*, 17, 205, 1970.

Woolsey, C. N., Erickson, T. C., and Gilson, W. E. Localization in somatic sensory and motor areas of human cerebral cortex as determined by direct recording of evoked potentials and electrical stimulation, *J. Neurosurg.*, 51, 476, 1979.

Wu, W. H., Beck, P. D., and Kaas, J. H. Ipsilateral cortical connections of S1 (3b) in prosimian primates: Evidence for five somatosensory areas, *Soc. Neurosci. Abstrs.*, 21, 112, 1995.

Wu, C. W., Beck, P. D., and Kaas, J. H. Cortical and thalamic connections of the second somatosensory area, S2, in the prosimian primate, *Galago Garnetti*, *Soc. Neurosci. Abstrs.*, 22, 107, 1996.

Wu, C. W. H., Bichot, N. P., and Kaas, J. H. Connections of the second (S2) and parietal ventral (PV) somatosensory areas with frontal motor cortex: A study combining electrorecording, microstimulation, cytoarchitecture, and connectivity, *Soc. Neurosci. Abstrs.*, 23, 1273, 1997.

Zhang, H. Q., Murray, G. M., Turman, A. B., Mackie, P. D., Coleman, G. T., and Rowe, M. J. Parallel processing in cerebral cortex of the marmoset monkey: effect of reversible SI inactivation on tactile responses in SII, *J. Neurophysiol.*, 76, 3633, 1996.

Zilles, K., Schlaug, G., Matelli, M., Luppino, G., Schleicher, A., Qu, M., Dabringhaus, A., Seitz, R., and Roland, P. E. Mapping of human and macaque sensorimotor areas by integrating architectonic, transmitter receptor, MRI, and PET data, *J. Anat.*, 187 (Pt 3), 515, 1995.

2 Cerebral Cortical Regions Devoted to the Somatosensory System: Results from Brain Imaging Studies in Humans

Harold Burton

CONTENTS

0-8493-2336-3/02/$0.00+$1.50
© 2002 by CRC Press LLC

2.1   INTRODUCTION

Studies in monkeys indicate that somatosensory processing of innocuous tactile stimuli occurs within an interconnected cortical network, which mostly resides in the parietal cortex. The major regions include multiple subdivisions of the primary somatosensory area (SI) in the postcentral gyrus, the secondary somatosensory area (SII) in the parietal operculum, additional lateral cortical areas buried within the Sylvian fissure, portions of the supramarginal gyrus, and granular prefrontal cortex. This review focuses on homologous somatosensory representations in the cortex of humans that are revealed with non-invasive recording and neuroimaging methods. Critical to the validation of these observations is confirming characteristic features known from direct neural recordings, especially in monkeys. Brain imaging studies like positron emission tomography (PET) and functional magnetic resonance imaging (fMRI) are especially appropriate in exposing the many cortical foci that respond during a single somatosensory stimulation paradigm. Non-invasive techniques have the great advantage of allowing replicate studies in conscious normal or patient populations performing a variety of tasks affecting tactile discrimination, sensorimotor integration, attention, and object recognition.

A discussion of different brain imaging techniques is beyond the scope of this review. However, some comments about underlying characteristics of these images are useful in evaluating the findings on somatosensory areas. PET and fMRI images are a consequence of local blood flow changes, which accompany immediate and focal neuronal activity. These images have low signal-to-noise ratios, are delayed in seconds from the millisecond intervals of neural activity by the slowness of hemodynamic responses, and are spatially constrained to the vascular distribution. However limited is the spatial resolution of PET or fMRI, the localization of activated areas is direct.

Magnetoencephalography (MEG) records evoked neural responses much like electroencephalography and, therefore, contribute explicit information about timing of neural activity. Signals have high reliability as they are based on averages of multiple, synchronized trials. Localizing the source of recorded electromagnetic fields requires inverse modeling to locate and orient the source and magnitude of responsible primary currents.[128] Although there are infinite solutions to the inverse problem, most recorded field patterns are dipolar, which suggests that source localizations can be confined to a single equivalent current dipole (ECD). Explicitly restricting the results to MR anatomical or functional foci in single subjects improves source localization models of these ECDs. However, modeled source localizations are always problematic, especially for long latency responses, in the presence of multiple, overlapping generators, and where stronger signals possibly mask weaker responses.[128]

2.2 PRIMARY SOMATOSENSORY CORTEX (SI)

2.2.1 DEFINING THE BOUNDARIES OF SI

2.2.1.1 Cytoarchitecture

In most descriptions, the SI portion of anterior parietal cortex encompasses four anterior to posterior subregions labeled, respectively, Brodmann areas 3a, 3b, 1, and 2.[17,58] Recent studies from autopsy data emphasize variable correlation among people between the location of cytoarchitectonic subareas and gross anatomical features.[84,209,230,231,243] Despite this caveat to interrelating brain imaging foci based on cross-subject averages, humans and monkeys show remarkably homologous subregions. Thus, using broad operational definitions that are applicable to humans and old-world monkeys, the following gross anatomical zones likely coincide with one cytoarchitectonic area: fundus of the central sulcus for 3a, nearly two-thirds of the posterior wall of the central sulcus for 3b, crown of the central sulcus and anterior half of the postcentral gyrus for 1, and posterior half of the postcentral gyrus and, possibly, the anterior wall of the postcentral sulcus for 2.[84,131] This cross-species homology is important to understanding SI in humans because findings from monkeys reveal not one but four relatively complete somatotopic maps whose boundaries correlate closely with the approximate borders of one of the four cytoarchitectonic subregions.[57,73,106,107,115,152,178,189] An important focus in describing brain imaging studies is the extent that these data confirm the four correlated functional/anatomical subdivisions included within SI.

2.2.1.2   Evidence for Medial to Lateral Somatotopy

Initial vetting of many brain imaging techniques derives from detecting somatosensory-driven activations attributable to anterior parietal cortex, and especially when results confirm medial to lateral sequences of activated foci from stimulating a small number of contralateral body regions (Table 2.1). For example, PET (reviewed in Reference 28) and fMRI (e.g., Reference 210) studies report a medial to lateral somatotopy for widely separate body representations (e.g., foot, hand, and face). Several studies describe activation of only a single peri-Rolandic site contralateral to stimulating the hand with high amplitude vibratory stimuli.[26,40,71,213] These observations thus corroborate earlier reports of a single distorted contralateral representation spanning the postcentral gyrus and having an approximate medial to lateral topography with respect to successive segmental sacral to cervical to trigeminal levels of the body.[181,233] Several studies also note ipsilateral activations even in the hand region of SI.[28,90,130,138]

Current imaging results must be tempered by recent findings of only 55% concordance between the centers of activated regions using fMRI and direct recordings in the same brains of anesthetized monkeys.[55] Thus, discrepancies from expected SI map features may only be a consequence of inherent technical limitations.[199] Further limiting possible findings is the crude spatial resolution of most imaging studies and the limited number of body representations studied.

2.2.1.3   SI Finger Representation

Despite limitations, brain imaging studies of the SI finger representation examine a number of features, such as overlapping representations from adjacent fingers, spatial extent of different finger representations, and the multiplicity of representations. Penfield's classic work illustrates a single hand representation with the thumb occupying a larger extent than that for any other finger, with all of the latter having nearly equal territories.[181] Later studies indicate larger representations for the thumb and index fingers in humans[233,236] and monkeys.[119,152,172]

Studies based on MEG or somatosensory evoked potential (SEP) measurements[8,221,239] indicate single and unique centers of mass for each finger. Electrically stimulating the proximal interphalangeal joint of each finger activated spatially separable magnetic or electrical field patterns for the initial 20 to 30 msec latency peak potentials. Source localizations, calculated from the equivalent current dipole (ECD) distributions, place non-overlapping foci for fingers 5 to 1, respectively, to a medial to lateral extent of parietal cortex in several studies.[8,10,221,239] All recordings reflect responses in or near the central sulcus. One study demonstrates distortions in the SI map by showing a ranking from largest to smallest for, respectively, thumb to finger 5 in the anterior to posterior lengths of ECD distributions.[221] In measurements from published figures, the average tangential distance between digits 5 and 1 is 18 mm in 4 subjects[8] and 23 mm between electric dipoles in 3 subjects.[221] This suggests approximately 3 to 4 mm sagittal width per digit, if each finger is assumed to have an equal representation. These data, however, do not show whether overlap exists at the boundaries between digit zones.

TABLE 2.1
SI Mapping Studies

Stimulation Sites	Type of Stimulus	Imaging Method	Authors
Fingers 1, 2, and 5 (thumb largest response)	Electrical	MEG	Reference 220
Hand	Large amplitude vibrator (130Hz)	PET	Reference 213
Median nerve	Electrical	MEG	Reference 92
Toes, hand, and lips	Large amplitude vibrator (130Hz)	PET	Reference 71
Ulnar nerve, digit 5 to digit 1, median nerve	Electrical	MEG, SEP	References 8 and 9
Ulnar nerve, digit 5 to digit 1, median nerve	Electrical	SEP	Reference 221
Fingers, hand, arm, and face	Electrical	MEG	Reference 239
Palm	Manual stroking	fMRI (1.5T)	Reference 89
Toes, fingertips, and tongue	Manual scrubbing	fMRI (1.5T)	Reference 210
Fingertips 5 to 2	Manual rubbing (~1 Hz)	fMRI (1.5T)	Reference 138
Fingertips 2 and 3	Controlled roughness	PET	Reference 28
Tongue, lips, fingers, arm, trunk, leg, and foot	Tactile	MEG	Reference 170
Fingers 1, 2, or 5	Vibration	fMRI (1.5T)	Reference 83
Left arm in a proximal-to-distal or distal-to-proximal direction	Air puffs	fMRI (1.5T)	Reference 214
Face, fingers, and toes	Air puffs	fMRI (1.5T)	Reference 98
Fingers 2 to 3 and 4 to 5	Brushing	fMRI (1.5T)	References 90 and 268
Fingertips 5 to 1	Piezoceramic stimulator (15 and 30 Hz)	fMRI (4T)	Reference 144
Distal-proximal direction over the volar surface of the right index finger and palm	Vibration (200 Hz)	MEG	Reference 96
Palm and finger	Manual punctate (~3Hz)	fMRI (3T)	Reference 164
Fingers 5 to 1	Electrical	fMRI (1.5T)	Reference 131
Fingertips 5 and 2	Piezoceramic stimulator (30 to 80 Hz)	fMRI (3T)	Reference 72

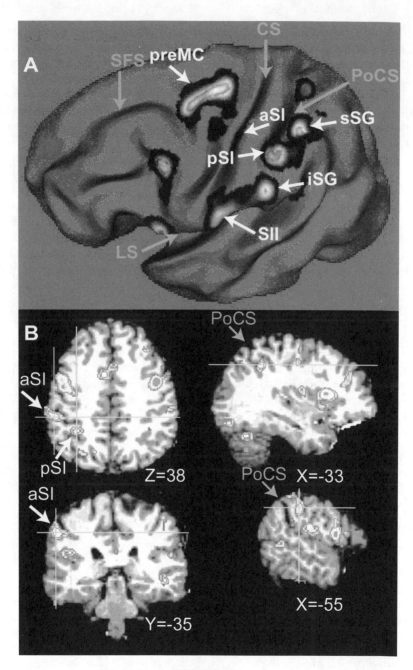

FIGURE 2.1

Some PET[28] and fMRI[32,72,130,138,144,164] studies reveal at least two finger foci occupying anterior and posterior portions of the postcentral gyrus. In a PET study, tactile stimulation of two fingertips activates contralateral foci extending into the posterior bank of the central sulcus, without reaching the fundus, and a second posterior focus covering all of the posterior postcentral gyrus and adjoining anterior bank of the postcentral sulcus.[28] Stimuli in this study were passively applied horizontal gratings that produced a roughness sensation. An earlier MEG study similarly illustrates one dipole source projected to the central sulcus and another to more posterior postcentral gyral cortex.[68] FMRI studies show more directly a single anterior focus within the central sulcus (Figure 2.1, aSI, and Reference 32). A completely separate posterior focus occupies the posterior half of the postcentral gyrus and anterior wall of the postcentral sulcus (Figure 2.1, pSI, and Reference 32). Complicating the identification of this posterior SI region is the presence of a nearby zone of increased activity on the posterior bank of the postcentral sulcus and adjoining superior aspect of the supramarginal gyrus (Figure 2.1, sSG). These distinctions of multiple foci in the brain imaging data when stimulating fingertips (see below and Figure 2.2) partially suggest the representation noted in monkeys: a rostrally pointing fingertip region within the posterior bank of the central sulcus and confined to area 3b, a second representation pointing posterior on the surface of the postcentral gyrus within area 1. The latter is possibly conjoined with a third site, which again has the distal finger pointing anterior, in area 2 (e.g., References 135, 146, 152, 172, 183, 185, and 238).

Overlap between foci for adjacent digits occurs in all of SI, but in monkeys the least occurs in anterior SI most likely associated with area 3b (Figure 2.2A). The question of overlap in humans remains unsettled because different studies report varying degrees of overlap from anterior to posterior across the postcentral gyrus. Comparisons of the peak source amplitudes for SI equivalent dipoles recorded to single compared to paired, simultaneous digit stimulation show greater overlap in the responses from adjacent digits 1 and 2 vs. 2 and 5.[10] In a fMRI study relying on electrical stimulation of single fingers, activations from adjacent digits overlap 41% in anterior foci centered in the central sulcus, 49% in the anterior half of the

---

FIGURE 2.1 (opposite) (See Color Figure 2.1 in color insert.) Z-score maps (P<.05) of MR signals in single subjects cued to attend to vibrotactile stimuli applied to fingertip 2 on the right hand with an MR compatible, computer-controlled vibrator. Results obtained using a general linear model, an assumed hemodynamic response, and a Boynton model.[15] Data shown in A are from trials when the subject selectively attended to changes in vibration frequency to detect whether the first or second sequentially presented stimulus was a higher frequency. Data shown in B are from another subject attending during different trials either selectively to frequency or to the duration of paired vibratory stimuli, or who divided attention between both frequency and duration features (see Reference 215). MR signals were collected and analyzed per trial (TR 3.68 seconds, 3.75mm voxel size, ~16 seconds per trial). A. Functional MR data superimposed on a three-dimensional, partially inflated view of left hemisphere for one subject. B. Slices at three orientations illustrate the location of MR signals in anterior and posterior portions of SI in another subject. Pink lines on each slice show registration of slices at orthogonal orientations. Coordinate values are referenced to the Talairach atlas.[222] Anatomical abbreviations (shown in red): CS, central sulcus; LS, lateral sulcus; PoCS, postcentral sulcus; SFS, superior frontal sulcus; functional abbreviations (shown in yellow): aSI, anterior primary somatosensory area; pSI, posterior primary somatosensory area; SII, secondary somatosensory area; preMC, premotor cortex; iSG, inferior supramarginal region; sSG, superior supramarginal region.

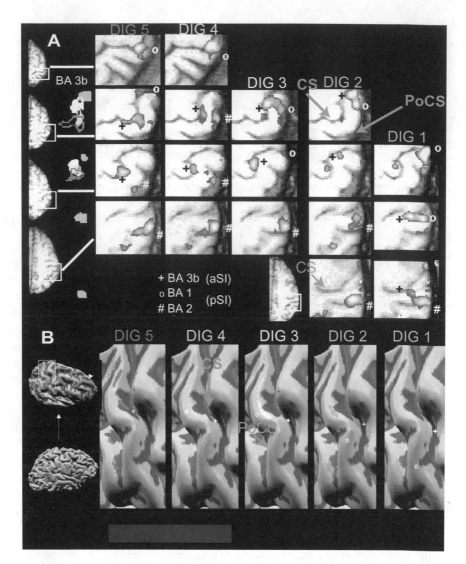

FIGURE 2.2 (See Color Figure 2.2 in color insert.) Activation sites in the left postcentral gyrus after electrical stimulation of single fingers (Digits 1 to 5) on the right hand. Results obtained from a correlation analysis in one subject. **A.** Distribution of activated pixels observed across a superior (top row) to inferior (bottom row) series of horizontal slices. Each column of magnified images shows the distribution of active foci noted from stimulating one of the digits. Symbols: +, o, and # demarcate, respectively, foci attributed to Brodmann areas 3b, 1, and 2. BA 3b occupies the posterior bank of the central sulcus (aSI in Figure 2.1) and BA 1 and 2 are, respectively, on the exposed postcentral gyral crown and adjoining anterior bank of the postcentral sulcus (combined as pSI in Figure 2.1). The column labeled BA 3b shows the outlines of foci activated at the same horizontal level (row of images to the right) from each of the digits. The colors for these outlines match the corresponding digit labels and foci shown in B. **B.** Expanded surface view of the postcentral gyrus presents distribution of active foci for each digit in one subject. [A is modified from Figure 1 and B from Figure 2 in Reference 131.]

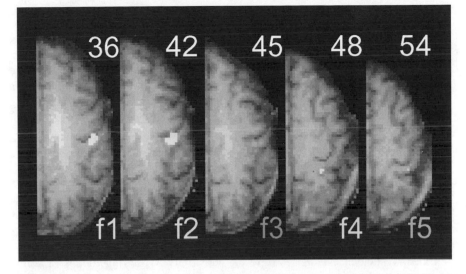

FIGURE 2.3 (See Color Figure 2.3 in color insert.) Group averaged Statistical Parameter Maps show maximum activated focus for each finger on a superior to inferior series of axial slices. Numbers are Z-axis coordinates in reference to Talairach atlas.[222] Stimulation of individual fingers obtained using piezoelectric vibration. [Modified from Reference 144.]

postcentral gyrus, and 68% in the vicinity of the postcentral sulcus. The anterior SI region also shows almost no overlap for foci from non-adjacent digits (see color coded insets in Figure 2.2A, Reference 131), while more posterior regions still exhibit overlaps of more than 25%.[131] Multiple digit representations occur within a single horizontal plane, but with clear medial to lateral separations in presumptive area 3b (Figure 2.2A). In addition, representations for different digits are separable in the superior to inferior axis (Figures 2.2 and 2.3). Another fMRI study reports considerable overlap between foci for fingers 2-3 vs. 4-5,[90] but a larger representation for fingers 2–3. This study, however, does not distinguish different anteroposterior sectors of the postcentral gyrus, and vein inflow effects contaminate the images. Yet another study describes nearly complete overlap in anterior foci from non-adjacent digits 2 and 5 and almost no overlap in more posterior postcentral regions.[72] Each fingertip was stimulated with individually mounted piezoelectric vibrators. Even group averaged images, obtained with higher field strength magnets and piezoelectric vibrators, show (Figure 2.3) distinct foci separated by >3 mm in the transverse axis for each vibrated fingertip.[144] However, three-dimensional reconstructions (Figure 2.2B, Reference 131) show some overlap between adjacent digits in anterior SI. The activated foci for even non-adjacent digits overlap in posterior SI (Figure 2.2B). Despite differences in imaging techniques, the fMRI studies agree with earlier MEG data in noting a 12–18 mm tangential length to the finger region.[131,144] These studies also confirm earlier observations of larger regions devoted to the thumb and index finger. Collectively, current brain imaging data provides a partial confirmation of maps obtained in monkeys. However, as techniques improve, this discrepancy will likely disappear.

2.2.1.4 Separate Anterior and Posterior Maps

Closely spaced mapping surveys in monkeys show a subsidiary anterior to posterior organization of two mirror representations for proximal parts of the fingers between the anterior and posterior distal tip regions, respectively, in areas 3b and 1.[152,153,172,183,185] Currently, no human study confirms this standard observation in monkeys of separate and sequential anterior to posterior foci from stimulating successive phalanges and palm.[96] However, one fMRI study, which utilized a novel technique of periodic, phase shifted, air puff stimulation of 5 sites from wrist to shoulder, describes a 10 mm strip of cortex within the central sulcus that sequentially responds to stimulation of proximal to distal surfaces on the ventral forearm.[214] Unlike prior investigations that considered a statistic based on MR signal magnitude, this study examined the mean phase of the fMRI signal as a function of cortical distance. Future applications of such phase analyses might demonstrate more precise maps in the rest of SI, and thereby show greater correspondence between SI in humans and monkeys.

Extensive evidence indicates differing sensitivity to varying modes of somatosensory stimulation from anterior to posterior aspects of SI.[57,73,106,107,115,152,178,189] Predominant responses in each map follow activation of different somatosensory receptors: muscle spindles in 3a, low threshold cutaneous receptors in 3b and 1, and both cutaneous and kinesthetic/proprioceptive receptors in 2. Although fMRI appears potentially capable of revealing three separate foci across the postcentral gyrus (Figure 2.2A), which presumptively represent activity in the Brodmann areas 3b, 1, and 2, only two brain imaging studies marginally support modality-based distinctions.[28,164]

In a PET study involving passive stimulation with tactile gratings, we describe two distinct but completely overlapping foci activated by two stimulation modes. In one mode, horizontal gratings directly rub against the skin surface of two fingertips; in a second mode, stimulation by the same surfaces is indirect and through a plectrum, which is attached to the finger and touches the gratings. However, skin mode stimulation correlates with significantly higher ($F_{(8,1)} = 4.63$, $P = .04$) blood flows in anterior SI.[28] This result concurs with likely greater stimulation of cutaneous receptors in the skin mode experiment, and therefore activation of likely area 3b in anterior SI.

An fMRI study[164] contrasts the distribution of SI responses to manually applied punctate tactile stimulation of the palm and fingers vs. flexion/extension movements of the fingers and palm. The first stimulation mode principally activates cutaneous receptors. The second, active mode engages a broad array of proprioceptive and skin stretch receptors together with activation of hand and distal arm muscles. By reconstructing the results from individual subjects, and flattening the cortical surface through the central sulcus, these investigators present distinct activation patterns across several SI subdivisions for the two tasks. Of greatest interest is the absence of activity across the fundus of the central sulcus, in presumed area 3a, during the tactile task (Figure 2.4B). Averaged data further shows distinct peaks in the gross anatomical zones marking areas 3b, 1, and 2 (Figure 2.4C). Responses stretch across all of the central sulcus during the kinesthetic/motor task (Figure 2.4B and 2.4C).

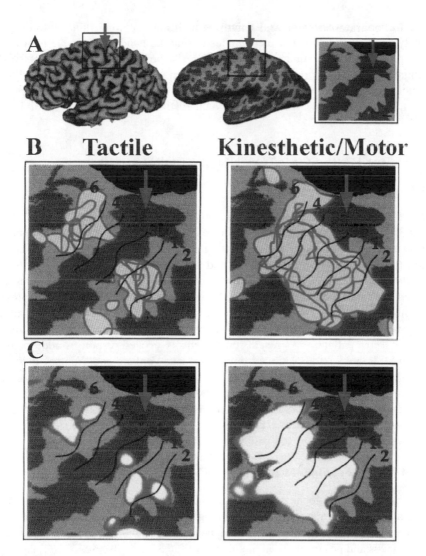

FIGURE 2.4 (See Color Figure 2.4 in color insert.) Distribution of MR signals across the central sulcus, hand area during tactile stimulation or a kinesthetic/motor protocol. Results are averaged maps from five subjects. **A.** *Left*: Three-dimensional reconstruction of the cortex. *Middle*: An inflated reconstruction of the same brain in which dark gray shows sulci and light gray illustrates gyri. Region enclosed in black box is over the hand area, which is shown enlarged on the *right* and below in B and C. Green arrow points to the fundal region of the central sulcus in all images. **B.** Distribution of activity from tactile stimulation (on left) or kinesthetic/motor protocol (on the right) from each of five subjects (red lines with yellow fill, t-test analysis) is projected onto the flattened reconstruction from one subject after transforming all distributions to a canonical representation. Solid black lines demarcate the presumed cytoarchitectonic Brodmann area borders, which are based on sulcal/gyral gross anatomy. **C.** Across subject average distribution of activity for the two tasks: tactile on the left and kinesthetic/motor on the right. [Modified from Figure 4 in Reference 164.]

By subtraction between the tasks, activity in the fundal region of the central sulcus is likely due to responsive muscle spindles during the hand/finger movement task.

Currently, no brain imaging study utilizes a precisely controlled, passively administered tactile or proprioceptive stimulation that activates selectively different subclasses of peripheral somatosensory receptors. The absence of such protocols limits prospects of assessing functional distinctions across the subdivisions of SI.

2.3 SECONDARY SOMATOSENSORY CORTEX (SII)

2.3.1 IDENTIFICATION OF SII

Initial evidence of a secondary somatosensory area (SII) in humans comes from 8 out of 350 patients who experienced somatic sensations, illusions of movements, or paresthesia in different body parts following electrical stimulation of surface cortex lateral to the face representation in SI, and near the posterior horizontal ramus of the Sylvian fissure.[179,181] Similar effects are reported from directly stimulating the exposed surface of the upper bank of the Sylvian fissure after temporal lobectomy[180] or through intracortical electrodes.[76] In addition, recordings from the upper bank of the Sylvian fissure (parietal operculum) of monkeys[234,235,237] and near the Sylvian fissure in a few patients[141,236] reveal somatosensory evoked responses additional to those in SI. In bold extrapolations from sparse data, Penfield and Woolsey independently summarize their findings as a single somatotopic representation in the parietal operculum, which consists of an upside-down body image with the head directed anterior and the distal fingers and toes pointing superficially toward the superior lip of the Sylvian fissure. However, the recording evidence for this SII map in humans remains fragmented and mixed results exist for the stimulation effects.[226]

2.3.1.1 Gross Anatomy of the Parietal Operculum

A precise anatomical localization of SII within the inferior parietal lobule must consider the sulcal arrangement surrounding the parietal operculum. On average, the anterior and posterior borders of the parietal operculum are, respectively, the central sulcus and posterior ascending ramus of the Sylvian fissure.[218] An imaginary lateral extension of the postcentral sulcus points to the middle of the parietal operculum, which is relatively co-extensive with the horizontal ramus of the Sylvian fissure (Figure 2.5). This horizontal segment, from inspection of 67 brains, is found in 99% and 70%, respectively, of left and right hemispheres.[232] It is twice as long on the left. Men with a consistent right hand preference show bilateral greater length horizontal rami; however, an asymmetry always exists, which is longer on the left.[232] An inferior portion of the supramarginal gyrus lies immediately posterior and superficial to the parietal operculum in most brains.[218] This gross anatomical location of the parietal operculum coincides with the position of SII in summary illustrations by Penfield[179,181] and occupies that portion of the parietal operculum included within area 40, by Brodmann[17] or area PF, of subsequent analyses (e.g., Reference 59).

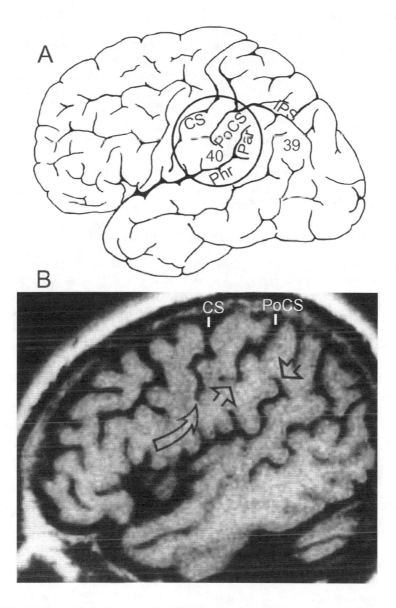

FIGURE 2.5 Sulcal pattern surrounding SII. **A**. Selective sulci of the lateral parietal cortex. Anatomical abbreviations: CS, central sulcus; LS, lateral sulcus; IPS, intraparietal sulcus; Par, posterior ascending ramus of the Sylvian fissure; Phr, posterior horizontal ramus of the Sylvian fissure; PoCS, postcentral sulcus; 39, 40 Brodmann's cytoarchitectonic areas. **B**. MR anatomical sagittal image illustrates the most common sulcal pattern surrounding the parietal operculum. Curved arrow marks CS, short left arrow points to PoCS, and right short arrow indicates Par. [A is modified from Figure 1 and B from Figure 2 in Reference 218.]

2.3.1.2   Cytoarchitecture

The cytoarchitecture in the parietal operculum consists of a prominent granule cell layer IV and small to medium sized pyramidal cells in supra- and infra-granular layers. A similar pattern prevails over all of the parietal operculum and adjoining anterior two-thirds of inferior aspects of the supramarginal gyrus in humans[59] and in the parietal operculum and inferior parietal lobule in animals.[21,191] Even a transitional area located close to the posterior ascending ramus of the Sylvian fissure, labeled PFG, still manifests a distinct layer IV.[59] This cytoarchitecture characterizes many paleocortical derivative, sensory dedicated cortical regions.[59,211] The connections to this lateral and inferior parietal cortex in monkeys are especially concerned with information from the somatosensory system.[20,27,74,75,114,116] Not too surprisingly, therefore, somatosensory stimulation evokes responses within and near the parietal operculum.[21]

2.3.1.3   Brain Imaging Evidence on SII Organization

Most studies of the parietal operculum with MEG, PET, or fMRI show bilateral distributions of increased regional activity (Table 2.2). All report larger responses contralateral to the stimulated site; however, the coordinates for responses from both sides are identical. FMRI findings of reduced or absent ipsilateral SII responses in patients with partial transections of the corpus callosum confirm anatomical studies in monkeys that ipsilateral responses in SII (and posterior portions of SI) are conducted by fibers crossing through the posterior third of the corpus callosum.[63] Studies in monkeys show these connections arising from SI and SII in the opposite hemisphere.[43,112,127,147,148] The average and standard deviation of stereotactic atlas coordinates[222] from eight studies (Table 2.2) are/± 49.3, ± 4.8, –21 ± 7.3, 17.4 ± 1.9 for the center of mass or peak activation. The small standard deviations for these averages are notable because the data include PET and fMRI images reconstructed to atlas space with different techniques. They demonstrate considerable consistency in the discovered centers of contralateral and ipsilateral responses during several types of somatosensory stimulation to different body regions. However, this spatial uniformity presents a problem to hypothesizing a somatotopic map with distinct zones for different body representations (see below).

2.3.1.3.1   Temporal Order of Processing through SII

The first brain imaging demonstration of bilateral somatosensory responses on the upper bank of the Sylvian fissure in humans comes from MEG recordings.[91,92] These, and several subsequent recordings, note long latency (~100ms), mostly positive somatosensory-evoked fields (SEF)[69,70,93-95,101,139,150] or SEPs.[2] Shorter latency SEPs are reported with intracortical electrodes implanted in the upper bank of (60ms negative cited in Reference 76) or surface cortex next to the Sylvian fissure (20 ms cited in Reference 141) in patients. Some MEG studies[121,128] emphasize short latency (20 to 26 msec), negative responses followed by additional, larger amplitude, mostly positive responses at the longer times noted by others.

Explanations for differences in reported response latencies require some comments on modeling dipole source localizations from the upper bank of the Sylvian

TABLE 2.2
SII Mapping Studies

Stimulation Sites	Type of Stimulus	Imaging	Active Regions	Coordinates (X, Y, Z)	Authors
Median nerve	Electrical	MEG	PO		References 91 and 92
Distal fingers 1 and 3	Electrical	MEG	PO		Reference 94
Hand	Large amplitude vibrator (130Hz)	PET	PO, Ri		References 207 and 213
Lower lip, wrist, and ankle	Electrical	MEG	PO		Reference 95
Toes, hand, and lips	Large amplitude vibrator (130Hz)	PET	PO, post. insula	44, −6.2, 18.2	Reference 26
	Vibration	PET	PO, post. insula	42, −21, 15	Reference 40
Hands, fingers	Roughness, length discriminations	PET	PO	55, −16, 17	Reference 133
Median nerves	Electrical, manual touch, movements	MEG	PO		Reference 101
Fingertips 5 to 2	Manual rubbing (~1 Hz)	fMRI (1.5T)	PO		Reference 138
	Electrical	MEG	PO		Reference 159
					Reference 150
Fingertips 2 and 3	Roughness discriminations	PET	PO	45, −23, 20	Reference 28
Fingers 1, 2, or 5	Vibration	fMRI	PO, insula		Reference 83
Face, fingers, and foot	Vibration (8 Hz) and air puff	fMRI	PO, Ri		Reference 98
Left radial nerve	Electrical, isometrics	MEG	PO		Reference 69
Fingertips 2 and 3	Tactile gratings	PET	PO, insula, Ri	52, −22, 20	Reference 30
Finger 3	Passive/active movement	PET	PO		Reference 161
Median nerve	Electrical	SEP	PO, insula		Reference 76
Left thumb pad	Piezoceramic stimulator (15 and 30 Hz)	fMRI	PO	55.5, −21, 17.5	Reference 145
Tibial nerve, fingers 1 and 3, upper and lower lips	Electrical	MEG	PO		Reference 143
Median nerve	Electrical	MEG	PO		Reference 121
Right median nerve		MEG, fMRI	AO, PO		Reference 128
Palm, fingers	Moving brush	PET	PO	50, −34, 16.5	Reference 13
Foot, trunk, hand, face, and lips	Manual touching	fMRI	PO/PV, other	50.8, −24.4, −14.7	Reference 54

fissure. This discussion, in turn, will suggest the possibility of multiple sources for the parietal opercular responses. The first issue is determining the orientation of the dipole. In MEG measurements, dipolar sources are modeled as arising within the depths of the fissure, and with a tangential orientation to the cortical surface and perpendicular to the axis of the Sylvian fissure.[91,92] In contrast, the polarity reversals for recorded intracortical SEPs are in the direction of the implanted electrodes, which suggests a radially oriented dipolar source, perpendicular to the cortical surface.[76] The underlying current source responsible for recorded SEPs and SEFs is synchronized responses from pyramidal cells in an area of cortex. The dipole orientation from these cells is theoretically perpendicular to the cortical surface. In a sulcus with a single infolding from the surface, these current sources would align tangential to the surface and be observed best by MEG, hence the success of this technique in detecting SII.[92] However, reports of different first response latencies from cortex deep within the Sylvian fissure may result from the anatomical complexity in this region, which is not completely accounted for in prior modeling of the source localization in MEG studies. Thus, models of deep dipolar sources, which are projected indirectly from surface recordings, fail to account for neuronal activity arising from potentially overlapping sources.[76] The inner and outer folds of the upper bank of that portion of the Sylvian fissure overlying the insula[113] probably have separate current sources oriented radially to their respective folds of cortex, and thus parallel to normal cortical functional columns. The current distributions from such closely spaced signals are likely inseparable using MEG,[121] or as noted in one study, might lead to unstable source localizations for equivalent current dipoles (ECDs) at 70–120 msec latencies at different anteroposterior positions.[128] Techniques that model multiple dipole sources using location constraints from subject-specific anatomical or functional MR images might be more sensitive to the actual neuronal source of the ECDs and, like direct intracortical SEPs,[76] are less subject to confounds of anatomical complexities. Studies using these methods report short and long latency SEFs in SII.[121,128] These data suggest the possibility of two adjacent cortical areas in the parietal operculum, with one responsible for short, and the other long, latency responses (see below and Reference 121).

Knowledge of the timing of parietal opercular responses relative to SI influences interpretations of processing sequences through somatosensory cortex. Thus, sequential timing of responses noted in MEG studies supports the hypothesis that processing proceeds serially from short latency responses in SI to later activity in SII.[70] Evidence supporting long latency SEPs and SEFs in SII is also consistent with the demonstration in macaque monkeys that SII fails to respond after destroying SI.[184,186] However, the time difference between SI and SII noted in earlier MEG studies is ~40 msec. This is incongruous with the ~12 –15 msec response latencies recorded from single SII neurons in monkeys to tactile stimuli.[25] Findings of short latency SII SEFs that equal or, in some subjects, precede activity in SI[121] are more compatible with evidence of parallel processing of somatosensory information in several animals.[23,82,224,241] The notion of parallel processing through private thalamic inputs also more suitably explains several additional observations: (1) preservation of responses in some SII cells with hand cutaneous receptive fields in monkeys that have lost the cutaneous SI finger/hand representations in areas 3b and 1;[24] (2) retention of tactile discrimination capabilities in infant[34] monkeys with near total SI lesions; and (3) persistence of ipsilateral SII responses in some patients with ischemic strokes that reduce or eliminate all contralateral SI and SII SEFs.[70] One explanation that accommodates both sets of findings is the existence of two SII areas. One SII might be responsible for long latency MEG recorded responses and is part of a serial processing network. A second, short latency component of SII processes somatosensory events nearly simultaneously with SI. (The evidence for two SII areas is discussed further below.)

2.3.1.3.2 One or Two Somatotopic Maps in SII

Only two brain imaging studies claim any evidence of a somatotopic map in SII. Most studies report complete overlap of activations evoked from stimulating different parts of the body. For example, ECD source localizations are coextensive for responses to electrically stimulating nerves from the hand (median) or foot (peroneal).[91,92,95] Similarly, distributions of responses in PET images[26] are indistinguishable to stimulating the hand or foot with large amplitude vibrations, or in fMRI images to applying air puffs to the face, fingers, or toes,[98] or lower amplitude vibrations to individual fingers 1, 2, or 5.[83] One MEG study,[143] though reporting no significant differences in group data for ECD locations from stimulating face, fingers and leg, shows ECD locations from stimulated fingers between anterior-lateral face and posterior-medial leg sites in a map from a single individual. One fMRI study[54] presents images, also from selected individual data, that might be construed as an anterior to posterior topography (Figure 2.6A1-3). These fMRI-determined maps illustrate a large, central domain for the hand representation, which also overlaps much of the activated foci for the foot or face (Figure 2.6A1-3).

At best, these brain imaging-determined maps of SII are preliminary because they do not present even the coarsest level of organization, such as separating the finger zone from those of the hand and arm. They especially lack information about the distal/proximal axis for the fingers and toes. Earlier views[179] claim that the zones for distal fingers and toes reside closer to the superior lip of the Sylvian fissure. Results from single neuron recordings in monkeys illustrate a map, such as those suggested for humans, with face to lower limb organized, respectively, from anterior-lateral to posterior-medial.[200] Later experiments, however, suggest the existence of two mirror image body maps with one representation anterior and oriented upside-down; the other, more posterior map, has an upright body orientation.[27,129] The junction between the two maps occupies the middle of the parietal operculum and falls along the finger and foot representations, which precludes earlier suggestions of any superficial orientation to the distal finger representation. The two subareas are labeled SII anterior (parietal ventral area in the terminology of Reference 129) and SII posterior. A major question is whether current brain imaging delimitations of SII in humans correctly defines it as a single functional entity.

A large paired finger/hand region dominates in the proposed dual maps in monkeys. A single, possibly similarly coupled foot region lies immediately juxtaposed to the posterior, inferior margin of the distal forelimb region. Single neuron recordings in both limb regions find neurons with multi-digit receptive fields, and few neurons with receptive fields on a single finger.[22,129,200] Earlier SEP studies in

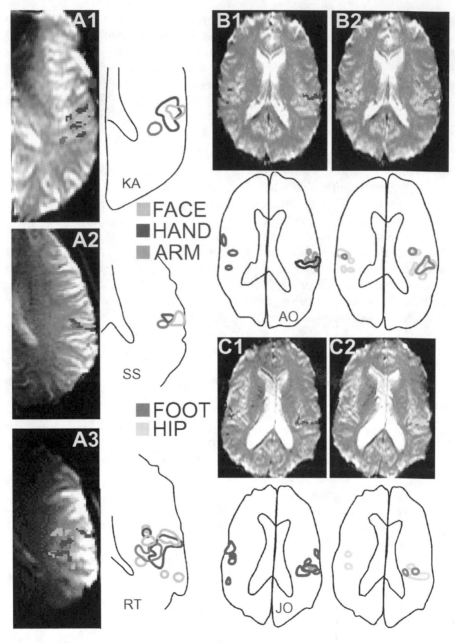

FIGURE 2.6

monkeys illustrate large receptive fields that include face and hand, multi-finger, hand and arm, or all of the foot, leg, and sacral/lumbar trunk.[237] Potential comparable receptive fields in humans would likely generate extensive spread of activity in fMRI or PET images when stimulating any finger, part of the hand, or lower leg, which might then eclipse separate representations from more selected body regions. An important identifying characteristic for two maps, however, is disjoint representations for proximal body regions that surround each fused distal limb region. One fMRI study shows activation of regions to stimulating the upper arm (Figure 2.6B1 and 2.6C1) or hip (Figure 2.6B2 and 2.6C2) that possibly surround foot and hand foci, but in only two subjects.[54] More extensive and accurate three-dimensional reconstructions are needed of images from several subjects stimulated identically in multiple body regions, before claiming these data confirm the presence of two maps in the parietal operculum of humans.

2.3.2 A ROLE FOR SII: SENSORIMOTOR EFFECTS IN SOMATOSENSORY AREAS

Activity in SII increases more than in SI during motor tasks with a tightly coupled sensory component that is topographically close to the body location of the contracting muscles.[101,139] For example, responses from the parietal operculum increase during trials with combined electrical stimulation of one median nerve and synchronized voluntary finger flexion movements only in the same hand.[101] Comparable topographical tuning effects occur in the absence of overt movements when isometric voluntary contractions of thenar or deltoid muscles in the same limb accompany electrical stimulation of the median nerve.[69,139] SII responses decrease when median nerve stimulation is combined with voluntary contractions of masseter or anterior tibial muscles, which are remote from the receptive territory of the stimulated nerve.[139] Additional examples of related findings come from PET data showing increases in regional cerebral blood flows (rCBF) in SI and SII during passive movements, but even greater rCBF changes, especially in contralateral SII, during active vs. passive movement of finger 3.[161] Similarly, bilateral MR signals in SII, but not SI, are greater when subjects must execute multiple manipulations in identifying, non-nameable objects vs. fewer movements when recognizing simpler shapes, such as a smooth sphere.[11,12]

These motor related modulations of SII activity possibly indicate mechanisms for enhancing sensory information from a limb as a guide to behavior involving that limb. A particular example of this behavior would be active manipulation of an object for purposes of identification. We previously hypothesized that SII provides

---

FIGURE 2.6 (opposite) (See Color Figure 2.6 in color insert.) Distribution of activated foci in the parietal operculum from tactile stimulation of a variety of body regions in individual subjects. Results obtained using a cross-correlation analysis. A1-3. Axial slices from three subjects depict foci from stimulating the foot, hand, and face. Outline drawings to the right of the MR images summarize the distribution of activity for each stimulated body region. These cases illustrate a medial to lateral organization for, respectively, the foot to face. In addition, the foot region occupies a more posterior position in two subjects. B1, C1. The location of activated foci from stimulating the upper arm-shoulder with respect to the hand in two subjects. In one subject (AO), the representation for the more proximal body region possibly surrounds that for the foci activated from the more distal hand. B2, C2. The location of activated foci from stimulating the upper leg-thigh with respect to the foot in two subjects. In both subjects, the representation for the more proximal body region appears to surround that for the foci activated from the more distal foot. [A is modified from Figure 7, and B, C are modified from Figure 8 in Reference 54.]

a conduit for information from cutaneous receptors to motor cortex given its considerable anatomical connections to ventral premotor frontal areas in monkeys.[21,149] Thus, during active manipulating, as in successive, targeted finger movements during haptic exploration, enhanced sensory messages processed in SII are available to direct and control integrated sequential touching.[11,12]

A potential confound in these sensorimotor studies is that the results might also reflect directing attention or vigilance to the stimulated part of the body as subjects contract particular muscles or move in selected ways. Discussed below is the effect of attention on increasing activity in SII and other somatosensory areas. Although the influence of attention cannot be ignored, subjects in the MEG studies had prior training with the electrical stimulation and some engaged in distracting tasks, such as reading.[69] The absence of responses from normally activated attention foci in medial cortex in most subjects[69,101,139] further limits the likelihood that attention was the sole basis of the observed modulations.

2.4 OTHER SOMATOSENSORY AREAS IN FRONTAL AND PARIETAL CORTEX

Several studies report somatosensory foci located at least 1 cm anterior to the parietal operculum.[26,40,54,76,98,111,128,207,213] These foci involve the superior and posterior parts of the insula and frontal operculum.[54,76,87,98,111] Even with the crudity of spatial resolution available in brain images, the location of these anterior foci clearly differ from responses attributable to posterior parietal operculum, SII.[26,40,54,98,128] SEFs in four out of five subjects trail, by 4 to 30 msec, those from the parietal operculum.[128] The coordinates for the centers of mass for these anterior foci have considerable variance irrespective of whether the same or different body sites are stimulated, which means somatotopic maps are improbable. An especially interesting observation is that activity in the insula also includes the nearby claustrum; and that this combined insula/claustrum focus might serve in the cross-modal transfer of shape information obtained through tactile or visual processing.[87]

2.4.1 SOMATOSENSORY ACTIVATION OF INSULA AND FRONTAL OPERCULUM

Finding somatosensory activated responses in the insula is entirely concordant with evidence in monkeys that this region connects with components of the cortical and thalamic somatosensory network[6,20,75,156,167,168] and responds to tactile stimuli, often to stimulation applied anywhere on the body.[201,212] The delayed timing of SEF responses recorded in this region also follows speculations that the somatosensory responses in the insula are part of a sequential processing cascade involving SII and corticolimbic pathways.[75,162,169] However, further study is needed to confirm the ancillary suggestion that these connections also are part of a network for remembering and learning new tactile discriminations.

Possible explanations for activity on the frontal operculum are confounded, however, by the known diversity in this region,[35,49,113,191,222] which includes language areas (e.g., Reference 77), premotor and primary motor regions for the head, and

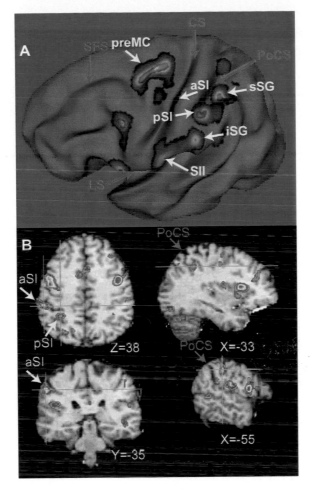

COLOR FIGURE 2.1 Z-score maps (P<.05) of MR signals in single subjects cued to attend to vibrotactile stimuli applied to fingertip 2 on the right hand with an MR compatible, computer-controlled vibrator. Results obtained using a general linear model, an assumed hemodynamic response, and a Boynton model.[15] Data shown in A are from trials when the subject selectively attended to changes in vibration frequency to detect whether the first or second sequentially presented stimulus was a higher frequency. Data shown in B are from another subject attending during different trials either selectively to frequency or to the duration of paired vibratory stimuli, or who divided attention between both frequency and duration features (see Reference 215). MR signals were collected and analyzed per trial (TR 3.68 seconds, 3.75mm voxel size, ~16 seconds per trial). **A.** Functional MR data superimposed on a three-dimensional, partially inflated view of left hemisphere for one subject. **B.** Slices at three orientations illustrate the location of MR signals in anterior and posterior portions of SI in another subject. Pink lines on each slice show registration of slices at orthogonal orientations. Coordinate values are referenced to the Talairach atlas.[222] Anatomical abbreviations (shown in red): CS, central sulcus; LS, lateral sulcus; PoCS, postcentral sulcus; SFS, superior frontal sulcus; functional abbreviations (shown in yellow): aSI, anterior primary somatosensory area; pSI, posterior primary somatosensory area; SII, secondary somatosensory area; preMC, premotor cortex; iSG, inferior supramarginal region; sSG, superior supramarginal region.

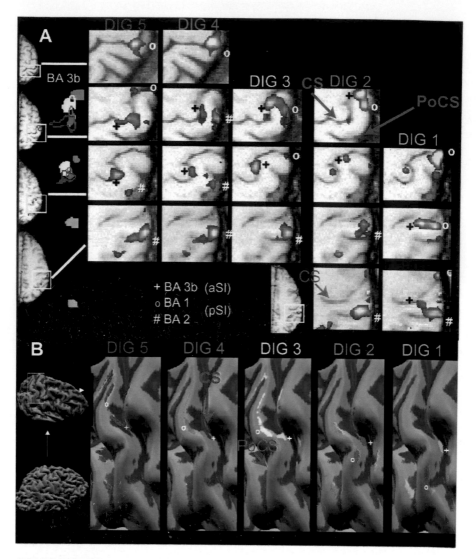

COLOR FIGURE 2.2 Activation sites in the left postcentral gyrus after electrical stimulation of single fingers (Digits 1 to 5) on the right hand. Results obtained from a correlation analysis in one subject. **A.** Distribution of activated pixels observed across a superior (top row) to inferior (bottom row) series of horizontal slices. Each column of magnified images shows the distribution of active foci noted from stimulating one of the digits. Symbols: +, o, #, demarcate, respectively, foci attributed to Brodmann areas 3b, 1, and 2. BA 3b occupies the posterior bank of the central sulcus (aSI in Figure 2.1) and BA 1 and 2 are, respectively, on the exposed postcentral gyral crown and adjoining anterior bank of the postcentral sulcus (combined as pSI in Figure 2.1). The column labeled BA 3b shows the outlines of foci activated at the same horizontal level (row of images to the right) from each of the digits. The colors for these outlines match the corresponding digit labels and foci shown in B. **B.** Expanded surface view of the postcentral gyrus presents distribution of active foci for each digit in one subject. [A is modified from Figure 1 and B from Figure 2 in Reference 131.]

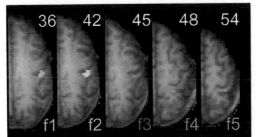

COLOR FIGURE 2.3 Group averaged Statistical Parameter Maps show maximum activated focus for each finger on a superior to inferior series of axial slices. Numbers are Z-axis coordinates in reference to Talairach atlas.[222] Stimulation of individual fingers obtained using piezoelectric vibration. [Modified from Reference 144.]

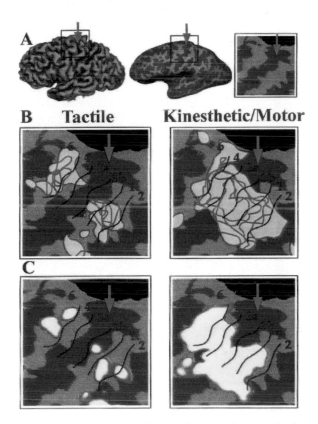

COLOR FIGURE 2.4 Distribution of MR signals across the central sulcus, hand area during tactile stimulation or a kinesthetic/motor protocol. Results are averaged maps from five subjects. **A.** *Left:* Three-dimensional reconstruction of the cortex. *Middle:* An inflated reconstruction of the same brain in which dark gray shows sulci and light gray illustrates gyri. Region enclosed in black box is over the hand area, which is shown enlarged on the *right* and below in B and C. Green arrow points to the fundal region of the central sulcus in all images. **B.** Distribution of activity from tactile stimulation (on left) or kinesthetic/motor protocol (on the right) from each of five subjects (red lines with yellow fill, t-test analysis) is projected onto the flattened reconstruction from one subject after transforming all distributions to a canonical representation. Solid black lines demarcate the presumed cytoarchitectonic Brodmann area borders, which are based on sulcal/gyral gross anatomy. **C.** Across subject average distribution of activity for the two tasks: tactile on the left and kinesthetic/motor on the right. [Modified from Figure 4 in Reference 164.]

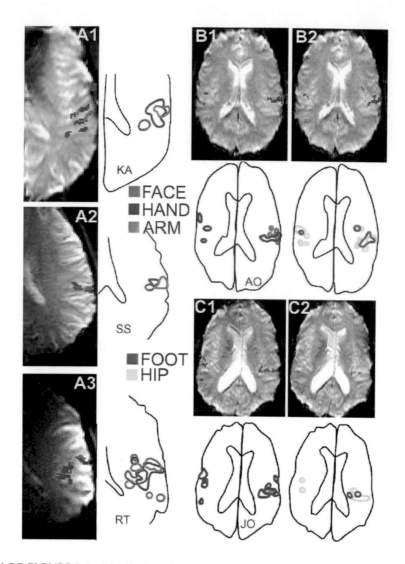

COLOR FIGURE 2.6 Distribution of activated foci in the parietal operculum from tactile stimulation of a variety of body regions in individual subjects. Results obtained using a cross-correlation analysis. **A1-3.** Axial slices from three subjects depict foci from stimulating the foot, hand, and face. Outline drawings to the right of the MR images summarize the distribution of activity for each stimulated body region. These cases illustrate a medial to lateral organization for, respectively, the foot to face. In addition, the foot region occupies a more posterior position in two subjects. **B1, C1.** The location of activated foci from stimulating the upper arm-shoulder with respect to the hand in two subjects. In one subject (AO), the representation for the more proximal body region possibly surrounds that for the foci activated from the more distal hand. **B2, C2.** The location of activated foci from stimulating the upper leg-thigh with respect to the foot in two subjects. In both subjects, the representation for the more proximal body region appears to surround that for the foci activated from the more distal foot. [A is modified from Figure 7, and B and C are modified from Figure 8 in Reference 54.]

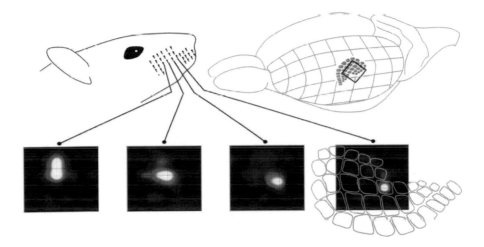

COLOR FIGURE 7.3 Topographically distributed "response maps" for four whiskers of the rat's snout (C_{1-4}) as revealed with large 10×10 microelectrode arrays inserted in somatosensory cortex. The four stimulus sites are shown on the drawing of the rat. To the right, the recording site is shown relative to the barrel field of the left hemisphere. The boxed area indicates the boundaries of the electrode array. At the bottom, spatial distributions of the four whiskers' representations are illustrated in relation to the barrel arrangement (right side). The center of each activated zone, where spike counts were highest, is given by the warmest color.

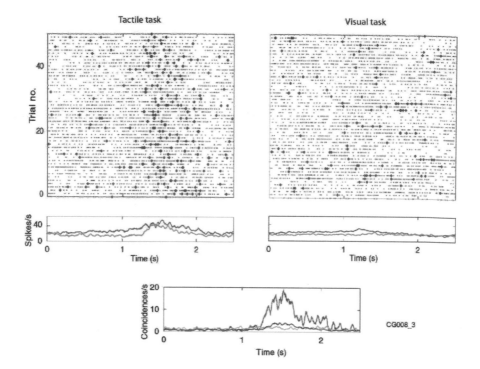

COLOR FIGURE 8.6 Effects of attention on synchrony. Response of a typical neuron pair (red and green) in monkey cortical area SII. The response rasters are triggered at the onsets of 50 tactile stimulus periods while the monkey performs the tactile letter discrimination task and the visual dim detection task. Each row in the top two rasters represents one stimulus period, 2.5 s long, corresponding to the presentation of one letter. Red and green dots represent the action potentials of the two neurons. Peristimulus time histograms are shown below each raster plot with corresponding colors. Synchronous events, defined as spikes from each neuron within 2.5 ms of each other, are represented as blue diamonds. This figure shows that the number of synchronous events is much higher when attention is directed towards the tactile stimuli. The number of coincident events is shown in the bottom figure — blue curve — tactile task, red curve — visual task, violet curve — coincidences expected by chance. [Adapted from Reference 46 with permission.]

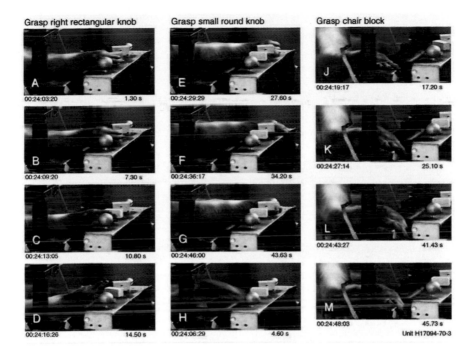

COLOR FIGURE 9.1 Digitized Hi-8 images of the hand kinematics during grasp of various objects. These pictures were cropped from full-frame views captured at the peak of bursts flagged in Figure 9.3; time code (left) and time in clip (right) are indicated below each image. Note that in D, the neuron failed to respond because the animal did not properly grasp the rectangular knob, and instead lifted it with the fingers extended.

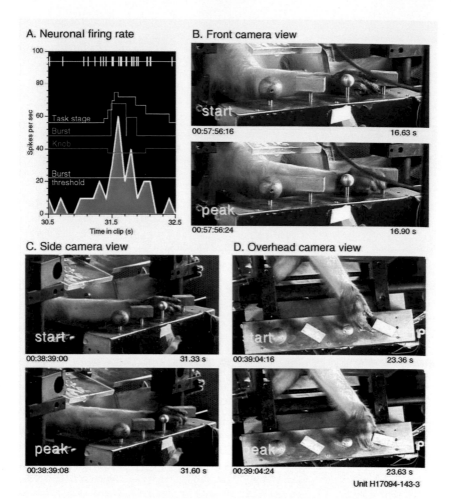

COLOR FIGURE 9.2 Burst analysis is used for objective evaluation of behaviors recorded with digital video. **A.** Spike train (top trace) and instantaneous firing rate (bottom, blue-and-white filled graph) during 2 s of continuous recording excerpted from a 2.5 minute duration video clip. **White threshold trace:** Burst threshold set 1 SD above the mean rate for the entire 2.5 min clip. **Green burst trace:** mean firing rate during the period of suprathreshold firing; bursts were offset by 48 spikes/s to aid visualization of the raw data. **Yellow task stage trace:** Upward deflections indicate the start of stages 1–4 (approach, contact, grasp, and lift); downward deflections mark the onset of stages 5–8 (hold, lower, relax, and release). **Red knob trace:** downward pulses span the contact through relax stages when the hand interacts with the object; the amplitude is proportional to the object's distance from the medial edge of the shape box. **B-D.** Video frames captured by three DV camcorders at the start of the burst, and at the moment of peak firing; images were cropped to highlight actions of the hand. Data from the three recorders were synchronized by matching the firing rates recorded on their audio channels. Spike trains in A were processed from recordings by the side camera. B and D show images captured with Sony DCR-TRV900 camcorders; images in C were cropped from a wider angle view recorded with the Canon XL1 instrument. Note the improved clarity and detail in all six panels compared to the Hi-8 images in Figure 9.1.

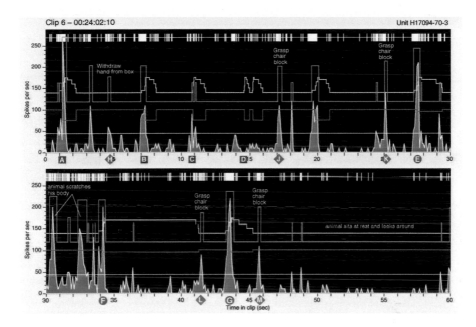

COLOR FIGURE 9.3   Burst analysis for the first minute of clip 6 from a neuron recorded in area 5 of PPC; same format as Figure 9.2A. Firing rates in the burst trace were offset by 120 spikes/s to aid visualization of the raw data. Symbols below the abscissa refer to the matching images in Figure 9.1 captured at the peak of these bursts. Bursts are correlated with grasping movements when the fingers flexed. **A-D.** Prehension of the right rectangular knob; note the difference in grip styles used in A-C, which evoked vigorous responses, and in D when no burst was detected. **E-G.** Prehension of the small round knob. H. Withdrawal of the hand from the box evoked a weak burst. **J-M.** Spontaneous grasp of a structural block on the chair during rest intervals evoked smaller amplitude, shorter duration bursts than grasp of the knobs tested in the task.

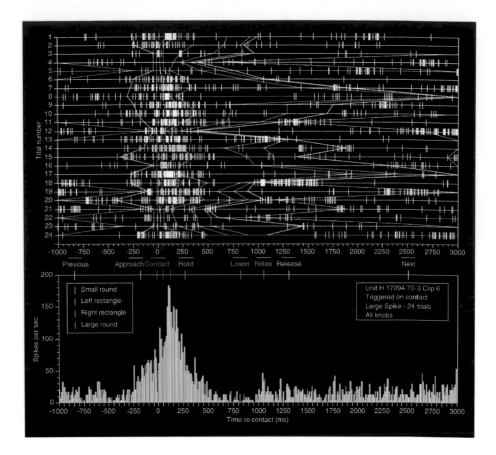

COLOR FIGURE 9.4 Rasters and PSTHs aligned to contact for an area 5 neuron. Trials 1–9 occurred during the first minute of the clip, and are shown in Figures 9.1 and 9.3; the remainder occurred in the subsequent 1.5 min. Spikes have been color-coded by the knob tested (see key); responses were similar for the four knobs tested. Color-coded lines spanning the raster mark the onset times of the task stages on each trial; the unlabelled magenta and dark blue lines indicate the grasp and lift stages, respectively. The PSTH averaged responses from all of these trials (binwidth = 10 ms); colored vertical bars mark the mean onset times of the task stages relative to contact in this clip.

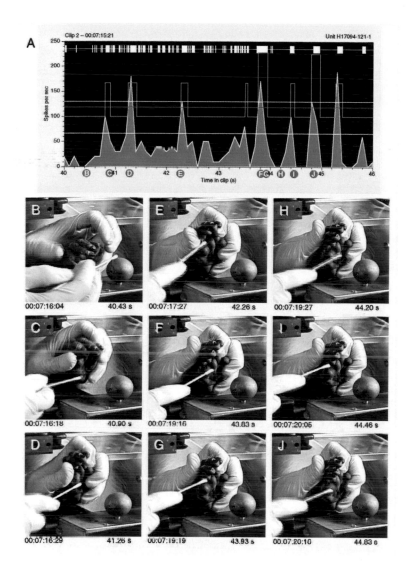

COLOR FIGURE 9.5 Mapping receptive fields on the palm with a cotton swab for a neuron recorded in anterior SI cortex. **A.** Burst analysis of firing patterns during a 6 s interval; mean firing rates during bursts are offset by 100 spikes/s. **B-J.** Cropped images of the hand coincident with the markers on the abscissa in A. The strongest responses occurred to touch applied to the interdigital pad below digit 5. DV images captured with Sony DCR-TRV900 camcorder.

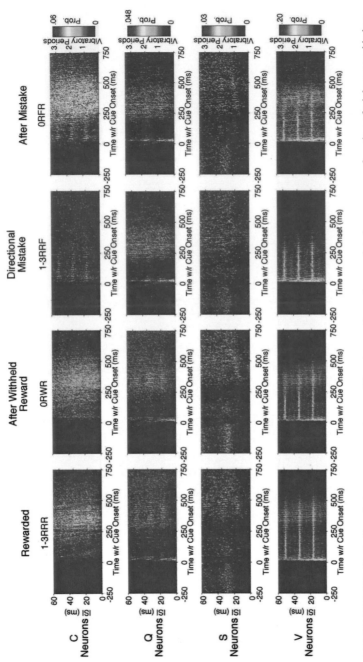

COLOR FIGURE 10.7 Population response patterns of four types of area 1 cue-related neurons as a function of trial type and behavioral time. Scatterplots constructed by binning all ISIs (5 ms × 1 ms) and normalizing by spikes/trial to give probability of spike occurrences in a bin. Probability indicated by attributed color. Tick marks to the right indicate time corresponding to the period of the 57 Hz vibratory stimulus. Response types as in Figure 10.6. Nomenclature for trial types as in text.

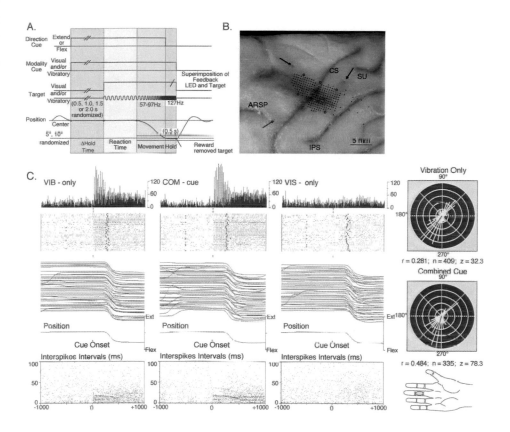

COLOR FIGURE 10.8 **A.** Schematic of the Guidance Task with the three variations tested. Animals held a steady position for a random time, detected the target, and then made wrist flexion or extension movements and held at the target zone. Targets in the visual-only (VIS-only) were LEDs signaling the position to which the animal moved. Approach to the target in the vibratory only (VIB-only) was signaled by increasing the vibratory frequency as the animal neared the target and decreasing frequency if he moved away from it. At the target, the vibratory frequency abruptly increased by 30 Hz. The third condition combined both visual and vibratory targets (COM-cue). All correct trials were rewarded. **B.** Dorsolateral view of the cortical surface of the most extensively studied monkey showing the locations of penetrations in the pre- and post-central cortices. Anterior (left); Medial (up). ARSP = Arcuate Spur; IPS = Intraparietal Sulcus; CS = Central Sulcus; SU = Superior Parietal Dimple. Open circles depict locations of electrolytic lesions. Arrows show postmortem pin marks placed relative to recording chamber coordinates. **C.** (below). A vibration-responsive neuron from primary somatosensory (SI) cortex with an RF on the third digit. The cycle distribution of the vibratory response became more bimodal in vibration-only trials and the mean vector length (r) was shorter. Increased bimodality or cycle distribution broadening occurred for most SI vibration-responsive neurons.

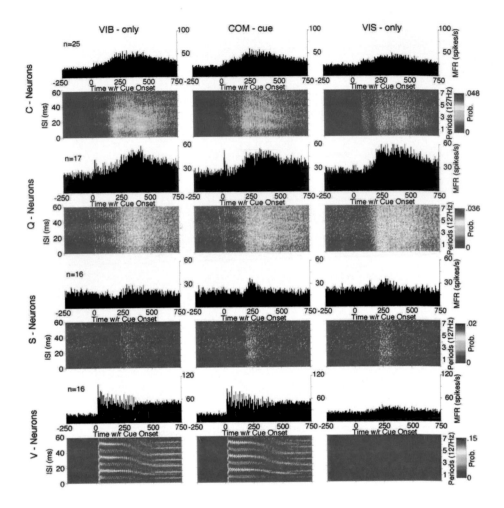

COLOR FIGURE 10.9 Response patterns of four populations of postcentral cortical cue-related neurons as a function of trial type and behavioral time. Scatterplots as described in Figure 10.7. Response types as in Figure 10.6. Initial vibratory cue frequency was 57 Hz, but increased linearly as animals moved the handle toward a target 10° from center. When the target was reached, the final stimulation frequency was 127 Hz. Thus, periods shown at the right of each scatterplot are for that frequency. Entrainment is evident in V neurons; weaker entrainment appears present in Q and C neurons.

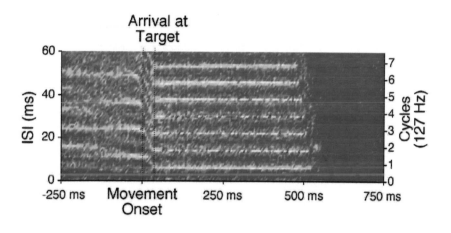

COLOR FIGURE 10.10 Scatterplot of the lower left panel of Figure 10.9 with data centered on wrist movement onset. When the target was reached, the final stimulation frequency was 127 Hz. Periods shown at the right of the scatterplot are for that frequency, which was presented while the animal held the handle 10° from the neutral position. Entrainment of this population of neurons is clearly seen for both the vibratory stimulus present before the animal moved and once he attained the target zone.

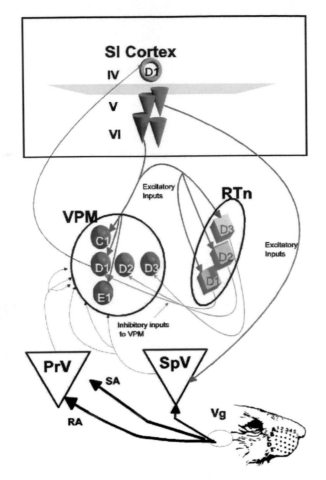

COLOR FIGURE 12.1 Schematic diagram of the rat trigeminal somatosensory system. Whiskers on the rat's snout are labeled according to the row and column in which they are located. Whisker columns are labeled from 1 to 5, caudal to rostral, while whisker rows are labeled A to E, dorsal to ventral. Peripheral nerve fibers innervating single whisker follicles have their cell bodies located in the trigeminal ganglion (Vg). Here, only the projections from Vg neurons to two main subdivisions of the trigeminal brainstem complex, the principal trigeminal nucleus (PrV) and the spinal trigeminal nucleus (SpV), are illustrated. Proponents of the feed-forward model of touch usually divide these projections into rapidly adapting (RA) and slowly adapting (SA) fibers, according to their physiological responses to tactile stimuli (see text). Each of these categories contains further subdivisions, which are not described here. Neurons located in these two brainstem nuclei give rise to parallel excitatory projections to the ventroposterior medial nucleus (VPM) of the thalamus. Neurons in VPM give rise to projections to layer IV of the primary somatosensory cortex (SI). A collateral of these thalamocortical projections reach the reticular nucleus (RT), whose neurons provide the main source of GABAergic inhibition to the VPM. Descending excitatory corticothalamic projections, originating in layer VI of the SI cortex, reach the VPM and the reticular nucleus of the thalamus (RTn). The assumed topographic arrangement of these projections in the VPM and the RT are illustrated in the scheme. Feedback corticofugal projections originated in layer V of the SI cortex also reach the trigeminal brainstem complex, targeting primarily the SpV subdivision.

lateral extensions of SI face areas (e.g., References 113 and 146). Currently, no brain imaging or single neuron recording data incorporates a somatosensory stimulation/behavioral paradigm that controls for the possible diversity of motor, cognitive, or sensory responses, and that consistently activates the insula or frontal operculum.

2.4.2 POSTERIOR PARIETAL CORTEX

2.4.2.1 Homologous Regions in Monkeys and Humans

Brain imaging evidence of additional activations related to somatosensory stimuli is present within posterior aspects of the postcentral sulcus, adjacent anterior portions of the superior parietal lobule, and superior aspects of the supramarginal gyrus (Figure 2.1, sSG). Interpreting the basis of these responses in posterior parietal cortex is difficult for two reasons. First, there is extreme variability in the gross anatomy of the cortex around the postcentral sulcus. Second, the expansion of the cortex associated with the postcentral sulcus creates a very different pattern from that in macaque brains because it displaces the intraparietal sulcus posterior into occipital cortex. The intraparietal sulcus in monkeys has greater continuity, extends to the posterior border of SI, and bisects posterior parietal cortex into superior and inferior lobules. These gross anatomical species distinctions make comparisons with results from monkeys ambiguous. Further, the Brodmann cytoarchitectonic definition of this region in primates adds uncertainty.[11] In monkeys, Brodmann assigned area 5 to all of the superior parietal lobule component and area 7 to inferior parietal cortex lateral to the intraparietal sulcus.[11] In humans, both areas 5 and 7 are assigned to the more medial cortex. In human atlases, Damasio reproduces Brodmann; however, the boundaries illustrated on various sections in the Talairach atlas denote a much broader definition of area 7 that includes superior and inferior parietal lobules.[49,222] A somewhat more consistent scheme by von Economo[58] labeled as PE, the posterior aspects of the postcentral sulcus and anterior portions of the superior parietal lobule. He subdivided the inferior parietal lobule (supramarginal gyrus) into areas PF and PG. Brodmann included all of the latter region in his area 40. Subsequent analyses suggest that the lateral-inferior portion of the supramarginal gyrus (Figure 2.1, iSG?) incorporates a subcomponent of cytoarchitectonic area PF[59] or area 7b.

It is doubtful that foci tied to somatosensory stimuli in superior and inferior parietal lobules are part of one functional entity. Functional heterogeneity is especially likely if the portion of the human inferior parietal lobule is homologous to cortex in monkeys that stretches from the fundus of the monkey intraparietal sulcus to the Sylvian fissure. This region of cortex in monkeys is now multiply subdivided according to unique anatomy and connections,[136,137] and with each part devoted to various aspects of visual-spatial functions.[5,42] Area 7b (PF), for example, has known connections with somatosensory, visual, and ventral premotor areas (e.g., References 4, 27, 149, 171, and 217), and responds to somatosensory and/or visual stimulation applied to relatively large, often inconclusively defined, receptive fields.[41,103,202] It is therefore possible that brain imaging evidence of activity in comparable cortex will reflect body spatial factors operating during active manipulation of objects, processing the three-dimensional properties of touched objects in tactile discrimination tasks and tactile attention (see below).

2.4.2.2 Identification of Active Foci

Different imaging studies report a varying extent of somatosensory-related activity throughout posterior parietal cortex. For example, Hodge and colleagues note activity from the fundus of the Sylvian fissure, lying posterior to the insula, and extending up to the posterior bank of the postcentral sulcus. They call this their retroinsular/parietal region.[98] Several investigations involving subjects engaged in tactile discrimination tasks describe activations that mostly occupy the posterior bank of the postcentral sulcus and adjacent surface cortex that lies lateral to, and overlaps the anterior tip of the intraparietal sulcus.[14,30,50,87,175,208] Similar coordinates and published axial scans identify a focus in the postcentral sulcus in an fMRI study including somatosensory exploration of complex objects.[11] This study labeled the active focus as an anterior intraparietal region. In Figure 2.1, the general location of these foci possibly includes the site labeled sSG. Roland and colleagues suggest that the role of this region is haptic processing of shapes,[87] which is consistent with results from other studies concerning active manipulation of objects.[11,50] However, this idea does not account for findings of corresponding activation patterns when non-shape tactile stimulation is applied passively.[30,32] A more general notion is that this higher level somatosensory region becomes active whenever somatosensory information must be attended and processed for object recognitions.

Activations in lateral/inferior supramarginal gyrus are posterior and superficial to the parietal operculum, and in probable parietal association cortex.[30,54,98] This region, labeled iSG in Figure 2.1, is adjacent to the posterior ascending ramus of the Sylvian fissure and is possibly related to area 7b.[59] The activity within this area has a separate focus from that in SII and is best noted when subjects receive discrete vibrotactile stimulation to a single fingertip, especially in the context of a challenging tactile discrimination task (Figure 2.1). A similar location and distinction from SII is identified in a re-evaluation of the stroke in a patient with tactile agnosia (see below).[197,198]

2.5 BRAIN REGIONS ACTIVE DURING TACTILE DISCRIMINATION TASKS

A notion of dedicated foci for different somatosensory features gains some support from prior claims that patients with lesions affecting the parietal operculum are unable to discriminate surface features, such as texture, while those with lesions of the postcentral gyrus, postcentral sulcus, and adjoining supramarginal gyrus cannot distinguish object shapes or sizes.[206] Implied is that SII is important for detecting roughness while posterior portions of SI and the nearby somatosensory association regions in the supramarginal gyrus are needed for detecting spatial features of objects. Lesion-behavioral studies in monkeys concur only partially. In agreement is the finding that lesions in monkeys involving posterior parts of the postcentral gyrus and adjoining medial bank of the intraparietal sulcus raise the threshold for distinguishing spatial features like object size and shape.[33] Interpreting backwards from recent studies, these lesions probably included the monkey equivalent of area 2 in posterior SI and parts of parietal somatosensory association cortex.[135] These cortical areas contain neurons especially responsive to the shape and curvature of touched objects.[105,107] These posterior parietal cortex neurons also respond selectively to hand positions, which possibly reflect selected responses to proprioceptive information from holding the fingers and hand around three-dimensional objects.[81] In disagreement with the thesis presented by Roland,[206] however, is that lesions in monkeys that affect anterior parts of SI, presumably confined to area 3b, disrupt detection thresholds for all types of tactile features including object sizes and shapes. In further contradistinction of the human clinical data which suggests a restricted role for SII in roughness detections,[206] lesions in monkeys that targeted area 1 in the middle of the postcentral gyrus affect mostly roughness detections,[33] and lesions of the parietal operculum in monkeys globally disrupts tactile discriminations for roughness, size, and shape.[34,79,169]

Results from several PET studies are consistent with conclusions from clinical studies in showing separate foci for selective processing of surface (e.g., roughness) as opposed to spatial (e.g., object length, curvature, or shape) features of touched objects.[14,133,175,208] In one study, subjects actively touched two successive plastic cylinders that differed in surface roughness or cylinder height. During different scans, subjects determined whether the second object was rougher or longer. In the control task, subjects executed similar movements through empty space while maintaining a pincher position with the same fingers. Differential activations spread across overlapping portions of anterior parietal cortex during both discrimination tasks contrasted to control. This coincides with the monkey results in which lesions to anterior SI[33] alter a common focus for processing surface and spatial features. However, in the PET images more extensive activity appears further posterior during the length task, which indicates possible distinctive processing of spatial features within the depths of the postcentral sulcus and adjoining supramarginal gyrus. Significant activations also appear posterior to posterior SI when comparing images based on discriminating differences in the curvature of objects passively placed in the hand[14] or differences in shapes vs. length or shapes vs. roughness of actively touched objects.[208] The same active touching paradigm results in greater activity in the parietal operculum during roughness discriminations.[133,208]

Fundamental to evaluating these results is recognizing that subjects use different exploratory procedures when distinguishing surface vs. spatial features of objects.[134,196] For example, one of the PET studies that reports task-specific foci, also notes that in the length discrimination task, relatively more time was spent exploring and defining the top edge, before the fingers were moved down the length of the stimulus (page 141 in Reference 175). Even without directly comparing sensory/perceptual distinctions between the length and roughness tasks, subjects positioned their fingers differently and attended to dissimilar physical properties of the cylinders. Thus, differences in haptic components of the two tasks confound acceptance of separate foci for the somatosensory properties of roughness and object length. Other PET studies that used passively applied tactile stimuli report fewer differences in the activation of SI and SII. One study reports no evidence of differences in activated SI and SII foci during roughness and length discrimination tasks when stimulation is always applied passively and identically.[30] Similarly, in another study, both the area 1 portion of SI and the parietal operculum respond during a discrimination task involving detection of differences in the velocity of a rotating brush across the volar surfaces of fingers 2 to 4.[13]

Undisputed is the observation that discriminating spatial features of touched objects leads to activity in a likely superior somatosensory parietal association region (Figure 2.1, sSG), which occupies the posterior wall of the postcentral sulcus and adjoining supramarginal gyrus.[14,208] This sSG region is posterior to area 2 and is, therefore, outside any traditional definitions of SI. In an fMRI study, an apparently similar region shows significant activity when subjects must recognize familiar three-dimensional objects through touch.[50] In one patient, an isolated infarct in more lateral aspects of possibly similar somatosensory association cortex in inferior supramarginal gyrus (Figure 2.1, iSG) produces true tactile agnosia, without affecting simple detections of touch.[197,198] Previously, this ventrolateral infarct was mistakenly described as an SII lesion,[37] which led to proposing SII as an area uniquely associated with tactile object recognition. This claim is incompatible with findings in MEG studies of ischemic stroke patients with reduced tactile abilities and dysfunctional SEFs in both SI and SII.[70] Collectively, these observations imply that somatosensory association areas within the supramarginal gyrus are specific for shape processing.[198] An alternative view is needed, however, because of the finding that presumably the same regions respond when subjects perform a vibrotactile discrimination task (Figure 2.1A). Thus, these parietal somatosensory regions potentially have a more general function for many kinds of tactile recognition processes. In all circumstances, responses appear where the recognition process necessitates comparing an internal representation (short or long-term memory?) of tactile properties to ones currently experienced. Thus, the supramarginal gyrus, somatosensory association region is possibly needed for comparing tactile properties of sequentially processed objects.

2.6 TACTILE ATTENTION

2.6.1 A CORTICAL NETWORK

In tactile and other forms of attention, targeted stimuli are processed selectively despite concurrent distracting stimuli competing for neural resources. Target stimuli may have specific attributes, such as spatial location on the body surface. Attention entails advantageous processing of targeted stimuli. As noted below, one underlying mechanism appears to be enhanced activity in somatosensory cortical areas during attention directed behavior. Absence of such heightened responses, possibly because strokes destroy the source for these enhancements in patients, often expresses as an inability to attend to stimulation presented simultaneously in different locations. A consequence for patients with tactile attention deficits is failure to acknowledge tactile stimulation in contralateral space (contra-lesion) when these occur simultaneously with stimulation in ipsilateral space (ipsi-lesion). Such symptoms mainly accompany right hemisphere infarcts in posterior parietal or pre-motor/pre-central regions.[225] These patients do not suffer sensory impairments as they usually detect isolated touch on contra-lesion parts of the body. Severely affected patients will ignore all somatosensory information arising from the contra-lesion half of the body irrespective of whether tactile stimulation simultaneously occurs on the ipsi-lesion side.[1,36,48,51,157,166] Some of the critical regions are not just sensory areas, are also activated during visual attention tasks, and may be involved in various motor behaviors. These regions possibly furnish internal representations of space that are the same spatial referents also needed when moving parts of the body to selected targets.[32,38]

2.6.1.1 Attention Effects in SI

Attention tasks alter activity in SI and SII. However, results from different studies show varying magnitudes of changes in these two regions (see below). The disparities largely rest on the question of attention effects in SI. Most neurophysiology studies in monkeys report fewer cells influenced by attention in SI.[31,100,102] A corresponding distinction exists for the visual system for which only some recent studies confirm attention effects in V1.[78,104,190,203,223] Underlying studies of visual attention is the hypothesis that processing through the attention network is hierarchical with the least attention-related activity in primary sensory cortex.[78,104,190,203,223] A similar conception of a hierarchical somatosensory attention network may be applicable.[29-32] Numerous brain imaging studies show attention related modulation of SI. For example, tasks requiring discriminating differences between touched objects demand attention and nearly always evoke increases relative to no stimulation/rest conditions in regional cerebral blood flows in peri-Rolandic cortex that is contralateral to the stimulated hand.[204] Several studies report blood flow increases more confined to just the contralateral postcentral gyrus during trials when subjects attend to vibratory stimuli compared to performing alternative visual tasks in the presence of the same tactile stimulation.[158,176] Relative increases in rCBF also occur bilaterally in SI when subjects selectively attend to different tactile stimulus attributes as compared to trials with the same stimuli, but a distracting counting task.[30] Blood flow changes also arise in the ipsilateral SI in the vicinity of the postcentral sulcus, but more prominently when ipsilateral is the right hemisphere. In fMRI studies, MR signal intensity increases on trials cued to detecting touches of the big toe compared to trials with attention to a visual display,[111] or the spatial extent of activations increases when contrasting responses during trials with a tactile vs. a distracting counting task.[88] Thus, activity in SI increases for attended tactile information. Unfortunately, there are currently no distinctions regarding attention effects in anterior vs. posterior parts of SI.

2.6.1.2 Attention Effects in SII

Brain imaging studies also document a role for SII in tactile attention. Examples from MEG recordings include greater magnitude SEF responses at 55 to 130 ms from SII when using an oddball paradigm that requires vigilance to detect variations in different stimulus attributes: spatial location,[94,151] stimulus intensity,[160] or both location and intensity.[99] Neither short (~20 ms) nor long (>35 ms) latency responses from SI show enhancements in the same tasks.[99] In contrast, several PET studies report rCBF increases in SI, but greater shifts in SII. This includes finding rCBF increases when subjects selectively attend to the roughness or length attributes of gratings passively rubbed against the skin[30] or when attending to roughness or the length of manipulated objects.[133] The distinctions regarding relative modulation in SI compared to SII also entails significantly larger spatial extent of MR signal distributions in SII (p<0.01) compared to a non-significant increase in SI when attending to tactile stimulation.[88] In further contradistinction with the MEG results, one fMRI study shows nearly equivalent increases in mean MR signal intensities in SI and SII when comparing responses during attention to tactile vs. visual stimuli.[111] Collectively, these studies show enhanced responses in SI and SII irrespective of the attended attribute (e.g., location, intensity, roughness, or object length).

2.6.1.3 Attention Effects in Posterior Parietal Cortex

Posterior parietal cortex displays substantial increases in rCBF in a variety of attention tasks involving tactile stimuli.[30,110,176,205] Confusing, however, is the variable definition of the affected region. Early studies indicate widespread activations that include portions of superior and inferior parietal lobules.[176,205] Subsequent descriptions primarily identify more focal activity. In a PET study involving passive stimulation with horizontal surface gratings, stimulation of digits two and three on the right hand, and selective cuing the attributes of surface roughness or length, we report an active focus in the depth and posterior wall of the postcentral sulcus.[30] The same region shows increases in MR signals during attention trials involving vibrotactile stimulation of a single fingertip (Figure 2.1, sSG). These fMRI images show activity associated with the posterior wall of the postcentral sulcus and extension of the region with increased MR signals to the superior portion of the supramarginal gyrus, just anterior and lateral to the intraparietal sulcus. Current results do not satisfactorily allow for the identification of two separate foci in this region: one within the postcentral sulcus and another on the exposed supramarginal gyrus. From published coordinates, a comparable superior supramarginal site is likely when contrasting attention to vibrotactile stimuli in an earlier PET study.[110] The critical somatosensory superior supramarginal region is anterior to much of the intraparietal sulcus and lateral to sites active during visual spatial attention tasks.[45-47,173] In reference to the Brodmann area designations in the Tailarach atlas,[222] the more anterior location appears to occupy cytoarchitectonic area 7, while the more posterior and inferior extensions of active cortex fall within a superior portion of area 40. This contrasts with visual attention foci within parietal-occipital/angular gyrus locations, some of which are within BA 39. However, one prior study identifies a focal increase in blood flow that just misses a .05 significance threshold in the superior part of the angular gyrus when contrasting attention or no attention to a vibrotactile stimulus.[110]

2.6.1.4 Attention Effects in Frontal Cortex

Two sites within frontal cortex routinely show activity changes during a variety of tasks including those involving attention. One is in area 6 of premotor cortex and mostly occupies the depths of medial precentral sulcus (Figure 2.1, preMC). A second is in the anterior cingulate cortex. Both regions show increased rCBF when subjects attend to selected attributes of tactile[30] or visual[44-46] stimuli. Motor tasks engage corresponding parts of premotor cortex, which as noted below, has prompted suggestions about common mechanisms between attention and intention to move.

2.6.2   Mechanisms for Tactile Attention Revealed in Brain Imaging

2.6.2.1   Response Amplification

Models of attention often incorporate the notion of amplified neural responses for selected, relevant stimuli, which improves probable preferential detection of desired target events.[52,187,188] In single neuron recording experiments, the most common expression of gain is enhanced firing rates to task relevant stimuli (reviewed in Reference 32). A potential expression of this process in brain imaging findings is the many examples of magnified responses in somatosensory cortical areas, and especially in SII, whenever attention focuses on some body part. Such amplifications might reflect a priming mechanism for any input applied to the receptive field of a neuron[29,99] or, more generally, to the collective responses of neurons representing an attended body region.[80]

2.6.2.2   Enlarged Pools of Active Neurons

Attention is also reported to increase synchronous firing patterns across populations of SII neurons in monkeys.[219] A potential effect of such synchronicity is expanding the pool of active neurons through improved synaptic efficacy that is a consequence of temporal summation. Thus, there would be further propagation of neural signals about stimulation at an attended location. This notion is consistent with reported increases in spatial extent of activated regions in brain imaging experiments.[88] The postulate is that attention amplifies signals through recruiting responses from more neurons. Attention thereby effects an expanded employment of the cortical network engaged by stimulation of a restricted part of the sensory surface. In the digit representation of SI, synchronization of responses might presumably enlarge the activated zones for the attended finger, and might potentially blur any underlying somatotopic organization. Just this effect is discussed in an MEG study in which the dipole distance between digits 1 and 5 increases when attention shifts to intervening, temporarily deafferented digits 2 to 4.[18] The precise source localization for a particular digit also changes depending on whether attention is directed to another digit on the same or opposite hand.[174] The problem of consistent somatotopic maps in most brain imaging studies might, therefore, be due to only minimal control of attentional variables and an absence of challenging and balanced somatosensory discrimination tasks when stimulating different fingertips.

2.6.2.3   Suppression of Non-essential Cortical Areas

Somewhat unique to brain imaging studies is the revelation of potential global shifts in relative activity depending on the focus of attention. For example, attending to tasks that involve non-tactile modalities downwardly modulates blood flows in SI.[65,97,124,142] A similar decrease in blood flow arises in the somatosensory cortical foci for one body representation whenever attention is directed to a distant, different part of the body.[56] Evidence from fMRI studies in the visual system also shows that attention directed to one stimulus counteracts competitive suppressions from multiple visual stimuli in nearby visual space.[123] These observations suggest that signals processed through an area of cortex utilized during attention are enhanced relative to depressed activity in non-essential regions, whether these are in other sensory cortical areas or the topographical representations for non-attended receptive fields. These suppressions might abet minimization of irrelevant information.

2.6.3 TACTILE ATTENTION AND INTENTION TO MOVE

The purpose of attention is in providing competitive capture of some extra-personal event in preparation and execution for some action.[3] This action might be to or projected from the spatial position of the detected event. In the visual system, for example, the location of an attended visual stimulus might coincide with preparing and executing eye movements to the same or a derivative location. In the somatosensory system, attention to stimulation of some body part might precede some intended movement with or to the coordinate location of that body part. Underlying both attention and intention to action is a mechanism dependent on cortical areas with corresponding responsibilities for coordinate space.[32,38,45,46] Brain imaging studies have uniquely contributed to this idea. First, there is evidence that similar cortical regions respond during attention or movement tasks. Second, there is evidence that related (possibly overlapping?) regions react during visual or tactile tasks that involve attention or movement.[32,38,45,46] This correspondence between visual and tactile coordinate space is reflected in recordings from single parietal cortex neurons that code hand/arm movements to selected visual and body centered spatial coordinates.[5,7]

An example of the likely similarity of foci active during attention and purposeful movements in the somatosensory system are parietal and frontal cortical areas affected during attention and tactile recognition tasks that involve object manipulation. Thus, MR signals increase near the postcentral sulcus and adjoining supramarginal gyrus or SII when subjects haptically explore an object.[11,12] Corresponding foci of rCBF increases are identified during attention to passively applied tactile stimuli to the fingers.[30,32] These regions are lateral to those associated with visually guided arm movements, described as activating cortex within the medial bank of the intraparietal sulcus.[125] Foci of rCBF increases in frontal pre-motor areas, especially within the depths of the pre-central sulcus and close to the posterior tip of the superior frontal sulcus (Figure 2.1, preMC), are present when subjects perform selective tactile attention tasks.[30] Nearby is evidence of MR signal increases during object manipulations.[11,12,32] Comparable analyses in the visual system show overlapping foci of MR signals in and near the intraparietal sulcus, and in frontal cortex, within the precentral sulcus and nearby middle frontal gyrus during covert visual spatial attention and overt eye movements.[38,45-47,182] Thus, tactile and visual attention activate corresponding parietal-frontal networks. The specific foci are not identical for these two modalities. In parietal cortex, those involved with visual processes are more extensive, reside further posterior within the intraparietal sulcus, and extend into occipital cortex. In frontal cortex, separations are less clear, although eye-movement regions are likely lateral to hand movement foci within the precentral sulcus. The relatively minimal extent of these discrepancies, however, further supports the notion that both spatial attention and motor intention requisition similar kinds of spatial coordinate systems to the external world.[32,38] The accumulated findings support the idea of similar parieto-premotor networks for tactile attention and tactile object recognition through haptic manipulations.[11,12,32]

2.7 PLASTICITY IN SI MAPS

2.7.1 EVIDENCE FROM ANIMALS SHOWS THAT SI MAPS CAN CHANGE

Related to difficulties in demonstrating detailed somatotopic maps in humans are possible effects due to the mechanisms responsible for cortical reorganization in adults after injuries (see reviews in References 53, 64, 120, 154, 155, and 193) or use and experience-dependent manipulations.[39,53,64,122,177,194,195,227,238] Of greatest significance are demonstrations that maps may be dynamic since reorganization can be immediate following local anesthesia[64] or temporally timed synchronization of peripheral inputs.[39,53] The expression of changes at any level will result in altered cortical somatotopy, whether the mechanisms responsible for short-term changes in maps are from alterations in synaptic weights in intrinsic cortical, subcortical, or combined circuits. Changes in maps, receptive fields, response latencies, and degrees of correlated activity all stem from likely alterations in the balance between excitatory and inhibitory network elements possibly operating through Hebbian mechanisms.[53,64,154,193] Thus, contrary to the precise boundaries for restricted body regions reported in studies with anesthetized animals, functional representations may be more general, rather than entirely dedicated to specific parts of the body. The brain imaging data discussed below is especially relevant in showing that cortical representations may expand or contract depending on behavioral circumstances.

2.7.2 EVIDENCE FROM HUMANS SHOWS THAT SI MAPS CAN CHANGE

Several brain imaging studies support the notion of a reorganized SI in adult subjects following peripheral or central injuries, sustained somatosensory stimulation through training, habituation to repetitive, simultaneous stimulation, or local anesthesia protocols. Much of this literature relies on observations from EEG or MEG recordings that are projected to be from the postcentral gyrus. However supportive these results are for findings of plasticity in animals, the data is sometimes contradictory and possibly open to alternative explanations. Brain imaging data by itself also cannot identify the locus responsible for observed changes. Instances of immediate, behaviorally mediated changes are likely due to unmasking existing subthreshold cortical connections through mechanisms, such as disinhibition.[117] However, patients with long-term amputations may show reorganizations based on more extensive anatomic changes, such as the axonal sprouting that has been observed in animals.[67,108,118] Changes in the connections between neighboring somatotopic representations in brainstem structures may be especially important (reviewed in Reference 165).

2.7.2.1   Effects Following Injuries

Amputees often report abnormal, unpleasant perceptual experiences of phantom limb sensations. These phantom sensations frequently arise from stimulation of skin areas whose cortical representations are immediately adjacent to the deafferented region. One hypothesis offered to explain phantom sensations is a remapping of SI where activity in cortical regions, with intact peripheral input, expands and takes over processing within the deafferented regions. These might best be described as somatotopic reorganizations.[165] For example, following amputations of the hand/arm, sequential stimulation of different facial regions elicits an orderly progression of sensations across the phantom (reviewed in Reference 193). Magnetic source imaging shows that these perceptual changes also correlate with extension of dipole source localizations to include the face and cortical region for the deafferented limb upon stimulating the face.[61,62,66,240] Direct evidence of augmented activity from neighboring cortex is presented in a PET study with two patients that experienced a supernumerary limb upon stimulating the dorsal or ventral trunk.[126] In this study, significant rCBF in SI expanded ventrally more than 20 mm into the deafferented hand/arm region from the normally evoked rCBF in the trunk region.[126] Similar effects are seen in patients with amputations restricted to a single finger. In those cases involving amputations of digit two, the dipole source locations from stimulating neighboring digits one and three invade the deafferented region within less than 10 days following surgery.[228,229] Referred phantom sensations also occur in spinal injury patients. In these cases, trigger zones that evoke phantoms may not necessarily be represented in adjacent sites in the SI map. A recent fMRI study describes displaced activation foci separated by more than 1.6 cm when such patients receive stimulation in trigger zones on the arm that also elicit the referred sensations to parts of the trunk represented below the segmental level of the spinal infarct.[165] An exciting aspect of these results is direct confirmation of earlier ideas that perceptual experiences are tied to dedicated parts of the SI map even when activating disparate parts of the map from one stimulation site. This study also proposes a more general hypothesis that SI map representations after longer-term injuries include the possibility of somatotopic reorganizations in subcortical structures.[165]

2.7.2.2   Effects Following Training and Experience

Numerous studies provide examples of a reorganized SI following use and training experience. These findings indicate that plasticity can develop from an existing network without hypothesizing effects induced from injuries. One early MEG study describes the changes following surgical separation of webbed fingers in two patients.[163] The SI hand area changed from a disorganized region before surgical separation of the fingers to a more normal map with distinct source localizations for each digit. A further example of dramatic dependence of the SI organization on training experience was first reported in MEG images from string players.[60] In individuals who learned to play as children, the right SI area devoted to the left fingering digits, 2 to 5, appears larger than in control subjects. The left SI hand area in the musicians had a normal spatial extent. A similar instance of an enlarged representation occurs in experienced readers of Braille, who read several hours a day.[177] Braille readers show a larger distribution of low threshold transcranial magnetic stimulation sites over the motor representation for the reading hand. An experimental confirmation of similar reorganized cortex from long-term experience shows multi-month long persistence of enlarged MR signal distribution in primary motor cortex for the fingers used in a practiced movement sequence.[122] These results suggest that prolonged training leads to an alteration in the proportion of cortex dedicated to processing information from the restricted part of the body engaged in the task.

2.7.2.2.1 Effects Manifested to Task Specific Parameters

Various factors appear responsible for such behaviorally induced changes in somatosensory maps. However, explanations for contrasting results in different studies are confusing because of the diversity of training, stimulation, and testing paradigms. Unpredicted from prior studies showing increases, one MEG study reports that the current dipole strengths decrease to stimulating skin surfaces following 3 to 4 weeks of practicing a same/different tactile discrimination task.[216] Contrary to the view of yet greater cortex for a practiced usage, one explanation for this reduction after intensive experience might be the efficient use of fewer neurons to perform a well-learned task.[216] An alternative, technical basis for these anomalous results is that plasticity effects are explicit to a task, and therefore not expressed when using different somatosensory stimuli during training and post-test, imaging trials, despite touching the same skin surfaces.[140]

In a study showing greater responses, subjects practice for three days and are tested with the same 21Hz vibration against the same digit surface when evaluating for changes in SEPs.[140] In this study, the test stimulus is always applied to digit 3. In two different training sessions, stimulation is simultaneous on digits 2, 3, and 4 or just on digits 2 and 4. Post-training testing shows smaller responses to stimulating the non-trained digit and greater responses when training involved all three digits. The results suggest that simultaneous stimulation serves to expand the cortical territory devoted to the trained areas of skin. Hence, the decreases in responses to testing digit 3 are due to shrinkage of its territory when training involved digits 2 and 4.

The demonstrated effects on map organization in SI may be even more dynamic in especially depending on behavioral context during testing. For example, after 4 weeks of training to discriminate the orientation of tactile stimuli applied synchronously to digits 1 and 5, the Euclidian distance between the source locations for these two digits expands nearly five-fold during recordings obtained while subjects actually perform the discrimination task, but the distance between the very same digits shrinks during passive stimulation when no attention is needed.[16] These results differ from findings in animals[109,194] in showing changes in SI representations only during task-relevant trials. Furthermore, similar shifts occur immediately in humans, without extensive training, merely by altering the focus of attention (see above References 18, 19, and 174). These latter findings imply that training engages similar mechanisms responsible for directing attention to the conditions of a task.

2.7.2.2.2   Effects Manifested through Co-Active Stimulation

The immediacy of these attention effects on map organization suggests that demonstrations of plasticity might depend on the flux of activity within existing networks. One possibility is the distribution of synchronized responses across adjacent cortex, because simultaneous stimulation is one factor present in all cited studies where training modified representations. Synchronized responses occur within cortical representations for attended stimulation (see above Reference 219). Thus, any procedures that lead to potential synchronized activity, without attention or behavioral training, might alter map representations, and even tactile discrimination performance.[86] Examples of transformed maps, without a specific behavioral task, occur in animals[53,85] and humans[242] from sustained passive and co-active stimulation of receptive fields. In the animal study, a fused and enlarged representation follows co-active stimulation of adjacent receptive areas. In the human study, shrinkage of Euclidian distance between the source localizations for the median and ulnar nerves follows 40 minutes of co-active, tactile stimulation of the fingertips for digits 1, 2, 4, and 5. An explanation consistent with both results is the expansion of the cortical territory for the co-activated digits 1, 2, 4, and 5 into the representations for the unstimulated digit 3. Hence, in the study with humans, when subsequently stimulating the original nerves, whose representation includes all five digits, there is an apparent shrinkage in the modeling of the source localizations because the expanded activity now collapses into the previously unstimulated representation, which results in dipole distributions that lie closer together.

2.7.3   CHANGES IN ACTIVE FOCI

The simple explanation that instances of enlarged dipole distributions reflect increases in the cortical zones devoted to overused somatosensory inputs requires some scrutiny. In animals, direct recordings from area 3b of SI illustrate map changes. MEG and EEG studies present indirect evidence collected from remote electrodes, and through inverse modeling, map characteristics are projected back to the underlying cortex. In these experiments, could the map changes expressed in the dipole data reflect shifts in dominant activity to different cortical areas? This concern may be unnecessary as most studies report no changes in dipole orientations after experimental manipulations. A change in dipole orientation would be expected if an active focus occupies a new location. However, we describe a shift in the site of maximal blood flow changes from anterior SI (3b?) to posterior SI upon repeated trials (just four runs) of the same stimulus.[28] The distance between peaks was under 10 mm and focal increases in blood flow still occur across multiple parts of SI. The change is in the location of maximum blood flow changes, which could appear as an enlarged SI representation in a dipole map, but without a change in dipole orientation. Other examples exist of even greater shifts in active foci following training (e.g., Reference 192). Many MEG and EEG studies demonstrating short-term manipulations of SI organization ignore the effects implied by these PET data and make no distinctions regarding plasticity effects in different parts of SI or other somatosensory regions. Direct evidence of expanded cortical activity is needed like that reported in patients with limb amputations[126] or spinal injury.[165]

2.8 SUMMARY

Brain imaging techniques offer a unique global overview of brain activity. Many experiments involving the somatosensory system have focused principally on the organization of the body representation in SI. Currently, the SI maps examined from brain images are relatively crude replicas of the more detailed representations studied previously with more direct recordings in animals. The explanations for differences between these two sets of observations are not yet fully known. Obvious are technical considerations related to spatial and temporal resolution. But functional distinctions not previously evident in the more focused animal studies may alter ideas about the likely rigidity of map organization. Recent studies in humans have uncovered remarkable dynamics in SI organization by manipulating tactile attention or through demonstrating alterations after training or injuries. Both attention and training effects may be based on manipulating synchronicity of inputs. These findings point toward a conception of SI maps as a neural resource within which the total pool of engaged neurons changes to meet behaviorally meaningful demands. In actuality, there may be no single SI map as originally conceived and refined by neurophysiology studies through much of the 20th Century. Missing still from most brain imaging data on SI is consistent evidence reflecting its four cytoarchitectonic subdivisions known for nearly 100 years. It will be difficult to extend theoretical predictions about the role of SI in tactile perceptions without such information.

Brain imaging has also provided substantial evidence that multiple cortical areas contribute to somatosensory perceptions. These have included the parietal operculum, posterior parietal cortex, frontal operculum, insula, and possibly portions of precentral frontal cortex. There are still other activated regions not discussed that include cingulate cortex and medial premotor regions. Intriguing has been converging data about overlapping functional domains for attention and intention for movements within some of these same regions. Although enticing, much of this literature is still preliminary and merely identifies the existence of several regions. It has yet to clarify specific roles for any region. Brain imaging studies, when coupled with well-designed behavioral protocols, will be particularly apt in helping define the spatial and temporal sequences of activation. This data will be needed in forming new hypotheses. One example already accessible is the relative interdependence between SI and SII or between anterior and posterior parts of the postcentral gyrus in tactile discrimination tasks.

ACKNOWLEDGMENTS

I thank the authors of several papers who generously sent me copies of their figures. My colleague, Dr. Robert Sinclair, provided especially valuable edits and comments. Work on this review was supported by funds from the National Institutes of Health, NS31005.

REFERENCES

1. Aglioti S., Smania N., and Peru A., Frames of reference for mapping tactile stimuli in brain-damaged patients, *J. Cogn. Neurosci.*, 11, 67, 1999.
2. Allison T., McCarthy G., Wood C. C., et al., Human cortical potentials evoked by stimulation of the II. Cytoarchitectonic areas generating long-latency, *J. Neurophysiol.*, 62, 711, 1989.
3. Allport A., Visual attention, in *Foundations of Cognitive Science*, Posner, M. I., (Ed), The MIT Press, Cambridge, 1989, 631.
4. Andersen R. A., Asanuma C., Essick G., et al., Corticocortical connections of anatomically and physiologically defined subdivisions within the inferior parietal lobule, *J. Comp. Neurol.*, 296, 65, 1990.
5. Andersen R. A., Multimodal integration for the representation of space in the posterior parietal cortex, *Philos. Trans. R. Soc. Lond. B Biol. Sci.*, 352, 1421, 1997.
6. Augustine J. R., Circuitry and functional aspects of the insular lobe in primates including humans, *Brain Res. Rev.*, 22, 229, 1996.
7. Battaglia-Mayer A., Ferraina S., Mitsuda T., et al., Early coding of reaching in the parietooccipital cortex, *J. Neurophysiol.*, 83, 2374, 2000.
8. Baumgartner C., Doppelbauer A., Deecke L., et al., Neuromagnetic investigation of somatotopy of human hand somatosensory cortex, *Exp. Brain Res.*, 87, 641, 1991.
9. Baumgartner C., Doppelbauer A., Sutherling W. W., et al., Human somatosensory cortical finger representation as studied by combined neuromagnetic and neuroelectric measurements, *Neurosci. Lett.*, 134, 103, 1991.
10. Biermann K., Schmitz F., Witte O. W., et al., Interaction of finger representation in the human first somatosensory cortex: a neuromagnetic study, *Neurosci. Lett.*, 251, 13, 1998.
11. Binkofski F., Buccino G., Posse S., et al., A fronto-parietal circuit for object manipulation in man: evidence from an fMRI-study, *Eur. J. Neurosci.*, 11, 3276, 1999.
12. Binkofski F., Buccino G., Stephan K. M., et al., A parieto-premotor network for object manipulation: evidence from neuroimaging, *Exp. Brain Res.*, 128, 210, 1999.
13. Bodegård A., Geyer S., Naito E., et al., Somatosensory areas in man activated by moving stimuli: cytoarchitectonic mapping and PET, *Neuroreport*, 11, 187, 2000.
14. Bodegård A., Ledberg A., Geyer S., et al., Object shape differences reflected by somatosensory cortical activation in human, *J. Neurosci.*, 20:RC51, 1, 2000.
15. Boynton G. M., Engel S. A., Glover G. H., et al., Linear systems analysis of functional magnetic resonance imaging in human V1, *J. Neurosci.*, 16, 4207, 1996.
16. Braun C., Schweizer R., Elbert T., et al., Differential activation in somatosensory cortex for different discrimination tasks, *J. Neurosci.*, 20, 446, 2000.
17. Brodmann K., *Brodmann's 'Localisation in the Cerebral Cortex,'* Smith-Gordon, London, 1994.
18. Buchner H., Reinartz U., Waberski T. D., et al., Sustained attention modulates the immediate effect of de-afferentation on the cortical representation of the digits: source localization of somatosensory evoked potentials in humans, *Neurosci. Lett.*, 260, 57, 1999.
19. Buchner H., Richrath P., Grunholz J., et al., Differential effects of pain and spatial attention on digit representation in the human primary somatosensory cortex, *Neuroreport*, 11, 1289, 2000.
20. Burton H. and Jones E. G., The posterior thalamic region and its cortical projection in new and old world monkeys, *J. Comp. Neurol.*, 168, 249, 1976.

21. Burton H., Second somatosensory cortex and related areas, in *Cerebral Cortex, Sensory-Motor Areas and Aspects of Cortical Connectivity*, Jones, E. G. and Peters, A. (Eds), Plenum, New York, 1986, 31.
22. Burton H. and Carlson M., Second somatic sensory cortical area (SII) in a prosimian primate, *Galago crassicaudatus*, *J. Comp. Neurol.*, 168, 200, 1986.
23. Burton H., Alloway K. D., and Rosenthal P., Somatotopic organization of the second somatosensory cortical area after lesions of the primary somatosensory area in infant and adult cats, *Brain Res.*, 448, 397, 1988.
24. Burton H., Sathian K., and Dian-Hua S., Altered responses to cutaneous stimuli in the second somatosensory cortex following lesions of the postcentral gyrus in infant and juvenile macaques, *J. Comp. Neurol.*, 291, 395, 1990.
25. Burton H. and Sinclair R. J., Second somatosensory area in macaque monkeys. I. Neuronal responses to controlled, punctate indentations of glabrous skin on the hand, *Brain Res.*, 520, 262, 1990.
26. Burton H., Videen T. O., and Raichle M. E., Tactile vibration activated foci in insular and parietal opercular cortex studied with positron emission tomography: mapping the second somatosensory area in humans, *Somatosen. & Mot. Res.*, 10, 297, 1993.
27. Burton H., Fabri M., and Alloway K., Cortical areas within the lateral sulcus connected to cutaneous representations in areas 3b and 1: A revised interpretation of the second somatosensory area in macaque monkeys, *J. Comp. Neurol.*, 355, 539, 1995.
28. Burton H., MacLeod A. M., Videen T. O., et al., Multiple foci in parietal and frontal cortex activated by rubbing embossed grating patterns across fingerpads: a positron emission tomography study in humans, *Cereb. Cortex*, 7, 3, 1997.
29. Burton H., Sinclair R. J., Hong S. Y., et al., Tactile-spatial and cross-modal attention effects in the second somatosensory and 7b cortical areas of rhesus monkeys, *Somatosen. & Mot. Res.*, 14, 237, 1997.
30. Burton H., Abend N., MacLeod A.-M. K., et al., Tactile attention tasks enhance activation in somatosensory regions of parietal cortex: A positron emission tomography study, *Cereb. Cortex*, 9, 662, 1999.
31. Burton H. and Sinclair R. J., Tactile-spatial and cross-modal attention effects in the primary somatosensory cortical areas 3b and 1-2 of rhesus monkeys, *Somatosen. & Mot. Res.*, 17, 213, 2000.
32. Burton H. and Sinclair R. J., Attending to and remembering tactile stimuli: a review of brain imaging data and single neuron responses, *J. Clin. Neurophysiol.*, 17, 575, 2000.
33. Carlson M., Characteristics of sensory deficits following lesions of Brodmann's areas 1 and 2 in the postcentral gyrus of macaca mulatta, *Brain Res.*, 204, 424, 1981.
34. Carlson M. and Burton H., Recovery of tactile function after damage to primary or secondary somatic sensory cortex in infant macaca mulatta, *J. Neurosci.*, 8, 833, 1988.
35. Carmichael S. T. and Price J. L., Architectonic subdivision of the orbital and medial prefrontal cortex in the macaque monkey, *J. Comp. Neurol.*, 346, 366, 1994.
36. Caselli R. J., Rediscovering tactile agnosia, *Mayo Clin. Proc.*, 66, 129, 1991.
37. Caselli R. J., Ventrolateral and dorsomedial somatosensory association cortex damage produces distinct somesthetic syndromes in humans, *Neurology*, 43, 762, 1993.
38. Chelazzi L. and Corbetta M., Cortical Mechanisms of Visuospatial Attention in the Primate Brain, in *The New Cognitive Neurosciences*, M.E. Gazzaniga, (Ed), MIT Press, Boston, 1999, 667.
39. Clark S. A., Allard T., Jenkins W. M., et al., Receptive fields in the body-surface map in adult cortex defined by temporally correlated inputs, *Nature*, 332, 444, 1988.

40. Coghill R. C., Talbot J. D., Evan A. C., et al., Distributed processing of pain and vibration by the human brain, *J. Neurosci.,* 14, 4095, 1994.

41. Colby C. L., Duhamel J. R., and Goldberg M. E., The analysis of visual space by the lateral intraparietal area of the monkey: the role of extraretinal signals, in *Progress in Brain Research,* Hicks, T. P., Molotchnikoff, S., and Ono, T. (Eds), Elsevier Science Publishers, New York, 1993, 307.

42. Colby C. L. and Goldberg M. E., Space and attention in parietal cortex, *Ann. Rev. Neurosci.,* 22, 319, 1999.

43. Conti F., Fabri M., and Manzoni T., Bilateral receptive fields and callosal connectivity of the body midline representation in the first somatosensory area of primates, *Somatosens. Res.,* 3, 273, 1986.

44. Corbetta M., Miezin F. M., Dobmeyer S., et al., Selective and divided attention during visual discriminations of shape, color, and speed: functional anatomy by positron emission tomography, *J. Neurosci.,* 11, 2383, 1991.

45. Corbetta M., Frontoparietal cortical networks for directing attention and the eye to visual locations: Identical, independent, or overlapping neural systems?, *Proc. Natl. Acad. Sci. (U.S.A.),* 95, 831, 1998.

46. Corbetta M., Akbudak E., Conturo T., et al., A common network of functional areas for attention and eye movements, *Neuron,* 21, 761, 1998.

47. Corbetta M., Kincade J. M., Ollinger J. M., et al., Voluntary orienting is dissociated from target detection in human posterior parietal cortex [published erratum appears in *Nat. Neurosci.* 2000 May; 3(5):521], *Nat. Neurosci.,* 3, 292, 2000.

48. Critchley M., The phenomenon of tactile inattention with special reference to parietal lesions, *Brain,* 72, 538, 1949.

49. Damasio H., *Human Brain Anatomy in Computerized Images,* Oxford University Press, New York, 1995.

50. Deibert E., Kraut M., Kremen S., et al., Neural pathways in tactile object recognition, *Neurology,* 52, 1413, 1999.

51. Denny-Brown D. and Chambers R. A., The parietal lobe and behavior, *Res. Pub. Ass. Nerv. and Ment. Dis.,* 36, 35, 1958.

52. Desimone R. and Duncan J., Neural mechanisms of selective visual attention, *Ann. Rev. Neurosci.,* 18, 193, 1995.

53. Dinse H. R., Godde B., Hilger T., et al., Short-term functional plasticity of cortical and thalamic sensory representations and its implication for information processing, *Adv. Neurol.,* 73, 159, 1997.

54. Disbrow E., Roberts T., and Krubitzer L., Somatotopic organization of cortical fields in the lateral sulcus of Homo sapiens: evidence for SII and PV, *J. Comp. Neurol.,* 418, 1, 2000.

55. Disbrow E. A., Slutsky D. A., Roberts T. P., et al., Functional MRI at 1.5 tesla: a comparison of the blood oxygenation level-dependent signal and electrophysiology, *Proc. Natl. Acad. Sci. (U.S.A.),* 97, 9718, 2000.

56. Drevets W. C., Burton H., Videen T. O., et al., Blood flow changes in human somatosensory cortex during anticipated stimulation, *Nature,* 373, 249, 1995.

57. Dreyer D. A., Loe P. R., Metz C. B., et al., Representation of head and face in postcentral gyrus of the macaque, *J. Neurophysiol.,* 38, 714, 1975.

58. Economo C. V., *The Cytoarchitectonics of the Human Cerebral Cortex,* Oxford University Press, London, 1929.

59. Eidelberg D. and Galaburda A. M., Inferior parietal lobule: divergent architectonic asymmetries in the human brain, *Arch. Neurol.,* 41, 843, 1984.

60. Elbert T., Pantev C., Wienbruch C., et al., Increased cortical representation of the fingers of the left hand in string players, *Science,* 270, 305, 1995.
61. Elbert T., Sterr A., Flor H., et al., Input-increase and input-decrease types of cortical reorganization after upper extremity amputation in humans, *Exp. Brain Res.,* 117, 161, 1997.
62. Elbert T. and Flor H., Magnetoencephalographic investigations of cortical reorganization in humans, *Electroencephalogr. Clin. Neurophysiol.,* 49, 284, 1999.
63. Fabri M., Polonara G., Quattrini A., et al., Role of the corpus callosum in the somatosensory activation of the ipsilateral cerebral cortex: an fMRI study of callosotomized patients, *Eur. J. Neurosci.,* 11, 3983, 1999.
64. Faggin B. M., Nguyen K. T., and Nicolelis M. A., Immediate and simultaneous sensory reorganization at cortical and subcortical levels of the somatosensory system, *Proc. Natl. Acad. Sci. (U.S.A.),* 94, 9428, 1997.
65. Fiez J. A., Raichle M. E., Miezin F. M., et al., PET studies of auditory and phonological processing: effects of stimulus characteristics and task demands, *J. Cogn. Neurosci.,* 7, 357, 1995.
66. Flor H., Elbert T., Muhlnickel W., et al., Cortical reorganization and phantom phenomena in congenital and traumatic upper-extremity amputees, *Exp. Brain Res.,* 119, 205, 1998.
67. Florence S. L., Taub H. B., and Kaas J. H., Large-scale sprouting of cortical connections after peripheral injury in adult macaque monkeys [see comments], *Science,* 282, 1117, 1998.
68. Forss N., Hari R., Salmelin R., et al., Activation of the human posterior parietal cortex by median nerve stimulation, *Exp. Brain Res,* 99, 309, 1994.
69. Forss N. and Jousmäki V., Sensorimotor integration in human primary and secondary somatosensory cortices, *Brain Res.,* 781, 259, 1998.
70. Forss N., Hietanen M., Salonen O., et al., Modified activation of somatosensory cortical network in patients with right-hemisphere stroke, *Brain,* 122, 1889, 1999.
71. Fox P. T., Burton H., and Raichle M. E., Mapping human somatosensory cortex with positron emission tomography, *J. Neurosurg.,* 67, 34, 1987.
72. Francis S. T., Kelly E. F., Bowtell R., et al., FMRI of the responses to vibratory stimulation of digit tips, *Neuroimage,* 11, 188, 2000.
73. Friedman D. P. and Jones E. G., Thalamic input to areas 3a and 2 in monkeys, *J. Neurophysiol.,* 45, 59, 1981.
74. Friedman D. P. and Murray E. A., Thalamic connectivity of the second somatosensory area and neighboring somatosensory cortical fields in the lateral sulcus of the monkey, *J. Comp. Neurol.,* 252, 348, 1986.
75. Friedman D. P., Murray E. A., O'Neill J. B., et al., Cortical connections of the somatosensory fields of the lateral sulcus of macaques: Evidence for a corticolimbic pathway for touch, *J. Comp. Neurol.,* 252, 323, 1986.
76. Frot M. and Mauguière F., Timing and spatial distribution of somatosensory responses recorded in the upper bank of the sylvian fissure (SII area) in humans, *Cereb. Cortex,* 9, 854, 1999.
77. Gabrieli J. D., Poldrack R. A., and Desmond J. E., The role of left prefrontal cortex in language and memory, *Proc. Natl. Acad. Sci. (U.S.A.),* 95, 906, 1998.
78. Gandhi S. P., Heeger D. J., and Boynton G. M., Spatial attention affects brain activity in human primary visual cortex, *Proc. Natl. Acad. Sci. (U.S.A.),* 96, 3314, 1999.
79. Garcha H. S. and Ettlinger G., The effects of unilateral or bilateral removals of the second somatosensory cortex (area SII): a profound tactile disorder in monkeys, *Cortex,* 14, 319, 1978.

80. García-Larrea L., Bastuji H., and Mauguière F., Mapping study of somatosensory evoked potentials during selective spatial attention, *Electroencephalogr. Clin. Neurophysiol.*, 80, 201, 1991.

81. Gardner E. P., Ro J. Y., Debowy D., et al., Facilitation of neuronal activity in somatosensory and posterior parietal cortex during prehension, *Exp. Brain Res.*, 127, 329, 1999.

82. Garraghty P. E., Florence S. L., Tenhula W. N., et al., Parallel thalamic activation of the first and second somatosensory areas in prosimian primates and tree shrews, *J. Comp. Neurol.*, 311, 289, 1991.

83. Gelnar P. A., Krauss B. R., Szeverenyi N. M., et al., Fingertip representation in the human somatosensory cortex: an fMRI study, *Neuroimage*, 7, 261, 1998.

84. Geyer S., Schleicher A., and Zilles K., Areas 3a, 3b, and 1 of human primary somatosensory cortex, *Neuroimage*, 10, 63, 1999.

85. Godde B., Spengler F., and Dinse H. R., Associative pairing of tactile stimulation induces somatosensory cortical reorganization in rats and humans, *Neuroreport*, 8, 281, 1996.

86. Godde B., Stauffenberg B., Spengler F., et al., Tactile coactivation-induced changes in spatial discrimination performance, *J. Neurosci.*, 20, 1597, 2000.

87. Hadjikhani N. and Roland P. E., Cross-modal transfer of information between the tactile and the visual representations in the human brain: A positron emission tomographic study, *J. Neurosci.*, 18, 1072, 1998.

88. Hämäläinen H., Hiltunen J., and Titievskaja I., fMRI activations of SI and SII cortices during tactile stimulation depend on attention, *Neuroreport*, 11, 1673, 2000.

89. Hammeke T. A., Yetkin F. Z., Mueller W. M., et al., Functional magnetic resonance imaging of somatosensory stimulation, *Neurosurgery*, 35, 677, 1994.

90. Hansson T. and Brismar T., Tactile stimulation of the hand causes bilateral cortical activation: a functional magnetic resonance study in humans, *Neurosci. Lett.*, 271, 29, 1999.

91. Hari R., Hämäläinen M., Kaukoranta E., et al., Neuromagnetic responses from the second somatosensory cortex in man, *Acta Neurol. Scand.*, 68, 207, 1983.

92. Hari R., Reinikainen K., Kaukoranta E., et al., Somatosensory evoked cerebral magnetic fields from SI and SII in man, *Electroencephalogr. Clin. Neurophysiol.*, 57, 254, 1984.

93. Hari R., Magnetic evoked fields of the human brain: basic principles applications, *Electroencephalogr. Clin. Neurophysiol.*, 41, 3, 1990.

94. Hari R., Hämäläinen H., Hämäläinen M., et al., Separate finger representations at the human second cortex, *Neuroscience*, 37, 245, 1990.

95. Hari R., Karhu J., Hämäläinen M., et al., Functional organization of the human first and second somatosensory cortices: a neuromagnetic study, *Eur. J. Neurosci.*, 5, 724, 1993.

96. Hashimoto I., Saito Y., Iguchi Y., et al., Distal-proximal somatotopy in the human hand somatosensory cortex: a reappraisal, *Exp. Brain Res.*, 129, 467, 1999.

97. Haxby J. V., Horwitz B., Ungerleider L. G., et al., The functional organization of human extrastriate cortex: A PET-rCBF study of selective attention to faces and locations, *J. Neurosci.*, 14, 6336, 1994.

98. Hodge C. J., Jr., Huckins S. C., Szeverenyi N. M., et al., Patterns of lateral sensory cortical activation determined using functional magnetic resonance imaging, *J. Neurosurg.*, 89, 769, 1998.

99. Hoechstetter K., Rupp A., Meinck H. M., et al., Magnetic source imaging of tactile input shows task-independent attention effects in SII, *Neuroreport*, 11, 2461, 2000.

100. Hsiao S. S., O'Shaughnesy D. M., and Johnson K. O., Effects of selective attention on spatial form processing in monkey primary and secondary somatosensory cortex, *J. Neurophysiol.*, 70, 444, 1993.
101. Huttunen J., Wikström H., Korvenoja A., et al., Significance of the second somatosensory cortex in sensorimotor integration: enhancement of sensory responses during finger movements, *Neuroreport*, 7, 1009, 1996.
102. Hyvärinen J., Poranen A., and Jokinen Y., Influence of attentive behavior on neuronal responses to vibration in primary somatosensory cortex of the monkey, *J. Neurophysiol.*, 43, 870, 1980.
103. Hyvärinen J., Posterior parietal lobe of the primate brain, *Physiol. Rev.*, 62, 1060, 1982.
104. Ito M. and Gilbert C. D., Attention modulates contextual influences in the primary visual cortex of alert monkeys, *Neuron*, 22, 593, 1999.
105. Iwamura Y. and Tanaka M., Postcentral neurons in hand region of area 2: Their possible role in the form discrimination of objects, *Brain Res.*, 150, 662, 1978.
106. Iwamura Y., Tanaka M., Sakamoto M., et al., Rostrocaudal gradients in the neuronal receptive field complexity in the finger region of the alert monkey's postcentral gyrus, *Exp. Brain Res.*, 92, 360, 1993.
107. Iwamura Y., Hierarchical somatosensory processing, *Curr. Opin. Neurobiol.*, 8, 522, 1998.
108. Jain N., Florence S. L., Qi H. X., et al., Growth of new brainstem connections in adult monkeys with massive sensory loss, *Proc. Natl. Acad. Sci. (U.S.A.)*, 97, 5546, 2000.
109. Jenkins W. M., Merzenich M. M., Ochs M. T., et al., Functional reorganization of primary somatosensory cortex in adult owl monkeys after behaviorally controlled tactile stimulation, *J. Neurophysiol.*, 63, 82, 1990.
110. Johannsen P., Jakobsen J., Bruhn P., et al., Cortical sites of sustained and divided attention in normal elderly humans, *Neuroimage*, 6, 145, 1997.
111. Johansen-Berg H., Christensen V., Woolrich M., et al., Attention to touch modulates activity in both primary and secondary somatosensory areas, *Neuroreport*, 11, 1237, 2000.
112. Jones E. G. and Powell T. P. S., Connexions of the somatic sensory cortex of the rhesus monkey. II. Contralateral cortical connexions, *Brain*, 92, 717, 1969.
113. Jones E. G. and Burton H., Areal differences in the laminar distribution of thalamic afferents in cortical fields of the insular, parietal, and temporal regions of primates, *J. Comp. Neurol.*, 168, 197, 1976.
114. Jones E. G., Coulter J. D., and Hendry S. H. C., Intracortical connectivity of architectonic fields in the somatic sensory, motor, and parietal cortex of monkeys, *J. Comp. Neurol.*, 181, 291, 1978.
115. Jones E. G. and Porter R., What is area 3a?, *Brain Res. Rev.*, 2, 1, 1980.
116. Jones E. G., *The Thalamus*, Plenum, New York, 1985.
117. Jones E. G., GABAergic neurons and their role in cortical plasticity in primates, *Cereb. Cortex*, 3, 361, 1993.
118. Jones E. G. and Pons T. P., Thalamic and brainstem contributions to large-scale plasticity of primate somatosensory cortex [see comments], *Science*, 282, 1121, 1998.
119. Kaas J. H. and Pons T. P., The somatosensory system of primates., in *Comparative Primate Biology, Volume 4: Neurosciences*, Steklis, H. P. (Ed), Alan R. Liss, New York, 1988, 421.
120. Kaas J. H., Plasticity of sensory and motor maps in adult mammals, *Ann. Rev. Neurosci.*, 14, 137, 1991.

121. Karhu J. and Tesche C. D., Simultaneous early processing of sensory input in human primary (SI) and secondary (SII) somatosensory cortices, *J. Neurophysiol.*, 81, 2017, 1999.

122. Karni A., Meyer G., Jezzard P., et al., Functional MRI evidence for adult motor cortex plasticity during motor skill learning, *Nature*, 377, 155, 1995.

123. Kastner S., De Weerd P., Desimone R., et al., Mechanisms of directed attention in the human extrastriate cortex as revealed by functional MRI [see comments], *Science*, 282, 108, 1998.

124. Kawashima R., O'Sullivan B. T., and Roland P. E., Positron-emission tomography studies of cross-modality inhibition in selective attentional tasks: closing the "mind's eye," *Proc. Natl. Acad. Sci. (U.S.A.)*, 92, 5969, 1995.

125. Kertzman C., Schwarz U., Zeffiro T. A., et al., The role of posterior parietal cortex in visually guided reaching movements in humans, *Exp. Brain Res.*, 114, 170, 1997.

126. Kew J. J., Halligan P. W., Marshall J. C., et al., Abnormal access of axial vibrotactile input to deafferented somatosensory cortex in human upper limb amputees, *J. Neurophysiol.*, 77, 2753, 1997.

127. Killackey H. P., Gould H. J., Cusick C. G., et al., The relation of corpus callosum connections to architectonic fields and body surface maps in sensorimotor cortex of new and old world monkeys, *J. Comp. Neurol.*, 219, 384, 1983.

128. Korvenoja A., Huttunen J., Salli E., et al., Activation of multiple cortical areas in response to somatosensory stimulation: combined magnetoencephalographic and functional magnetic resonance imaging, *Hum. Brain Mapp.*, 8, 13, 1999.

129. Krubitzer L., Clarey J., Tweedale R., et al., A redefinition of somatosensory areas in the lateral sulcus of macaque monkeys, *J. Neurosci.*, 15, 3821, 1995.

130. Kurth R., Villringer K., Mackert B. M., et al., fMRI assessment of somatotopy in human Brodmann area 3b by electrical finger stimulation, *Neuroreport*, 9, 207, 1998.

131. Kurth R., Villringer K., Curio G., et al., fMRI shows multiple somatotopic digit representations in human primary somatosensory cortex, *Neuroreport*, 11, 1487, 2000.

132. LaMotte R. H. and Mountcastle V. B., Disorders in somesthesis following lesions of parietal lobe, *J. Neurophysiol.*, 42, 400, 1979.

133. Ledberg A., O'Sullivan B. T., Kinomura S., et al., Somatosensory activations of the parietal operculum of man. A PET study, *Eur. J. Neurosci.*, 7, 1934, 1995.

134. Lederman S. J. and Klatzky R. L., Hand movements: A window into haptic object recognition, *Cognit. Psychol.*, 19, 342, 1987.

135. Lewis J. W., Burton H., and Van Essen D. C., Anatomical evidence for the posterior boundary of area 2 in the macaque monkey, *Somatosen. & Mot. Res.*, 16, 382, 1999.

136. Lewis J. W. and Van Essen D. C., Corticocortical connections of visual, sensorimotor, and multimodal processing areas in the parietal lobe of the macaque monkey, *J. Comp. Neurol.*, 428, 112, 2000.

137. Lewis J. W. and Van Essen D. C., Mapping of architectonic subdivisions in the macaque monkey, with emphasis on parieto-occipital cortex, *J. Comp. Neurol.*, 428, 79, 2000.

138. Lin W., Kuppusamy K., Haacke E. M., et al., Functional MRI in human somatosensory cortex activated by touching textured surfaces, *J. Magn. Reson. Imaging*, 6, 565, 1996.

139. Lin Y. Y., Simões C., Forss N., et al., Differential effects of muscle contraction from various body parts on neuromagnetic somatosensory responses, *Neuroimage*, 11, 334, 2000.

140. Liu L. C., Gaetz W. C., Bosnyak D. J., et al., Evidence for fusion and segregation induced by 21 Hz multiple-digit stimulation in humans, *Neuroreport*, 11, 2313, 2000.

141. Lüders H., Lesser R. P., Dinner D. S., et al., The second sensory area in humans: evoked potential and electrical stimulation studies, *Ann. Neurol.*, 17, 177, 1985.

142. Macaluso E., Frith C., and Driver J., Selective spatial attention in vision and touch: unimodal and multimodal mechanisms revealed by PET, *J. Neurophysiol.*, 83, 3062, 2000.

143. Maeda K., Kakigi R., Hoshiyama M., et al., Topography of the secondary somatosensory cortex in humans: a magnetoencephalo-graphic study, *Neuroreport*, 10, 301, 1999.

144. Maldjian J. A., Gottschalk A., Patel R. S., et al., The sensory somatotopic map of the human hand demonstrated at 4 Tesla, *NeuroImage*, 10, 55, 1999.

145. Maldjian J. A., Gottschalk A., Patel R. S., et al., Mapping of secondary somatosensory cortex activation induced by vibrational stimulation: an fMRI study, *Brain Res.*, 824, 291, 1999.

146. Manger P. R., Woods T. M., and Jones E. G., Representation of face and intra-oral structures in area 3b of macaque monkey somatosensory cortex, *J. Comp. Neurol.*, 371, 513, 1996.

147. Manzoni T., Conti F., and Fabri M., Callosal projections from area SII to SI in monkeys: anatomical organization and comparison with association projections, *J. Comp. Neurol.*, 252, 245, 1986.

148. Manzoni T., Barbaresi P., Conti F., et al., The callosal connections of the primary somatosensory cortex and the neural bases of midline fusion, *Exp. Brain Res.*, 76, 251, 1989.

149. Matelli M., Camarda R., Glickstein M., et al., Afferent and efferent projections of the inferior area 6 in the macaque monkey, *J. Comp. Neurol.*, 251, 281, 1986.

150. Mauguière F., Merlet I., Forss N., et al., Activation of a distributed somatosensory cortical network in the human brain. A dipole modelling study of magnetic fields evoked by median nerve stimulation. Part I: Location and activation timing of SEF sources, *Electroencephalogr. Clin. Neurophysiol.*, 104, 281, 1997.

151. Mauguière F., Merlet I., Forss N., et al., Activation of a distributed somatosensory cortical network in the human brain: a dipole modelling study of magnetic fields evoked by median nerve stimulation. Part II: Effects of stimulus rate, attention and stimulus detection, *Electroencephalogr. Clin. Neurophysiol.*, 104, 290, 1997.

152. Merzenich M. M., Kaas J. H., Sur M., et al., Double representation of the body surface within cytoarchitectonic areas 3b and 1 in "S1" in the owl monkey *(Aotus trivirgatus)*, *J. Comp. Neurol.*, 181, 41, 1978.

153. Merzenich M. M., Kaas J. H., Wall J., et al., Topographic reorganization of somatosensory cortical areas 3b and 1 in adult monkeys following restricted deafferentation, *Neuroscience*, 8, 33, 1983.

154. Merzenich M. M. and Jenkins W. M., Reorganization of cortical representations of the hand following alterations of skin inputs induced by nerve injury, skin island transfers, and experience, *J. Hand Ther.*, 6, 89, 1993.

155. Merzenich M. M. and Sameshima K., Cortical plasticity and memory, *Curr. Opin. Neurobiol.*, 3, 187, 1993.

156. Mesulam M. M. and Mufson E. J., Insula of the old world monkey III: efferent cortical output and comments on function, *J. Comp. Neurol.*, 212, 38, 1982.

157. Mesulam M. M., A cortical network for directed attention and unilateral neglect, *Ann. Neurol.*, 10, 309, 1981.

158. Meyer E., Ferguson S., Zatorre R., et al., Attention modulates somatosensory cerebral blood flow response to vibrotactile stimulation measured by positron emission tomography, *Ann. Neurol.*, 29, 440, 1991.

159. Mima T., Ikeda A., Nagamine T., et al., Human second somatosensory area: subdural and magnetoencephalographic recording of somatosensory evoked responses, *J. Neurol. Neurosurg. Psychiatry,* 63, 501, 1997.

160. Mima T., Nagamine T., Nakamura K., et al., Attention modulates both primary and secondary somatosensory cortical activities in humans: a magnetoencephalographic study, *J. Neurophysiol.,* 80, 2215, 1998.

161. Mima T., Sadato N., Yazawa S., et al., Brain structures related to active and passive finger movements in man, *Brain,* 122, 1989, 1999.

162. Mishkin M., Analogous neural models for tactual and visual learning, *Neuropsychologia,* 17, 139, 1979.

163. Mogilner A., Grossman J. A., Ribary U., et al., Somatosensory cortical plasticity in adult humans revealed by magnetoencephalography, *Proc. Natl. Acad. Sci. (U.S.A.),* 90, 3593, 1993.

164. Moore C. I., Stern C. E., Corkin S., et al., Segregation of somatosensory activation in the human rolandic cortex using fMRI, *J. Neurophysiol.,* 84, 558, 2000.

165. Moore C. I., Stern C. E., Dunbar C., et al., Referred phantom sensations and cortical reorganization after spinal cord injury in humans, *Proc. Natl. Acad. Sci. (U.S.A.),* 97, 14703, 2000.

166. Moscovitch M. and Behrmann M., Coding of spatial information in the somatosensory system: Evidence from patients with neglect following parietal lobe damage, *J. Cogn. Neurosci.,* 6, 151, 1994.

167. Mufson E. J. and Mesulam M. M., Thalamic connections of the insula in the rhesis monkey and comments on the paralimbic connectivity of the medial pulvinar nucleus, *J. Comp. Neurol.,* 227, 109, 1984.

168. Mufson E. J. and Mesulam M. M., Insula of the old world monkey. II. Afferent cortical input and comments on the claustrum, *J. Comp. Neurol.,* 212, 23, 1982.

169. Murray E. A. and Mishkin M., Relative contributions of SmII and area 5 to tactile discrimination of monkeys, *Beh. Brain Res.,* 11, 67, 1984.

170. Nakamura A., Yamada T., Goto A., et al., Somatosensory homunculus as drawn by MEG, *Neuroimage,* 7, 377, 1998.

171. Neal J. W., Pearson R. C. A., and Powell T. P. S., The cortico-cortical connections of area 7b, PF, in the parietal lobe of the monkey, *Exp. Brain Res.,* 419, 341, 1987.

172. Nelson R. J., Sur M., Felleman D. J., et al., Representations of the body surface in postcentral parietal cortex of *Macaca fascicularis, J. Comp. Neurol.,* 192, 611, 1980.

173. Nobre A. C., Sebestyen G. N., Gitelman D. R., et al., Functional localization of the system for visuospatial attention using positron emission tomography, *Brain,* 120, 515, 1997.

174. Noppeney U., Waberski T. D., Gobbele R., et al., Spatial attention modulates the cortical somatosensory representation of the digits in humans, *Neuroreport,* 10, 3137, 1999.

175. O'Sullivan B. T., Roland P. E., and Kawashima R., A PET study of somatosensory discrimination in man: microgeometry vs. macrogeometry, *Eur. J. Neurosci.,* 6, 137, 1994.

176. Pardo J. V., Fox P. T., and Raichle M. E., Localization of a human system for sustained attention by positron emission tomography, *Nature,* 349, 61, 1991.

177. Pascual-Leone A. and Torres F., Plasticity of the sensorimotor cortex representation of the reading finger in Braille readers, *Brain,* 116, 39, 1993.

178. Paul R. L., Merzenich M., and Goodman H., Representation of slowly and rapidly adapting cutaneous mechanoreceptors of the hand in Brodman's areas 3 and 1 of *Macaca mulatta, Brain Res.,* 36, 229, 1972.

179. Penfield W. and Rasmussen T., *Secondary Sensory and Motor Representation*, Macmillan, New York, 1950.
180. Penfield W. and Jasper H., *Epilepsy and the Functional Anatomy of the Human Brain*, Churchill, London, 1954.
181. Penfield W. G. and Boldrey E., Somatic motor and sensory representation in the cerebral cortex of man as studied by electrical stimulation, *Brain*, 60, 389, 1937.
182. Petit L. and Haxby J. V., Functional anatomy of pursuit eye movements in humans as revealed by fMRI, *J. Neurophysiol.*, 82, 463, 1999.
183. Pons T. P., Garraghty P. E., Cusick C. G., et al., The somatotopic organization of area 2 in macaque monkeys, *J. Comp. Neurol.*, 241, 445, 1985.
184. Pons T. P., Garraghty P. E., Friedman D. P., et al., Physiological evidence for serial processing in somatosensory cortex, *Science*, 237, 417, 1987.
185. Pons T. P., Wall J. T., Garraghty P. E., et al., Consistent features of the representation of the hand in area 3b of macaque monkeys, *Somatosen. Res.*, 4, 309, 1987.
186. Pons T. P., Garraghty P. E., and Mishkin M., Serial and parallel processing of tactual information in somatosensory cortex of rhesus monkeys, *J. Neurophysiol.*, 68, 518, 1992.
187. Posner M. I. and Petersen S. E., The attention system of the human brain, *Ann. Rev. Neurosci.*, 13, 25, 1990.
188. Posner M. I. and Dehaene S., Attentional networks, *Trends Neurosci.*, 17, 75, 1994.
189. Powell T. P. S. and Mountcastle V. B., Some aspects of the functional organization of the cortex of the postcentral gyrus of the monkey: A correlation of findings obtained in a single unit analysis with cytoarchitecture, *Bull. Johns Hopkins Hosp.*, 105, 133, 1959.
190. Press W., Knierim J., and Van Essen D., Effects of spatial attention on macaque V1, (in preparation), 2000.
191. Preuss T. M. and Goldman-Rakic P. S., Architectonics of the parietal and temporal association cortex in the strepsirhine primate *Galago* compared to the anthropoid primate *Macaca*, *J. Comp. Neurol.*, 310, 475, 1991.
192. Raichle M. E., Fiez J. A., Videen T. O., et al., Practice-related changes in human brain functional anatomy during nonmotor learning, *Cereb. Cortex*, 4, 8, 1994.
193. Ramachandran V. S. and Hirstein W., The perception of phantom limbs. The D. O. Hebb lecture, *Brain*, 121, 1603, 1998.
194. Recanzone G. H., Merzenich M. M., Jenkins W. M., et al., Topographic reorganization of the hand representation in cortical area 3b of owl monkeys trained in a frequency-discrimination task, *J. Neurophysiol.*, 67, 1031, 1992.
195. Recanzone G. H., Merzenich M. M., and Schreiner C. E., Changes in the distributed temporal response properties of SI cortical neurons reflect improvements in performance on a temporally based tactile discrimination task, *J. Neurophysiol.*, 67, 1071, 1992.
196. Reed C. L., Lederman S. J., and Klatzky R. L., Haptic integration of planar size with hardness, texture, and planar contour, *Can. J. Psychol.*, 44, 522, 1990.
197. Reed C. L. and Caselli R. J., The nature of tactile agnosia: a case study, *Neuropsychologia*, 32, 527, 1994.
198. Reed C. L., Caselli R. J., and Farah M. J., Tactile agnosia. Underlying impairment and implications for normal tactile object recognition, *Brain*, 119, 875, 1996.
199. Roberts T. P., Disbrow E. A., Roberts H. C., et al., Quantification and reproducibility of tracking cortical extent of activation by use of functional MR imaging and magnetoencephalography, *AJNR Am. J. Neuroradiol.*, 21, 1377, 2000.
200. Robinson C. J. and Burton H., Somatotopographic organization in the second somatosensory area of M. fascicularis, *J. Comp. Neurol.*, 192, 43, 1980.

201. Robinson C. J. and Burton H., Organization of somatosensory receptive fields in cortical areas 7b, retroinsula, postauditory and granular insula of *M. fascicularis*, *J. Comp. Neurol.*, 192, 69, 1980.

202. Robinson C. J. and Burton H., Somatic submodality distribution within the second somatosensory (SII), 7b, retroinsular, postauditory, and granular insular cortical areas of *M. fascicularis*, *J. Comp. Neurol.*, 192, 93, 1980.

203. Roelfsema P. R., Lamme V. A., and Spekreijse H., Object-based attention in the primary visual cortex of the macaque monkey, *Nature*, 395, 376, 1998.

204. Roland P. E., Somatotopical tuning of postcentral gyrus during focal attention in man. A regional cerebral blood flow study, *J. Neurophysiol.*, 46, 744, 1981.

205. Roland P. E., Cortical regulation of selective attention in man. A regional cerebral blood flow study, *J. Neurophysiol.*, 48, 1059, 1982.

206. Roland P. E., Somatosensory detection of miocrogeometry, macrogeometry, and kinesthesia after localized lesions of the cerebral hemispheres in man, *Brain Res. Rev.*, 12, 43, 1987.

207. Roland P. E. and Seitz R. J., Positron emission tomography studies of the somatosensory system in man, *Ciba Found. Symp.*, 163, 113, 1991.

208. Roland P. E., O'Sullivan B., and Kawashima R., Shape and roughness activate different somatosensory areas in the human brain, *Proc. Natl. Acad. Sci. (U.S.A.)*, 95, 3295, 1998.

209. Roland P. E. and Zilles K., Structural divisions and functional fields in the human cerebral cortex, *Brain Res. Rev.*, 26, 87, 1998.

210. Sakai K., Watanabe E., Onodera Y., et al., Functional mapping of the human somatosensory cortex with echo-planar MRI, *Magn. Reson. Med.*, 33, 736, 1995.

211. Sanides F., Comparative architectonics of the neocortex of mammals and their evolutionary interpretation, *Ann. NY Acad. Sci.*, 167, 404, 1969.

212. Schneider R. J., Friedman D. P., and Mishkin M., A modality-specific somatosensory area within the insula of the rhesus monkey, *Brain Res.*, 621, 116, 1993.

213. Seitz R. J. and Roland P. E., Vibratory stimulation increases and decreases the regional cerebral blood flow and oxidative metabolism: a positron emission tomography (PET) study, *Acta Neurol. Scand.*, 86, 60, 1992.

214. Servos P., Zacks J., Rumelhart D. E., et al., Somatotopy of the human arm using fMRI, *Neuroreport*, 9, 605, 1998.

215. Sinclair R. J., Kuo J. J., and Burton H., Effects on discrimination performance of selective attention to tactile features, *Somatosen. & Mot. Res.*, 17, 145, 2000.

216. Spengler F., Roberts T. P., Poeppel D., et al., Learning transfer and neuronal plasticity in humans trained in tactile discrimination, *Neurosci. Lett.*, 232, 151, 1997.

217. Stanton G. B., Cruce W. L. R., Goldberg M. E., et al., Some ipsilateral projections to areas PF and PG of the inferior parietal lobule in monkeys, *Neurosci. Lett.*, 6, 243, 1977.

218. Steinmetz H., Ebeling U., Huang Y. X., et al., Sulcus topography of the parietal opercular region: an anatomic and MR study, *Brain Lang.*, 38, 515, 1990.

219. Steinmetz P. N., Roy A., Fitzgerald P. J., et al., Attention modulates synchronized neuronal firing in primate somatosensory cortex, *Nature*, 404, 187, 2000.

220. Suk J., Ribary U., Cappell J., et al., Anatomical localization revealed by MEG recordings of the somatosensory system, *Electroencephalogr. Clin. Neurophysiol.*, 78, 185, 1991.

221. Sutherling W. W., Levesque M. F., and Baumgartner C., Cortical sensory representation of the human hand: size of finger regions and nonoverlapping digit somatotopy, *Neurology*, 42, 1020, 1992.

222. Talairach J. and Tournoux P., *Coplanar Stereotaxic Atlas of the Human Brain*, Thieme Medical, New York, 1988.
223. Tootell R. B., Hadjikhani N., Hall E. K., et al., The retinotopy of visual spatial attention, *Neuron*, 21, 1409, 1998.
224. Turman A. B., Ferrington D. G., Ghosh S., et al., Parallel processing of tactile information in the cerebral cortex of the cat: effect of reversible inactivation of SI on responsiveness of SII neurons, *J. Neurophysiol.*, 67, 411, 1992.
225. Vallar G., Rusconi M. L., Bignamini L., et al., Anatomical correlates of visual and tactile extinction in humans: a clinical CT scan study, *J. Neurol. Neurosurg. Psychiatry*, 57, 464, 1994.
226. Van Buren J. M., Sensory responses from stimulation of the inferior Rolandic and Sylvian regions in man, *J. Neurosurg.*, 59, 119, 1983.
227. Wang X., Merzenich M. M., Sameshima K., et al., Remodelling of hand representation in adult cortex determined by timing of tactile stimulation, *Nature*, 378, 71, 1995.
228. Weiss T., Miltner W. H., Dillmann J., et al., Reorganization of the somatosensory cortex after amputation of the index finger, *Neuroreport*, 9, 213, 1998.
229. Weiss T., Miltner W. H., Huonker R., et al., Rapid functional plasticity of the somatosensory cortex after finger amputation, *Exp. Brain Res.*, 134, 199, 2000.
230. White L., Andrews T., Hulette C., et al., Structure of the human sensorimotor system. II: Lateral symmetry, *Cereb. Cortex*, 7, 31, 1997.
231. White L., Andrews T., Hulette C., et al., Structure of the human sensorimotor system. I: morphology and cytoarchitecture of the central sulcus, *Cereb. Cortex*, 7, 18, 1997.
232. Witelson S. F. and Kigar D. L., Sylvian fissure morphology and asymmetry in men and women: bilateral differences in relation to handedness in men, *J. Comp. Neurol.*, 323, 326, 1992.
233. Woolsey C. N., Marshall W. H., and Bard P., Representation of cutaneous tactile sensibility in the cerebral cortex of the monkey as indicated by evoked potentials, *Bull. Johns Hopkins Hosp.*, 70, 399, 1942.
234. Woolsey C. N., "Second" somatic receiving areas in the cerebral cortex of cat, dog, and monkey, *Fed. Proc.*, 2, 55, 1943.
235. Woolsey C. N. and Fairman D., Contralateral, ipsilateral, and bilateral representation of cutaneous receptors in somatic areas I and II of the cerebral cortex of the pig, sheep, and other mammals, *Surgery*, 19, 684, 1946.
236. Woolsey C. N., Erickson T. C., and Gilson W. E., Localization in somatic sensory and motor areas of human cerebral cortex as determined by direct recording of evoked potentials and electrical stimulation, *J. Neurosurg.*, 51, 476, 1979.
237. Woolsey C. N. and Walzl E. M., Cortical auditory area of *Macaca mulatta* and its relation to the second somatic sensory area (Sm II): determation by electrical excitation of auditory nerve fibers in the spiral osseous lamina and by click stimulation, in *Cortical Sensory Organization. Volume 3: Multiple Auditory Areas*, Woolsey, C. N. (Ed), Humana Press, Clifton, NJ, 1982.
238. Xerri C., Merzenich M. M., Peterson B. E., et al., Plasticity of primary somatosensory cortex paralleling sensorimotor skill recovery from stroke in adult monkeys, *J. Neurophysiol.*, 79, 2119, 1998.
239. Yang T. T., Gallen C., Schwarz B., et al., Non-invasive somatosensory homunculus mapping in humans by using a large-array biomagnetometer, *Proc. Natl. Acad. Sci. (U.S.A.)*, 90, 3098, 1993.
240. Yang T. T., Gallen C. C., Ramachandran V. S., et al., Noninvasive detection of cerebral plasticity in adult human somatosensory cortex, *Neuroreport*, 5, 701, 1994.

241. Zhang H. Q., Murray G. M., Turman A. B., et al., Parallel processing in cerebral cortex of the marmoset monkey: effect of reversible SI inactivation on tactile responses in SII, *J. Neurophysiol.*, 76, 3633, 1996.
242. Ziemus B., Huonker R., Haueisen J., et al., Effects of passive tactile co-activation on median ulnar nerve representation in human SI, *Neuroreport*, 11, 1285, 2000.
243. Zilles K., Schlaug G., Matelli M., et al., Mapping of human and macaque sensorimotor areas by integrating architectonic, transmitter receptor, MRI, and PET data, *J. Anat.*, 187, 515, 1995.

3 Neural Mechanisms of Tactile Form and Texture Perception

Kenneth O. Johnson and Takashi Yoshioka

CONTENTS

3.1   OVERVIEW

Information about the external world is analyzed and subdivided into separate processing streams in each of the sensory systems. That division begins at the very first stage of sensory processing within the somatosensory system: nociceptors, thermoreceptors, proprioceptors, and cutaneous mechanoreceptors transduce different stimulus properties and channel their information into separate, parallel streams. The same principle applies to the four cutaneous mechanoreceptors that are responsible for tactile perception. Evidence from three decades of psychophysical and neurophysiological research shows that each mechanoreceptive system (the mechanoreceptors of a single type and the pathways that convey their information to perception) serves a distinctly different function and that, taken together, these functions account for tactile perception (reviewed in Reference 1). In this chapter, we review briefly the functions of the mechanoreceptors, but we concentrate on the peripheral and cortical neural mechanisms of functions served by the slowly adapting type 1 (SA1) afferents (i.e., form and texture perception).

3.2 PERIPHERAL NEURAL MECHANISMS

3.2.1 MECHANORECEPTION

The four cutaneous mechanoreceptive afferent neuron types innervating the glabrous skin comprise SA1 afferents, which end in Merkel cells; rapidly adapting (RA) afferents, which end in Meissner corpuscles; Pacinian (PC) afferents, which end in Pacinian corpuscles; and slowly adapting type 2 (SA2) afferents, which are thought to end in Ruffini endings. Two of the four, the RA and PC afferents, respond only to skin motion; they are classed as rapidly adapting because they respond only transiently to sudden, steady indentation. The other two, the slowly adapting SA1 and SA2 afferents are classed as slowly adapting because they respond to sustained skin deformation with a sustained discharge that declines slowly, but they (particularly SA1 afferents) are much more sensitive to skin movement than to static deformation. The neural response properties of these cutaneous afferents have been studied extensively in both human and nonhuman primates and, except for the SA2 afferents, which are not found in nonhuman primates, there are no interspecies differences. We use the terms SA1, SA2, RA, and PC systems throughout this chapter.[2] By SA1 system, for example, we mean the SA1 receptors (the Merkel-neurite complex), the SA1 afferent nerve fiber population, and all the central neuronal pathways that convey the SA1 signal to memory and perception. We do not mean to imply that there is no central convergence between these systems or that the systems do not overlap.

SA1 afferent fibers branch repeatedly before they lose their myelin and end in the basal layer of the epidermis where specialized (Merkel) epidermal cells enfold the unmyelinated endings.[3] Although there are synapse-like junctions between the Merkel cells and the axon terminals, the transduction of local tissue deformation appears to arise as the result of the activation of mechanosensitive ion channels in the bare nerve endings.[4,5] SA1 afferents innervate the skin of the fingerpad densely and have small receptive fields. Consequently, they transmit high resolution spatial neural images of stimuli contacting the fingerpad. A striking response property is surround suppression,[6] which gives SA1 afferents response properties like those attributed to surround inhibition in the central nervous system. In the central nervous system, inhibition surrounding an excitatory center makes neurons sensitive to local curvature and, depending on the balance between excitation and inhibition, insensitive to a uniform stimulus field. Surround suppression in the responses of tactile peripheral afferents confers similar response properties but instead of being based on synaptic mechanisms it is based entirely on mechanoreceptor sensitivity to a specific component of tissue strain near the nerve ending (strain energy density or a closely related component of strain[7,8]). As a consequence, SA1 afferents respond strongly to points, edges, and curvature and these responses are suppressed by the presence of stimuli in the surrounding skin. Also, because of this surround suppression, SA1 afferents are minimally responsive to uniform skin indentation. Therefore, local spatial features such as edges, and curves are represented strongly in the neural image conveyed by the peripheral SA1 population response (illustrated in Figure 3.3). Combined psychophysical and neurophysiological experiments, reviewed below, show that the SA1 system is responsible for form and texture perception.

RA afferent fibers also branch repeatedly as they near the epidermis, ending in 30–80 Meissner's corpuscles.[1] Meissner's corpuscles occur in dermal pockets between the sweat-duct and adhesive ridges,[9,10] which puts them as close to the surface of the epidermis as possible within the dermis.[11] This may account, in part, for the RA's greater sensitivity to minute skin deformation compared with SA1 afferents, whose receptors are on the tips of the deepest epidermal ridges. RA afferents innervate the skin of the fingerpad more densely than do the SA1 afferents. Based on the available studies of innervation density,[12-14] Johnson et al.[1] have concluded that innervation densities in humans and monkeys are not significantly different and that the best estimates at the fingertip are 100 SA1 and 150 RA afferents/cm^2. Although RA afferents have a greater potential for transmitting spatial information because of their greater innervation density, they resolve spatial detail poorly. Their receptive field sizes depend strongly on stimulus intensity and are much larger than SA1 receptive fields at indentation levels that occur in ordinary tactile experience.[15] The striking feature of RA responses is their sensitivity to minute skin motion. The effective operating range for RAs is about 4–400 µm indentation; the comparable SA1 range is about 15–1500 µm or more.[15,16-18] The SA1 and RA response properties are complementary. The RA and SA1 systems are, in some ways, like the scotopic and photopic systems in vision. The RA system, like the scotopic system, has greater sensitivity but poorer spatial resolution and limited dynamic range. The SA1 system, like the photopic system, is less sensitive but has higher spatial resolution and operates over a wider dynamic range. The neural response properties of RA afferents make them ideally suited for motion perception. In fact, combined psychophysical and neurophysiological studies show that the RA system is responsible for the perception of events that produce low-frequency, low-amplitude skin motion. That includes the detection of microscopic surface features, the detection of low frequency vibration, and the detection of slip, which is critical for grip control (reviewed in Reference 1).

PC afferent fibers end in single Pacinian corpuscles, which occur in the dermis or the deeper tissues. Bell, Bolanowski and Holmes[19] have provided an extensive review of the history, structure, and electrophysiological properties of this receptor. The PC's most striking property is its extreme sensitivity, which derives from mechanosensitive ion channels in the afferent's unmyelinated ending. The most sensitive Pacinian corpuscles respond with action potentials to vibratory amplitudes as small as 3 nm applied directly to the corpuscle[20] and 10 nm applied to the skin.[21]

Pacinian corpuscles comprise multiple layers of fluid-filled sacs; these sacs act as a cascade of high-pass filters that shield the unmyelinated ending from the large, low frequency deformations that accompany most manual tasks.[22,23] If it was not for this intense filtering, the transducer, which is two orders of magnitude more sensitive than any of the other mechanoreceptive transducers, would be overwhelmed by most cutaneous stimuli. Because of their extreme sensitivity, receptive field boundaries are difficult to define. The most sensitive PCs have receptive fields that encompass an entire hand or even an entire arm; a less sensitive PC may have a receptive field restricted to a single phalanx. There are about 2500 Pacinian corpuscles in the human hand and they are about twice as numerous in the fingers as in the palm (about 350 per finger and 800 in the palm; reviewed in Reference 21). Because of the small number of PC afferents and their very large receptive fields, the PC population transmits little, if any useful information about the spatial properties of a stimulus. Instead, it transmits information communicated by vibrations in objects, probes, or tools held in the hand (reviewed in Reference 1).

SA2 afferents are distinguished from SA1 afferents by four properties: (1) their receptive field areas are about five times larger and their receptive field borders are not clearly demarcated;[24] (2) they are about six times less sensitive to cutaneous indentation;[25,26] (3) they are 2–4 times more sensitive to skin stretch;[27] and (4) their interspike intervals are more uniform.[27,28] SA2 afferents are thought to end in Ruffini complexes,[3] although the association of afferents with these response properties with a specific receptor is not as secure as with the other three cutaneous mechanorecep-tors. Both SA1 and SA2 afferents respond to forces orthogonal and parallel to the skin surface, but between them the SA1 afferents are biased toward responsiveness to orthogonal forces and SA2 afferents to parallel forces.[29] The minimal SA2 responses to raised dot patterns (e.g., Braille patterns in Figure 3.4) and to curved surfaces[30] suggest that they play no role in form perception.[31] Based on their responses to curved surfaces, Goodwin et al. conclude that "SA2 responses are unlikely to signal information to the brain about the local shape of an object" (page 2887 in Reference 30). Because of their sensitivity to skin stretch, SA2s are well suited to signal lateral forces such as active forces pulling on an object held in the hand. A more interesting possibility is that they send a neural image of skin stretch that plays a significant, or possibly even the dominant, role in our perception of hand conformation (reviewed in Reference 1) and of the direction of motion of an object moving across the skin.[32]

3.2.2 FORM AND TEXTURE PERCEPTION

Form and texture perception have in common only that both depend on surface structure. Form perception is perception of the specific geometric structure of a surface or object; texture perception corresponds to the subjective feel of a surface and it depends on its distributed, statistical properties. Form perception has many dimensions; texture perception has only two or possibly three dimensions (see below). Form perception can be studied with objective methods (i.e., the subject's responses can be classified as right or wrong); texture perception cannot. If a subject

TABLE 3.1
Afferent Types and Their Functions

Afferent Type	Receptor	Adaptation to Steady Deformation	RF Size	Spatial Resolution	Temporal Sensitivity (Hz)	Function
SA1	Merkel	Slow	Small	0.5 mm	0–100	Form, texture
RA	Meissner	Rapid	Small	3–5 mm	2–100	Motion perception, grip control
PC	Pacinian	Rapid	Large	2 cm	10–1000	Transmitted vibration, tool use
SA2	Ruffini	Slow	Large	1 cm	0–20	Lateral force, hand shape, motion direction

is asked whether the dot or ridge spacing of one surface is greater than another the response is a judgment about the surface and it provides a clue to the subject's capacity for form perception (e.g., Reference 33 and 34); if the same subject is asked whether the second surface feels rougher (or softer) than the first, the response is a description of his or her experience and it provides a clue to the subject's perception of texture.

3.2.2.1 Form Perception

The ability to discriminate object or surface features and the capacity for pattern recognition at the fingertip are the same whether the object is contacted by active touch or is applied to the passive hand.[35] Form perception is affected only marginally by whether the object is stationary or moving relative to the skin; it is unaffected by scanning speed up to 40 mm/s; it is unaffected by contact force, at least over the range from 0.2–1 N; and it is affected only marginally by the heights (relief) of spatial features over a wide range of heights.[36-39]

Three psychophysical studies of the limit of tactile spatial acuity are illustrated in Figure 3.1. In all three studies, the element width that resulted in performance midway between chance and perfect discrimination was between 0.9 and 1.0 mm, which is close to the theoretical limit set by the density of SA1 and RA primary afferents at the fingertip.[40] Acuity declines progressively from the index to the fifth finger[41] and it declines progressively with age.[42-44] Whether these differences in acuity are due to differences in innervation density is not known. Spatial acuity at the fingertip is the same in man and monkey.[45] Spatial acuity at the lip and tongue is significantly better than at the fingertip.[46-48]

3.2.2.1.1 Spatial Acuity

Tactile spatial resolution of about 1 mm requires an innervation density of (at least) about one afferent per square mm and it requires that individual afferents resolve the spatial details at least as well as human subjects. Neither the PC nor the SA2 system comes close on either score.[13,49] Note that the human performance illustrated

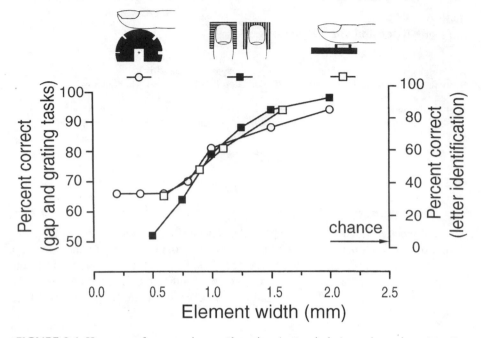

FIGURE 3.1 Human performance in gap detection (open circles), grating orientation discrimination (filled squares), and letter recognition (open squares) tasks. The abscissa represents the fundamental element width for each task, which was gap size for the gap detection task, bar width (half the grating period) for the grating orientation discrimination task, and the average bar and gap width within letters (approximately one fifth the letter height) for the letter recognition task. Threshold is defined as the element size producing performance midway between chance (50% correct for the gap and grating tasks, 1/26 for letter recognition) and perfect performance. Adapted from Johnson, K.O. and Phillips, J.R., *J. Neurophysiol.* 46, 1177–1191, 1981, with permission.

in Figure 3.1 begins to rise above chance at element sizes around 0.5 mm, which means that either the SA1 or RA system must begin to resolve spatial detail at 0.5 mm or less. Evidence that only the SA1 afferents account for the spatial resolution illustrated in Figure 3.1 comes from neurophysiological experiments in which SA1 and RA afferents were studied with the same periodic gratings used in the psychophysical experiments. SA1 responses to a periodic grating convey information about spatial structure when the groove and ridge widths are 0.5 mm wide (e.g., see Figure 3.2). When the grooves and ridges are 1 mm wide, SA1s provide a robust neural image of the stimulus. In contrast, RAs require grooves that are at least 3 mm wide before their responses begin to distinguish a grating from a flat surface; most RAs fail even to register grooves 3 mm wide.[7] The RA response illustrated in Figure 3.2 was the most sensitive to spatial detail of all the RAs studied. Kops and Gardner[50] obtained nearly identical results with an Optacon, a dense array of vibrotactile probes designed as a reading aid for the blind.[51] PC afferents were unable to resolve grooves that were 5 mm wide (Figure 3.2).

3.2.2.1.2 Pattern Recognition

The relationship between SA1 response properties and pattern recognition behavior in a letter recognition experiment is illustrated in Figure 3.3. In the study illustrated in Figure 3.3, Vega-Bermudez et al.[35] showed that there was no detectable difference in human performance between active and passive touch and that the confusion matrix shown in Figure 3.3 is characteristic of human letter recognition performance across a wide range of stimulus conditions. Recognition behavior is highly pattern specific; recognition accuracy differs significantly between letters (ranging from 15% for the letter N to 98% for the letter I) and more than 50% of the confusions are confined to 7% of all possible confusion pairs (22 out of 325 possible confusion pairs), which are enclosed in boxes in Figure 3.3. The confusions in all but 5 of those 22 pairs are highly asymmetric ($p < 0.001$). Analysis of the hit rates and false positive rates suggests that this recognition behavior bears no relationship to cognitive bias. The frequency of occurrence of letters in English bears no relationship to the rates of correct responses, false positives, or total responses. Further, if the recognition behavior illustrated in Figure 3.3 was related to cognitive biases, the hit rates and false positive rates should be related but they are not.[35]

The responses of SA1 afferents to the same letters scanned across their receptive fields (Figure 3.3) seem to explain the recognition behavior. For example, B is rarely identified as B; instead, it is called D more often than it is called B. Conversely, D is virtually never called B (Figure 3.3 top). The reason for this response bias can be explained by the SA1 surround suppression mechanism discussed earlier, which suppresses the response to the central, horizontal bar of the B: the neural representation of the B does, in fact, resemble a D more than it does a B (see Figure 3.3). For another example, C is often called G or Q, but G and Q are almost never called C. An explanation is that many of the features that discriminate the letters are missing in the neural representations, so a lack of the features that distinguish a G or Q from a C in the representation of the C is not a strong reason to not respond G or Q. Conversely, the strong representation of the distinctive features of the G and Q make confusion with a C unlikely. The performance illustrated in Figure 3.3 is for naive subjects in their first testing session. Performance improves steadily on repeated testing.[35] One explanation for this improvement is that subjects learn the idiosyncracies of the neural representations (e.g., when a subject recognizes the distinctive feature of the G in the neural representation he or she is less likely to mistake a C for a G).

The responses of typical human cutaneous afferents to Braille symbols (top row) scanned over their receptive fields are illustrated in Figure 3.4. The human SA1, RA, and PC responses to these raised-dot patterns are indistinguishable from the responses of monkey SA1, RA, and PC afferents to similar patterns.[36] SA1 afferents provide a sharp, isomorphic representation of the Braille patterns, RA afferents provide a less sharply defined isomorphic representation, and PCs and SA2s provide no useful spatial information.

3.2.2.1.3 Studies with the Optacon

Psychophysical and neurophysiological studies with the Optacon provide a unique window on tactile perception in the absence of activity in the SA1 system. Neurophysiological studies show that the Optacon activates the RA and PC systems

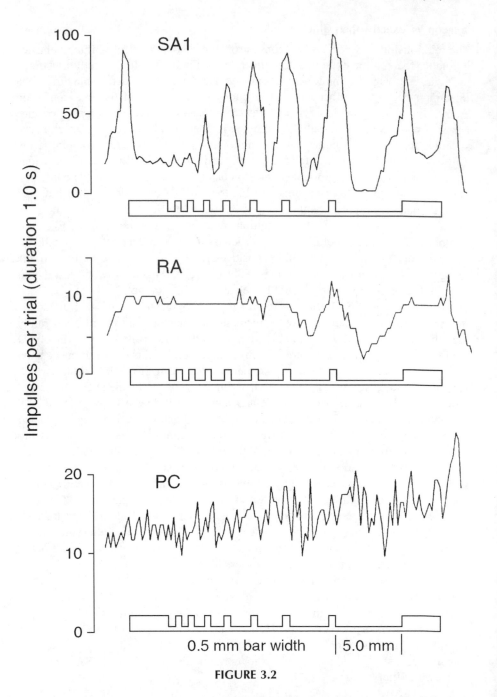

FIGURE 3.2

well, that it fails to activate the SA1 system, and that the PC system cannot account for spatial pattern recognition performance with the Optacon;[52,53] therefore, the many psychophysical studies employing the Optacon[54] are studies of the sensory capacity provided by the RA system. The human ability to resolve spatial patterns with the

Optacon is exactly that which would be predicted based on the RA responses illustrated in Figure 3.2. For example, humans cannot discriminate the orientation of an Optacon grating pattern until the grooves in the grating exceed 5 mm width.[50] This implies that the SA1 system is solely responsible for the limits of resolution illustrated in Figure 3.1.

3.2.2.1.4 Curvature Perception

Combined psychophysical and neurophysiological studies of curvature perception provide evidence of the neural mechanisms of form perception not based on the limits of spatial acuity.[55-58] These studies by Goodwin, Wheat, and their colleagues show that estimates of curvature are unaffected by changes in contact area and force and, conversely, estimates of force are unaffected by changes in curvature. This latter finding is particularly surprising considering that SA1 firing rates are strongly affected by curvature.[58,59,60] These psychophysical observations (that curvature perception is unaffected by changes in contact area or force) suggest that the spatial profile of neural activity in one or more of the afferent populations is used for the perception of curvature and that a different neural code (e.g., total discharge rate) is used for the perception of force. Only the SA1 population response provides a veridical representation of curvature that can account for the psychophysical observations.[58,61] The SA1 population responses to a wide range of curvatures are shown in Figure 3.5. RAs respond poorly to such stimuli and provide no signal that might account for the ability of humans to discriminate curvature.[58,62,63]

3.2.2.2 Texture Perception

Our knowledge of texture perception and its neural mechanisms has changed dramatically in the last decade. A major step is the demonstration that texture perception involves two strong dimensions, roughness and softness, and a weaker third dimension described as something like stickiness. Multidimensional scaling studies have shown that texture perception includes soft–hard and smooth–rough as independent perceptual dimensions, surface hardness and roughness can occur in almost any combination, and that they account for most or all of texture perception.[64,65] A third,

---

FIGURE 3.2 (opposite) Responses of SA1, RA, and PC afferents to a grating pressed into the skin. The grating is shown in cross section beneath each response profile. The bars are 0.5 mm wide; the grooves are deeper than illustrated (2.0 mm deep) and are 0.5, 0.5, 0.75, 1.0, 1.5, 2.0, 3.0, and 5.0 mm wide. The responses displayed in each profile were obtained by indenting the skin to a depth of 1 mm, holding the indentation for 1 second, raising the grating, and then moving it laterally by 0.2 mm before the next indentation. The horizontal dimension of the response profile represents the location of the center of the receptive field relative to the grating; for example, the left peak in the SA1 response profile (approximately 95 imp/s) occurred when the center of the SA1 receptive field was directly beneath the left edge of the grating. The RA illustrated here was the most sensitive to the spatial structure of the grating of all RAs studied. Some RAs barely registered the presence of the 5 mm gap even though they responded vigorously at all grating positions. Adapted from Phillips, J.R. and Johnson, K.O., *J. Neurophysiol.* 46, 1192–1203, 1981, with permission.

The Somatosensory System

FIGURE 3.3 Confusion matrix of responses obtained from humans in a letter recognition task (top) and responses of a monkey SA1 afferent to the same letter stimuli (A-Z), which approximate the neural images of letter stimuli conveyed to the brain (bottom). The confusion matrix is derived from the pooled results of 64 subjects who performed either the active or passive letter identification task. The letters were raised 0.5 mm above the background and were 6 mm high. Matrix entries represent the frequencies of all possible responses to each letter (e.g., the letter A was called N on 8% of presentations). The numbers in the bottom row represent column sums. The numbers in the right-most column represent the number of presentations of each letter. Boxes around entries represent letter pairs whose mean confusion rates exceed 8%. For example, the mean confusion rate for B and G is 8% because G is called B on 11% of trials and B is called G on 5% of trails. The neural image (bottom) was derived from action potentials recorded from a single SA1 afferent fiber in a monkey. The stimuli consisted of the same embossed letters as in the letter recognition task scanned repeatedly from right to left across the receptive field of the neuron (equivalent to finger motion from left to right). Each black tick in the raster represents the occurrence of an action potential (see legend of Figure 3.4 for details). Adapted from Vega-Bermudez, F., Johnson, K.O., and Hsiao, S.S., *J. Neurophysiol.* 65, 531–546, 1991, with permission.

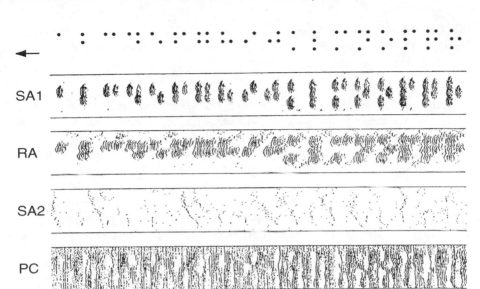

FIGURE 3.4 Responses of human SA1, RA, SA2, and PC afferent fibers to Braille symbols corresponding to the letters A–R. The Braille symbols were scanned (60 mm/s) repeatedly from right to left over the afferent fibers' receptive fields (equivalent to finger motion from left to right). Each black tick in the impulse rasters represents the occurrence of an action potential. Each row of the raster represents the response to a single scan. After each scan, the Braille pattern was shifted 0.2 mm at right angles to the scanning direction. The receptive fields were all on the distal fingerpads. Adapted from Phillips, J.R., Johansson, R.S., and Johnson, K.O., *Exp. Brain Res.* 81, 589–592, 1990, with permission.

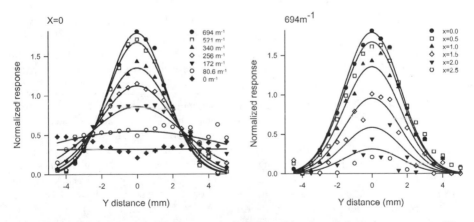

FIGURE 3.5 SA1 population response to indentation with spheres of varying curvature. The left plot shows the mean responses of SA1 afferents as a function of proximal–distal distance from the center of indentation. Data is shown for seven curved surfaces with radii ranging from 1.44 mm (curvature = 694 m^{-1}) to a flat surface (curvature = 0 m^{-1}). The right plot shows population response profiles in proximal–distal slices at varying distances from the center of indentation. From Goodwin, A.W., Browning, A.S., and Wheat, H.E., *J. Neurosci.* 15, 798–810, 1995, with permission.

weak dimension (sticky–slippery) improves the multidimensional scaling fit in some subjects. The discussion here is restricted to the neural mechanisms of the subjective sense of roughness and softness.

3.2.2.2.1 Roughness

The subjective sense of roughness has been studied extensively.[65-75] These studies show that roughness perception is unidimensional (the test of unidimensionality being the ability to assign numbers on a unidimensional continuum and to make greater-than and less-than judgments); that it depends on element height, diameter, shape, compliance, and density; and that the relationship between roughness and the different element parameters is complex and nonlinear. Lederman and her colleagues showed that scanning velocity, contact force, and friction between the finger and a surface have minor or no effects on roughness magnitude judgments.[68,69,76]

A series of studies (reviewed in Reference 1) has examined neural codes based on mean impulse rates, temporal measures of firing, and spatial measures of firing in each of the afferent systems as possible bases for the subjective sense of roughness. These studies were based explicitly on the idea of falsification.[77] The test of each neural coding hypothesis was consistency and it was implemented by plotting subjects' mean roughness judgments against each putative neural measure. A putative neural coding measure was rejected only when there was no consistent (one-to-one) relationship between the two. Those studies showed that only spatial variation in SA1 firing rates accounts for roughness perception over a wide range of stimuli and stimulus conditions.[71-73,75] All other measures failed the consistency test in one or more studies. Specifically, spatial variation was computed as the mean absolute difference in firing rates between SA1 afferents with receptive field centers separated by 2–3 mm. The correlation between subject's roughness judgments and this measure was 0.97 or greater in all studies (Figure 3.6).

A clue to the actual neural mechanism is provided by recent studies, reviewed below, that show that neurons in area 3b of primary somatosensory cortex (S1) have receptive fields composed of spatially separated regions of excitation and inhibition and that they compute spatial variation in the afferent discharge as hypothesized in

---

FIGURE 3.6 (opposite) Perceived roughness and spatial variation in SA1 firing rates in four studies with different textured surfaces. The left ordinate, filled circles, and solid lines in each graph represents the mean reported roughness for each surface (first normalized to a grand mean of 1.0 for each subject). The right ordinate, open circles, and dashed lines represent a measure of spatial variation in SA1 firing rates. The same measure of spatial variation was used for all four studies. The surface pattern used in each study is illustrated below the data to which it applies. The top row (A) shows results from Connor et al.,[71] who used 18 raised dot patterns with different dot spacings and diameters. The middle row (B) shows results from Blake et al.,[73] who used 18 raised dot patterns with different dot heights and diameters. The two left graphs in the bottom row (C) show results from Connor and Johnson,[72] who varied pattern geometry to distinguish temporal and spatial neural coding mechanisms. The right graph shows data from Yoshioka et al.,[75] who used fine gratings with spatial periods ranging from 0.1 to 2.0 mm. The lines connect stimulus patterns with constant spatial periods. From Yoshioka, T., Dorsch, A.K., Hsiao, S.S., and Johnson, K.O. *J. Neuroscience*, in press, 2001, with permission.

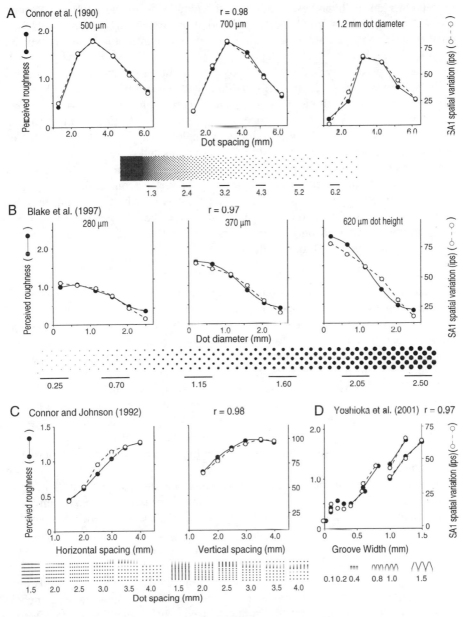

FIGURE 3.6

the studies described above. The mean firing rate of a population of such neurons with excitatory and inhibitory receptive fields separated by 2–3 mm would correspond closely to a subject's roughness judgments.[75] Such a mechanism has several advantages. Like roughness perception, it is unidimensional and it is affected only secondarily by factors such as scanning velocity and contact force.[78] Whether or not neurons in area 3b are responsible for the computations that underlie roughness perception requires hypothesis testing of the kind done in the studies described above. Nonetheless, a strong hypothesis has to be that roughness perception is based on the mean firing rate of a population of area 3b neurons that compute spatial variation in the SA1 population response.

A recent psychophysical study[79] sheds some light on the cortical mechanisms underlying roughness perception. If roughness is computed by neurons with receptive fields that span multiple digits (e.g., in S2 cortex) then roughness judgments should be affected by the spatial properties of surfaces contacting multiple fingers but they are not. Subjects either scanned a single sandpaper with one finger or scanned two adjacent sandpapers with adjacent fingers and were asked to base their subjective response on a single finger. The result was that the roughness judgments were completely unaffected by the roughness of the surface contacting the adjacent, unattended finger.[79] This suggests that if roughness is based on the mean impulse rate of a set of neurons that compute spatial variation in the SA1 population response then that rate is affected only by the surface contacting a single finger (i.e., the receptive field is restricted to a single finger).

3.2.2.2.2 Softness

The nature of the texture dimension, soft–hard, can be appreciated by pressing the keys of a computer keyboard and then other, softer objects. Perceived softness does not depend on the relationship between force and object displacement, the fact that the space bar gives way easily does not make it soft. A soft object conforms to the finger or hand as it is manipulated, but conformation is not sufficient; the keyboard keys that are molded to conform to the skin of a fingertip feel as hard as flat keys. The essence of softness is progressive conformation to the contours of the fingers and hand in proportion to contact force. The degree of softness is signaled by the rate of growth of contact area with contact force and by the uniformity of pressure across the contact area. Conversely, the essence of hardness is invariance of object form with changes in contact force. Resistance to deformation signals the hardness of an object or surface.

The neural mechanisms of the subjective sense of softness have not been studied systematically. Except for a study by Harper and Stevens,[80] most psychophysical studies have focused on the objective ability to discriminate compliance. Harper and Stevens showed that subjective softness judgments were related to the compliance of their test objects by a power function (exponent 0.8) and that hardness and softness judgments were reciprocally related. The most extensive study of the ability of humans to discriminate compliance is by Srinivasan and LaMotte,[81] who used cutaneous anesthesia and various modes of stimulus contact to show that cutaneous information alone is sufficient to discriminate the compliance of objects with deformable surfaces. Subjects discriminate softness when an object is applied to the passive, immobile finger as accurately as when they actively palpate the object. Moreover, they showed that this ability is unaffected when the velocity and force of application are randomized.

There are no combined psychophysical and neurophysiological experiments that systematically address the neural mechanisms of hardness perception, but the likely mechanism can be inferred from what we know about the response properties of each of the afferent types. As in roughness perception, the possible coding mechanisms are intensive, temporal, or spatial codes in one or more of the cutaneous afferent populations. Intensive codes are unlikely because random changes in velocity and force, which do not affect discrimination performance, have strong effects on afferent impulse rates.[82] Purely temporal codes seem unlikely because perceived softness (or hardness) is based on perceived changes in object form with changing contact force. Consequently, the system responsible for the perception of object form (i.e., the SA1 system) seems to be responsible for hardness/softness perception. Therefore, it is likely that the SA1 system is responsible for tactile perception in both of the principal dimensions of texture (smooth–rough and soft–hard) as well as for the perception of form.

3.2.2.3 Texture Perception with a Probe

When we use a tool or probe, we perceive distant events almost as if our fingers were present at the working surfaces of the tool or probe. An early demonstration of this was by Katz,[83] who showed that we can discriminate the texture of a surface as well with a probe as with a finger applied directly to the surface. He showed further that this capacity is lost when vibrations in the probe are damped. A recent study[21] has shown that when subjects grasp a probe, some subjects can detect transmitted vibrations with amplitudes less than 10 nm (the mean is 30 nm, Figure 3.7). Only the PC system can account for this capacity.

The hypothesis that the PC system is responsible for the perception of vibrations transmitted through an object held in the hand supposes that Pacinian receptors detect the transmitted vibration and that the PC population transmits a neural representation of the vibratory signal sufficient to account for this perceptual capacity. Work on the human ability to detect and discriminate complex vibratory stimuli supports this idea by showing that we are sensitive to the temporal structure of high frequency stimuli that only activate PC afferents.[84-86] For example, humans can discriminate the frequency with which a 250-Hz carrier stimulus is modulated for modulation frequencies as high as 60 Hz.[86] In contrast, Bensmaïa and Hollins[87] have shown that the discrimination of complex waveforms composed of high frequencies is poor and they have suggested that RA afferents play a significant role in temporal coding.

3.3 CORTICAL MECHANISMS

The clear segregation of function between the SA1, RA, SA2, and PC systems and the widely accepted idea that these afferent systems remain segregated in the central nervous system suggests a distinct segregation of function within the central pathways. However, considering how distinctly separate the functions of the four afferent systems are, it is surprising how little concrete evidence there is of functional segregation paralleling the functions of the four afferent systems. One of the few definite claims that can be made is that area 3b is critical for form perception. In the remainder of this chapter, we review the mechanisms of form processing in area 3b.

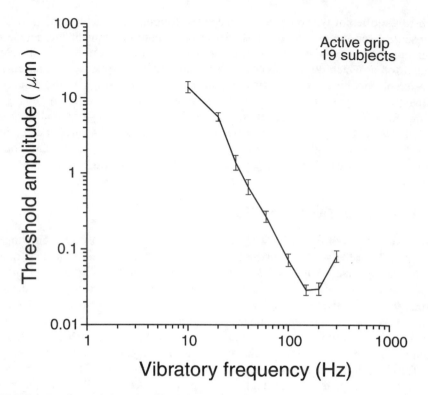

FIGURE 3.7 Vibratory detection for subjects grasping a 35 mm diameter rod lightly. A shaker embedded within the rod induced sinusoidal vibrations parallel to the axis of the rod. The rod motion was monitored by a sensitive, three-dimensional accelerometer. The ordinate represents the mean threshold (half the peak-to-peak excursion). Adapted from Brisben, A.J., Hsiao, S.S., and Johnson, K.O., *J. Neurophysiol.* 81, 1548–1558, 1999, with permission.

3.3.1 FUNCTION OF AREA 3B

The psychophysical studies reviewed earlier show that the limit of spatial acuity (about 1 mm) lies at the theoretical limit imposed by the peripheral innervation density. That implies central mechanisms specialized for the preservation of spatial information in the face of growing receptive field size within the central pathways. Neurophysiological studies have shown that area 3b has smaller receptive fields[88-90] and a higher proportion of cells responding to static skin indentation[88,91] than other cortical areas, but those observations are not especially indicative of a role in spatial information processing. The smaller receptive fields in area 3b are almost certainly a consequence of the fact that area 3b lies at an earlier stage of processing within pathways leading to distributed representations of spatial form.[92,93] A continued response to sustained, steady indentation implies input from slowly adapting afferents but the converse is not true, a transient response may result from lack of slowly adapting input or, just as likely, delayed inhibition that shuts off the response to a sustained input.[91,94-97] Ablation studies shed little light on the function of area 3b. Removal of area 3b produces profound behavioral deficits in all somatosensory tasks tested, while removal of other S1 areas appears to produce more specific deficits in the tactile discrimination of textures (area 1) and three-dimensional forms (area 2).[98] A clue that area 3b is specialized for processing spatial information comes from the relative expansions of the representations of the digits in areas 3b and 1. The cortical magnification factors in areas 3b and 1 are approximately equal over most of the postcentral gyrus except in the finger region, where the magnification in area 3b climbs to approximately five times that in area 1 (at the representation of the fingertips[89]).

Another clue comes from the response properties of neurons in area 3b. Almost all area 3b neurons appear to have homogeneous, excitatory receptive fields when probed with punctate stimuli.[89,99] The literature on area 3b is striking for the small number of neurons with apparent inhibitory subfields and for the small number of neurons with even mildly complex response properties like direction, orientation, or velocity selectivity when probed manually.[99-104] Hyvärinen and Poranen,[105] for example, reported that only 4 of 122 neurons in area 3b of the alert monkey had mildly complex response properties; no neurons in their study were orientation or directionally selective. These observations match our own experience when exploring receptive fields in area 3b manually. However, when neurons with receptive fields on the fingerpads are stimulated with scanned, complex spatial stimuli almost all yield responses that are more complex than can be accounted for by simple, excitatory receptive fields and it is evident that the neurons are responding to specific features of the stimuli.[92,93,106,107] Responses of two SA neurons in area 3b are illustrated in Figure 3.8. The neuron illustrated in the top two rows of Figure 3.8 is clearly responding to the orientations of letter segments. The neuron illustrated in the bottom two rows has a complex response that is not easily interpreted; for example, the overall response to the letter is suppressed by the horizontal bars within the B, E, and F, but not the H. Thus, the neuronal responses in area 3b suggest specialization for form processing rather than, for example, motion processing.

3.3.2 RECEPTIVE FIELDS IN AREA 3B

A recent series of quantitative studies with controlled, scanned stimuli has confirmed that there is little or no directional selectivity in area 3b and that area 3b neuronal discharge rates are affected only mildly by changes in stimulus velocity across the skin.[78,97,108] But they also show that all neurons in area 3b are selectively responsive to particular spatial patterns of stimulation, that they are sensitive to the orientation of these patterns, and that this selectivity is shaped as much by inhibition as excitation. The reason that previous studies have generally failed to identify this inhibition is that it is manifested only as a reduction of the response to a stimulus in the excitatory part of the receptive field. Ninety-five percent of area 3b neuronal receptive fields have three components: (1) a single, central excitatory region of short duration (10 ms at most), (2) one or more inhibitory regions that are adjacent to and synchronous with the excitation, and (3) a larger inhibitory region that overlaps the excitation partially or totally and is delayed with respect to the first two components (by 30 ms on average). The remaining 5% have two or more regions of excitation.

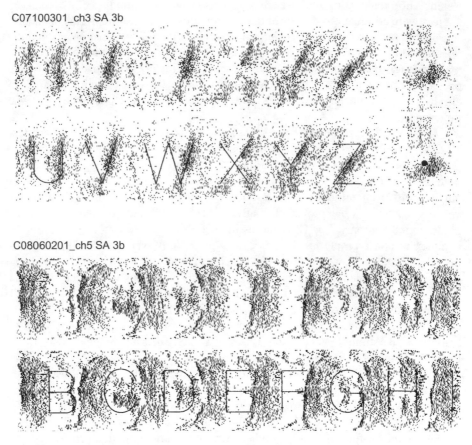

FIGURE 3.8 S1 cortical neuronal responses to letters, 8 mm high, scanned at 50 mm/sec. Both neurons were recorded from area 3b and were classed as SA, i.e., each responded to sustained skin indentation with a sustained discharge. The bottom set of rasters in each panel illustrates the neuron's response with the stimulus superimposed to show how the neuron's activity related to the spatial features of the stimulus. From Johnson, K.O., Hsiao, S.S., and Twombly, I.A., in *The Cognitive Neurosciences*, M. S. Gazzaniga, Ed., pp. 235–268, The MIT Press, Cambridge, MA, 1995, with permission.

The receptive fields of 247 area 3b neurons mapped with scanned, random-dot stimuli are illustrated in Figure 3.9, which shows that nearly all receptive fields are characterized by a single central region of excitation with inhibition on one, two, or three sides. Surround inhibition occurred rarely. The inhibitory area was, on average, about 30% larger than the excitatory area (means were 18 and 14 mm²) and, like the excitatory area, varied greatly (from 1 to 47 mm²). The inhibitory mass (absolute value of inhibition integrated over the entire inhibitory field) like the excitatory mass (comparable definition), varied by 50 to 1 between neurons (125 to 6830 mass units; mean 1620 mass units). There was no evidence of clustering into distinct receptive field types. The distributions of excitatory and inhibitory areas and masses were all Gaussian in logarithmic coordinates (i.e., lognormal); the excitatory and inhibitory masses were more closely correlated ($r = 0.56$) than were the areas ($r = 0.26$). Receptive fields mapped in this way predict neuronal responses to stimulus features such as orientation accurately.[97]

3.3.3 DELAYED INHIBITION

Area 3b neurons have two striking response properties that are not evident in Figure 3.9. The first is that the spatiotemporal structure of their neuronal responses and the spatial structures of their receptive fields are virtually unaffected by the velocity with which a stimulus moves across the skin or, conversely, how rapidly a finger is scanned over a surface for velocities up to at least 80 mm/sec;[78] that is, the spatial structure of raster plots like the ones illustrated in Figure 3.8 are unaffected by changes in scanning velocity. Increasing scanning velocity causes a marked increase in the intensities of the excitatory and inhibitory subfields without affecting their geometries. This results in increased firing rates without any loss of the response selectivity conferred by the receptive field geometry. The mechanism of this increased responsiveness with increased velocity lies in an interaction between the excitation and the delayed inhibition.[78]

The delayed inhibition confers a second property, which is that the receptive field geometry depends strongly on scanning direction.[97] A typical example from a study in which eight scanning directions were used is shown in Figure 3.10. The receptive field at the left of each group of three receptive field diagrams is the receptive field obtained directly from the neuron's responses to random-dot stimuli scanned in one of the eight directions. By inspecting each receptive field in Figure 3.10, it can be seen that, regardless of scanning direction, there is a fixed region of inhibition distal to and left of the excitation. It can also be seen that there is a region of inhibition displaced in the scanning direction (opposite to the finger motion) from the region of excitation. To visualize this more clearly, each neuron's response to scanning in multiple directions was fitted with a three-component receptive field model comprising a Gaussian excitatory region and two Gaussian inhibitory regions, one to simulate the region of fixed, synchronous inhibition and one to simulate the region of delayed inhibition. This model is illustrated by the simulated receptive field in the central panel in each group of three panels and by the corresponding diagram in the right panel. The degree to which the model description accounts for the observed receptive fields can be seen by comparing the model receptive field in the central panel and the actual receptive field in the left panel. This comparison shows that the three-component model explains the effect of scanning direction on receptive field shape well. The correlation between the model and observed receptive fields averaged 0.81 in 62 neurons studied with four or more scanning directions. Neurons with lower correlations all had a third, fixed region of inhibition not accounted for by the three-component model (i.e., they would have been described by the three components enumerated above if the model had allowed for more than one region of fixed inhibition).

Area 3b

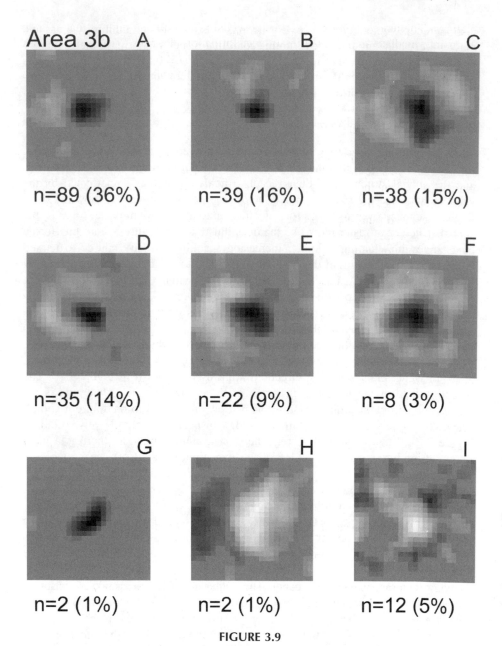

FIGURE 3.9

3.3.4 FUNCTIONAL IMPLICATIONS

The wide range of receptive field geometries and the wide range of responses to complex, scanned stimuli found in area 3b shows that the initial, isomorphic neural representation of spatial form that prevails in the periphery gives way to an altered form of representation in which neuronal responses represent the presence of specific features. The more complex responses observed in S2 cortex[93] suggest that area 3b is an intermediate step in a series of transformations leading to a more complex form of representation.[92,93,108]

The fixed inhibitory components of each neuron's receptive field interact with the central excitation to act as a spatial filter, conferring selectivity for particular spatial features or patterns independent of scanning direction and velocity. For example, when the fixed inhibition lies on two adjacent sides, the neuron is more responsive to corners that protrude into the excitatory subfield without activating the inhibitory subregions. When the fixed inhibitory subfield occupies a single location on one side of the excitatory subfield, both tend to be elongated and to lie parallel to one another; as a result, the neuron is more responsive to edges oriented orthogonal to the displacement between the two.[97,108]

The delayed inhibitory component serves three functions. First, it confers sensitivity to stimulus gradients in the scanning direction, whatever that direction. The delayed inhibition suppresses the response to uniform surfaces and thereby emphasizes the effects of spatial or temporal novelty. When scanning the finger over a surface, elevated features trigger the excitation and fixed inhibition first and then the lagged inhibition 30 ms later. Second, when the delayed inhibition is centered on the excitation, it produces a progressive increase in discharge rate with increasing scanning velocity. The acquisition of tactile spatial information by scanning one's finger over a surface compensates for the very limited field of view provided by a single fingerpad. It is clearly an advantage to be able to scan one's fingers over an object or a surface rapidly without loss of information. Psychophysical experiments show that there is no loss of performance in pattern recognition as scanning rate increases to 40 mm/s and then only a small loss as the rate is increased to 80 mm/s.[35] Without a compensatory mechanism, rapid scanning has a substantial cost. As scanning velocity increases, each stimulus element spends less time within the receptive field (reduced dwell time) and the element is represented by fewer action

---

FIGURE 3.9 (opposite) Receptive fields in area 3b of the alert monkey. Each panel illustrates an example of a receptive field type, the total number of receptive fields of that type, and their percentage of the total receptive field sample (n = 247, all on distal fingerpads). The gray scale represents the grid of excitation and inhibition (25 × 25 bins = 10 × 10 mm) that best described the neuron's response to a random stimulus pattern. Dark regions represent excitatory regions; lighter regions represent inhibitory regions. The uniform background gray level represents the region where stimuli had no effect. The types are shown in decreasing order of frequency: **A.** A single inhibitory region located on the trailing (distal) side of the excitatory region (left). **B.** A region of inhibition located on one of the three nontrailing sides of the excitatory region. **C.** Two regions of inhibition on opposite sides of the excitatory region. **D.** Inhibition on three sides of the excitatory region. **E.** Inhibition on two contiguous sides of the excitatory region. **F.** A complete inhibitory surround. **G.** An excitatory region only. **H.** Receptive field dominated by inhibition. **I.** Receptive fields not easily assigned to one of the preceding categories. This categorization is meant only to show the range and relative frequency of these receptive field types. In fact, the receptive fields formed a continuum spanning these shapes. From DiCarlo, J.J., Johnson, K.O., and Hsiao, S.S., *J. Neurosci.* 18, 2626–2645, 1998, with permission.

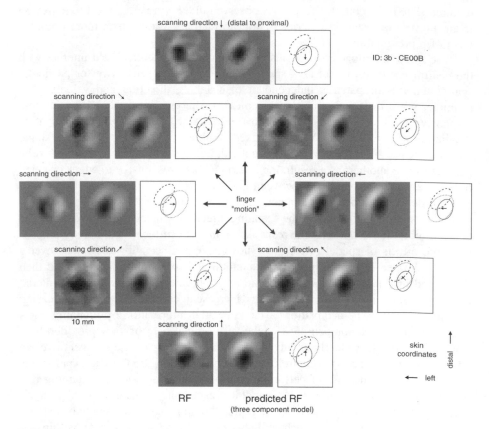

FIGURE 3.10 Receptive fields from a single area 3b neuron determined in eight scanning directions. The three squares in each group display the receptive field estimated from the raw data (on the left), the receptive field predicted by the three-component model (middle), and the positions of the model Gaussian components (on the right). The ellipses in the right square in each group are iso-amplitude contours at 1.5 SD. The scanning direction is shown above each group. Each receptive field is plotted as if it were viewed through the dorsum of the finger with the finger pointed toward the top of the figure; the effect of relative motion between the finger and the stimulus pattern on the receptive field can be visualized by placing a fingerpad in the center of the figure and sliding it along the arrow labeled "finger motion" toward the receptive field of interest. Note how the locations of the model's excitatory (solid ellipse) and fixed inhibitory components (dashed ellipse) are unaffected by scanning direction and, similarly, how the lagged inhibitory component (dotted ellipse) trails the lag center by a fixed distance in each direction (the lag distance is the same in all directions because the scanning velocity was constant, 40 mm/s). The arrow in each right-hand square shows the displacement of the lagged inhibitory component due to scanning. The degree to which the model accounts for receptive field structure in each direction can be seen by comparing the left and middle panels in each group. From DiCarlo, J.J. and Johnson, K.O., *J. Neurosci.* 20, 495–510, 2000, with permission.

potentials. The delayed inhibition provides a compensatory mechanism that increases firing rate with increasing scanning velocity. As velocity increases, the delayed inhibition lags progressively to expose more excitation. Consequently, the excitatory and inhibitory components of the receptive field grow rapidly in intensity with no effect on receptive field geometry. The result is a representation of spatial form that is invariant with scanning velocity and is more intense than it would be without this mechanism.[78] Third, and least significant in area 3b, selectivity for motion direction results when the center of the delayed inhibition is displaced from the center of excitation.[96,97,109,110] Motion in the direction of the displacement exposes progressively more excitation, which produces a progressively greater discharge rate. Motion in the opposite direction shifts the delayed inhibition over the center of excitation, thereby reducing the discharge rate. The center of the delayed inhibition in area 3b is, with few exceptions, close to the center of excitation, which explains why so few neurons in area 3b exhibit directional selectivity. When the center of the delayed inhibition is displaced from the center of excitation, it predicts the neuron's directional selectivity accurately.[97]

3.4 CONCLUSION

In this chapter, we have reviewed the peripheral and cortical neural mechanisms of tactile form and texture perception. Three decades of combined psychophysical and neurophysiological experiments suggest a sharp division of function among the four cutaneous afferent systems that innervate the human hand. The SA1 system provides a high-quality neural image of the spatial structure of objects and surfaces that contact the skin and this is the basis of form and texture perception. The RA system provides a neural image of motion signals from the whole hand. From this, the brain extracts information that is critical for grip control and also information about the motion of objects contacting the skin. The PC system provides a neural image of vibrations transmitted to the hand from objects contacting the hand or, more frequently, objects grasped in the hand. The SA2 system provides a neural image of skin stretch over the whole hand. The evidence is less secure but the most likely hypothesis is that the brain extracts information about hand conformation and the direction of motion of objects moving across the skin. S1 cortex transforms this information to an altered form of representation that is, as yet, not well understood. The available evidence suggests that area 3b is specialized for processing spatial information. Evidence of that is its magnification of body parts with high spatial resolution compared with area 1 and its neuron's response properties, which appear to be specialized for processing spatial information.

REFERENCES

1. Johnson, K.O., Yoshioka, T., and Vega-Bermudez, F., Tactile functions of mechanoreceptive afferents innervating the hand. *J. Clin. Neurophysiol.* 17, 539, 2000.
2. Johnson, K.O. and Hsiao, S.S., Neural mechanisms of tactual form and texture perception. *Annu. Rev. Neurosci.* 15, 227, 1992.

3. Iggo, A. and Andres, K.H., Morphology of cutaneous receptors. *Annu. Rev. Neurosci.* 5, 1, 1982.

4. Diamond, J., Mills, L.R., and Mearow, K.M., Evidence that the Merkel cell is not the transducer in the mechanosensory Merkel cell-neurite complex. *Prog. Brain Res.* 74, 51, 1988.

5. Ogawa, H., The Merkel cell as a possible mechanoreceptor cell. *Prog. Neurobiol.* 49, 317, 1996.

6. Vega-Bermudez, F. and Johnson, K.O., Surround suppression in the responses of primate SA1 and RA mechanoreceptive afferents mapped with a probe array. *J. Neurophysiol.* 81, 2711, 1999.

7. Phillips, J.R. and Johnson, K.O., Tactile spatial resolution: II. Neural representation of bars, edges, and gratings in monkey primary afferents. *J. Neurophysiol.* 46, 1192, 1981.

8. Srinivasan, M.A. and Dandekar, K., An investigation of the mechanics of tactile sense using two-dimensional models of the primate fingertip. *J. Biomech. Eng.* 118, 48, 1996.

9. Munger, B.L. and Ide, C., The structure and function of cutaneous sensory receptors. *Arch. Histol. Cytol.* 51, 1, 1988.

10. Guinard, D., Usson, Y., Guillermet, C., and Saxod, R., PS-100 and NF 70-200 double immunolabeling for human digital skin Meissner corpuscle 3D imaging. *J. Histochem. Cytochem.* 48, 295, 2000.

11. Quilliam, T.A., The structure of finger print skin. In: *Active Touch*, G. Gordon, Ed., 1, Pergamon Press, Oxford, 1978.

12. Johansson, R.S. and Vallbo, Å.B., Skin mechanoreceptors in the human hand: an inference of some population properties. In: *Sensory Functions of the Skin in Primates*, Y. Zotterman, Ed., 171, Pergamon Press, Oxford, 1976.

13. Johansson, R.S. and Vallbo, Å.B., Tactile sensibility in the human hand: relative and absolute densities of four types of mechanoreceptive units in glabrous skin. *J. Physiol. (Lond.)* 286, 283, 1979.

14. Darian-Smith, I. and Kenins, P., Innervation density of mechanoreceptive fibers supplying glabrous skin of the monkey's index finger. *J. Physiol. (Lond.)* 309, 147, 1980.

15. Vega-Bermudez, F. and Johnson, K.O., SA1 and RA receptive fields, response variability, and population responses mapped with a probe array. *J. Neurophysiol.* 81, 2701, 1999.

16. Mountcastle, V.B., Talbot, W.H., and Kornhuber, H.H., The neural transformation of mechanical stimuli delivered to the monkey's hand. In: *Touch, Heat, Pain and Itch*, A.V. de Reuck and J. Knight, Eds., 325, Churchill, London, 1966.

17. Johansson, R.S., Tactile afferent units with small and well demarcated receptive fields in the glabrous skin area of the human hand. In: *Sensory Function of the Skin in Humans*, D.R. Kenshalo, Ed., 129, Plenum, New York, 1979.

18. Blake, D.T., Johnson, K.O., and Hsiao, S.S., Monkey cutaneous SAI and RA responses to raised and depressed scanned patterns: effects of width, height, orientation, and a raised surround. *J. Neurophysiol.* 78, 2503, 1997.

19. Bell, J., Bolanowski, S.J., and Holmes, M.H., The structure and function of Pacinian corpuscles: A review. *Prog. Neurobiol.* 42, 79, 1994.

20. Bolanowski, S.J. and Zwislocki, J.J., Intensity and frequency characteristics of Pacinian corpuscles. I. Action potentials. *J. Neurophysiol.* 51, 793, 1984.

21. Brisben, A.J., Hsiao, S.S., and Johnson, K.O., Detection of vibration transmitted through an object grasped in the hand. *J. Neurophysiol.* 81, 1548, 1999.

22. Hubbard, S.J., A study of rapid mechanical events in a mechanoreceptor. *J. Physiol. (Lond.)* 141, 198, 1958.

23. Loewenstein, W.R. and Skalak, R., Mechanical transmission in a Pacinian corpuscle. An analysis and a theory. *J. Physiol. (Lond.)* 182, 346, 1966.

24. Johansson, R.S. and Vallbo, Å.B., Spatial properties of the population of mechanoreceptive units in the glabrous skin of the human hand. *Brain Res.* 184, 353, 1980.

25. Johansson, R.S. and Vallbo, Å.B., Detection of tactile stimuli. Thresholds of afferent units related to psychophysical thresholds in the human hand. *J. Physiol. (Lond.)* 297, 405, 1979.

26. Johansson, R.S., Vallbo, Å.B., and Westling, G., Thresholds of mechanosensitive afferents in the human hand as measured with von Frey hairs. *Brain Res.* 184, 343, 1980.

27. Edin, B.B., Quantitative analysis of static strain sensitivity in human mechanoreceptors from hairy skin. *J. Neurophysiol.* 67, 1105, 1992.

28. Chambers, M.R., Andres, K.H., von Duering, M., and Iggo, A., The structure and function of the slowly adapting type II mechanoreceptor in hairy skin. *Q.J. Exp. Physiol.* 57, 417, 1972.

29. Macefield, V.G., Hager-Ross, C., and Johansson, R.S., Control of grip force during restraint of an object held between finger and thumb: responses of cutaneous afferents from the digits. *Exp. Brain Res.* 108, 155, 1996.

30. Goodwin, A.W., Macefield, V.G., and Bisley, J.W., Encoding of object curvature by tactile afferents from human fingers. *J. Neurophysiol.* 78, 2881, 1997.

31. Phillips, J.R., Johansson, R.S., and Johnson, K.O., Representation of Braille characters in human nerve fibers. *Exp. Brain Res.* 81, 589, 1990.

32. Olausson, H., Wessberg, J., and Kakuda, N., Tactile directional sensibility: peripheral neural mechanisms in man. *Brain Res.* 866, 178, 2000.

33. Lamb, G.D., Tactile discrimination of textured surfaces: psychophysical performance measurements in humans. *J. Physiol. (Lond.)* 338, 551, 1983.

34. Morley, J.W., Goodwin, A.W., and Darian-Smith, I., Tactile discrimination of gratings. *Exp. Brain Res.* 49, 291, 1983.

35. Vega-Bermudez, F., Johnson, K.O., and Hsiao, S.S., Human tactile pattern recognition: Active vs. passive touch, velocity effects, and patterns of confusion. *J. Neurophysiol.* 65, 531, 1991.

36. Johnson, K.O. and Lamb, G.D., Neural mechanisms of spatial tactile discrimination: neural patterns evoked by Braille-like dot patterns in the monkey. *J. Physiol. (Lond.)* 310, 117, 1981.

37. Loomis, J.M., On the tangibility of letters and Braille. *Percept. Psychophys.* 29, 37, 1981.

38. Phillips, J.R., Johnson, K.O., and Browne, H.M., A comparison of visual and two modes of tactual letter resolution. *Percept. Psychophys.* 34, 243, 1983.

39. Loomis, J.M., Tactile recognition of raised characters: a parametric study. *Bull. Psychon. Soc.* 23, 18, 1985.

40. Johnson, K.O. and Phillips, J.R., Tactile spatial resolution: I. Two-point discrimination, gap detection, grating resolution, and letter recognition. *J. Neurophysiol.* 46, 1177, 1981.

41. Vega-Bermudez, F. and Johnson, K.O. Differences in spatial acuity between digits. *Neurology* in press. 2001.

42. Stevens, J.C. and Choo, K.K., Spatial acuity of the body surface over the life span. *Somatosens. Mot. Res.* 13, 153, 1996.

43. Wohlert, A.B., Tactile perception of spatial stimuli on the lip surface by younger and older adults. *J. Speech Hear. Res.* 39, 1191, 1996.

44. Sathian, K., Zangaladze, A., Green, J., Vitek, J.L., and DeLong, M.R., Tactile spatial acuity and roughness discrimination: impairments due to aging and Parkinson's disease. *Neurology* 49, 168, 1997.

45. Hsiao, S.S., O'Shaughnessy, D.M., and Johnson, K.O., Effects of selective attention of spatial form processing in monkey primary and secondary somatosensory cortex. *J. Neurophysiol.* 70, 444, 1993.

46. Van Boven, R.W. and Johnson, K.O., The limit of tactile spatial resolution in humans: Grating orientation discrimination at the lip, tongue, and finger. *Neurology* 44, 2361, 1994.

47. Sathian, K. and Zangaladze, A., Tactile spatial acuity at the human fingertip and lip: Bilateral symmetry and interdigit variability. *Neurology* 46, 1464, 1996.

48. Essick, G.K., Chen, C.C., and Kelly, D.G., A letter-recognition task to assess lingual tactile acuity. *J. Oral Maxillofac. Surg.* 57, 1324, 1999.

49. Phillips, J.R., Johansson, R.S., and Johnson, K.O., Responses of human mechanoreceptive afferents to embossed dot arrays scanned across fingerpad skin. *J. Neurosci.* 12, 827, 1992.

50. Kops, C.E. and Gardner, E.P., Discrimination of simulated texture patterns on the human hand. *J. Neurophysiol.* 76, 1145, 1996.

51. Bliss, J.C., A relatively high-resolution reading aid for the blind. *IEEE Trans. Man-Machine Sys.* 10, 1, 1969.

52. Gardner, E.P. and Palmer, C.I., Simulation of motion on the skin. I. Receptive fields and temporal frequency coding by cutaneous mechanoreceptors of Optacon pulses delivered to the hand. *J. Neurophysiol.* 62, 1410, 1989.

53. Palmer, C.I. and Gardner, E.P., Simulation of motion on the skin. IV. Responses of Pacinian corpuscle afferents innervating the primate hand to stripe patterns on the OPTACON. *J. Neurophysiol.* 64, 236, 1990.

54. Craig, J.C. and Rollman, G.B., Somesthesis. *Annu. Rev. Psychol.* 50, 305, 1999.

55. Goodwin, A.W., John, K.T., and Marceglia, A.H., Tactile discrimination of curvature by humans using only cutaneous information from the fingerpads. *Exp. Brain Res.* 86, 663, 1991.

56. Goodwin, A.W. and Wheat, H.E., Human tactile discrimination of curvature when contact area with the skin remains constant. *Exp. Brain Res.* 88, 447, 1992.

57. Goodwin, A.W. and Wheat, H.E., Magnitude estimation of force when objects with different shapes are applied passively to the fingerpad. *Somatosens. Mot. Res.* 9, 339, 1992.

58. Goodwin, A.W., Browning, A.S., and Wheat, H.E., Representation of curved surfaces in responses of mechanoreceptive afferent fibers innervating the monkey's fingerpad. *J. Neurosci.* 15, 798, 1995.

59. Srinivasan, M.A. and LaMotte, R.H., Tactile discrimination of shape: responses of slowly and rapidly adapting mechanoreceptive afferents to a step indented into the monkey fingerpad. *J. Neurosci.* 7, 1682, 1987.

60. LaMotte, R.H. and Srinivasan, M.A., Responses of cutaneous mechanoreceptors to the shape of objects applied to the primate fingerpad. *Acta Psychol. (Amst.)* 84, 41, 1993.

61. Dodson, M.J., Goodwin, A.W., Browning, A.S., and Gehring, H.M., Peripheral neural mechanisms determining the orientation of cylinders grasped by the digits. *J. Neurosci.* 18, 521, 1998.

62. Khalsa, P.S., Friedman, R.M., Srinivasan, M.A., and LaMotte, R.H., Encoding of shape and orientation of objects indented into the monkey fingerpad by populations of slowly and rapidly adapting mechanoreceptors. *J. Neurophysiol.* 79, 3238, 1998.

63. LaMotte, R.H., Friedman, R.M., Lu, C., Khalsa, P.S., and Srinivasan, M.A., Raised object on a planar surface stroked across the fingerpad: Responses of cutaneous mechanoreceptors to shape and orientation. *J. Neurophysiol.* 80, 2446, 1998.

64. Hollins, M., Faldowski, R., Rao, S., and Young, F., Perceptual dimensions of tactile surface texture: A multidimensional-scaling analysis. *Percept. Psychophys.* 54, 697, 1993.

65. Hollins, M., Bensmaïa, S.J., Karlof, K., and Young, F., Individual differences in perceptual space for tactile textures: Evidence from multidimensional scaling. *Percept. Psychophys.* 62, 1534, 2000.

66. Meenes, M. and Zigler, M.J., An experimental study of the perceptions of roughness and smoothness. *Am. J. Psychol.* 34, 542, 1923.

67. Stevens, S.S. and Harris, J.R., The scaling of subjective roughness and smoothness. *J. Exp. Psychol.* 64, 489, 1962.

68. Lederman, S.J., Tactile roughness of grooved surfaces: The touching process and the effects of macro- and micro-surface structure. *Percept. Psychophys.* 16, 385, 1974.

69. Lederman, S.J., Tactual roughness perception: Spatial and temporal determinants. *Can. J. Psychol.* 37, 498, 1983.

70. Sathian, K., Goodwin, A.W., John, K.T., and Darian-Smith, I., Perceived roughness of a grating: correlation with responses of mechanoreceptive afferents innervating the monkey's fingerpad. *J. Neurosci.* 9, 1273, 1989.

71. Connor, C.E., Hsiao, S.S., Phillips, J.R., and Johnson, K.O., Tactile roughness: neural codes that account for psychophysical magnitude estimates. *J. Neurosci.* 10, 3823, 1990.

72. Connor, C.E. and Johnson, K.O., Neural coding of tactile texture: comparisons of spatial and temporal mechanisms for roughness perception. *J. Neurosci.* 12, 3414, 1992.

73. Blake, D.T., Hsiao, S.S., and Johnson, K.O., Neural coding mechanisms in tactile pattern recognition: the relative contributions of slowly and rapidly adapting mechanoreceptors to perceived roughness. *J. Neurosci.* 17, 7480, 1997.

74. Meftah, E.M., Belingard, L., and Chapman, C.E., Relative effects of the spatial and temporal characteristics of scanned surfaces on human perception of tactile roughness using passive touch. *Exp. Brain Res.* 132, 351, 2000.

75. Yoshioka, T., Dorsch, A. K., Hsiao, S. S., and Johnson, K. O. Neural coding mechanisms underlying perceived roughness of finely textured surfaces. *J. Neuroscience*, in press, 2001.

76. Taylor, M.M. and Lederman, S.J., Tactile roughness of grooved surfaces: A model and the effect of friction. *Percept. Psychophys.* 17, 23, 1975.

77. Popper, K., *The Logic of Scientific Discovery*, Basic Books, New York, 1959.

78. DiCarlo, J.J. and Johnson, K.O., Velocity invariance of receptive field structure in somatosensory cortical area 3b of the alert monkey. *J. Neurosci.* 19, 401, 1999.

79. Dorsch, A. K., Hsiao, S. S., Johnson, K. O., and Yoshioka, T. Tactile attention: subjective magnitude estimates of roughness using one or two fingers. *Soc. Neuroscience Abstr.* 27, 2001.

80. Harper, R. and Stevens, S.S., Subjective hardness of compliant materials. *Q. J. Exp. Psychol.* 16, 204, 1964.

81. Srinivasan, M.A. and LaMotte, R.H., Tactual discrimination of softness. *J. Neurophysiol.* 73, 88, 1995.

82. Srinivasan, M.A. and LaMotte, R.H., Tactual discrimination of softness: Abilities and mechanisms. In: *Somesthesis and the Neurobiology of the Somatosensory Cortex*, O. Franzén, R.S. Johansson, and L. Terenius, Eds., 123, Birkhäuser Verlag, Basel, 1996.

83. Katz, D., *The World of Touch*, Erlbaum (Krueger, L.E., translator; published originally in 1925), Hillsdale, NJ, 1925.

84. Weisenberger, J.M., Sensitivity to amplitude-modulated vibrotactile signals. *J. Acoust. Soc. Am.* 80, 1707, 1986.

85. Lamore, P.J.J., Muijser, H., and Keemink, C.J., Envelope detection of amplitude-modulated high-frequency sinusoidal signals by skin mechanoreceptors. *J. Acoust. Soc. Am.* 79, 1082, 1986.

86. Formby, C., Morgan, L.N., Forrest, T.G., and Raney, J.J., The role of frequency selectivity in measures of auditory and vibrotactile temporal resolution. *J. Acoust. Soc. Am.* 91, 293, 1992.

87. Bensmaïa, S.J. and Hollins, M., Complex tactile waveform discrimination. *J. Acoust. Soc. Am.* 108, 1236, 2000.

88. Paul, R.L., Goodman, H., and Merzenich, M.M., Alterations in mechanoreceptor input to Brodmann's areas 1 and 3 of the postcentral hand area of Macaca mulatta after nerve section and regeneration. *Brain Res.* 39, 1, 1972.

89. Sur, M., Merzenich, M.M., and Kaas, J.H., Magnification, receptive-field area, and hypercolumn size in areas 3b and 1 of somatosensory cortex in owl monkeys. *J. Neurophysiol.* 44, 295, 1980.

90. Sur, M., Garraghty, P.E., and Bruce, C.J., Somatosensory cortex in macaque monkeys: laminar differences in receptive field size in areas 3b and 1. *Brain Res.* 342, 391, 1985.

91. Sur, M., Wall, J.T., and Kaas, J.H., Modular distribution of neurons with slowly adapting and rapidly adapting responses in area 3b of somatosensory cortex in monkeys. *J. Neurophysiol.* 51, 724, 1984.

92. Bankman, I.N., Johnson, K.O., and Hsiao, S.S., Neural image transformation in the somatosensory system of the monkey: Comparison of neurophysiological observations with responses in a neural network model. *Cold Spring Harb. Symp. Quant. Biol.* 55, 611, 1990.

93. Johnson, K.O., Hsiao, S.S., and Twombly, I.A., Neural mechanisms of tactile form recognition. In: *The Cognitive Neurosciences*, M.S. Gazzaniga, Ed., 235, The MIT Press, Cambridge, MA, 1995.

94. Andersson, S.A., Intracellular postsynaptic potentials in the somatosensory cortex of the cat. *Nature* 205, 297, 1965.

95. Innocenti, G.M. and Manzoni, T., Response patterns of somatosensory cortical neurones to peripheral stimuli. An intracellular study. *Arch. Ital. Biol.* 110, 322, 1972.

96. Gardner, E.P. and Costanzo, R.M., Temporal integration of multiple-point stimuli in primary somatosensory cortical receptive fields of alert monkeys. *J. Neurophysiol.* 43, 444, 1980.

97. DiCarlo, J.J. and Johnson, K.O., Spatial and temporal structure of receptive fields in primate somatosensory area 3b: effects of stimulus scanning direction and orientation. *J. Neurosci.* 20, 495, 2000.

98. Randolph, M. and Semmes, J., Behavioral consequences of selective ablations in the postcentral gyrus of Macaca mulatta. *Brain Res.* 70, 55, 1974.

99. Mountcastle, V.B. and Powell, T.P.S., Neural mechanisms subserving cutaneous sensibility, with special reference to the role of afferent inhibition in sensory perception and discrimination. *Bull. Johns Hopkins Hosp.* 105, 201, 1959.

100. Whitsel, B.L., Roppolo, J.R., and Werner, G., Cortical information processing of stimulus motion on primate skin. *J. Neurophysiol.* 35, 691, 1972.

101. Hyvärinen, J. and Poranen, A., Movement-sensitive and direction- and orientation-selective cutaneous receptive fields in the hand area of the post-central gyrus in monkeys. *J. Physiol. (Lond.)* 283, 523, 1978.

102. Pubols, L.M. and LeRoy, R.F., Orientation detectors in the primary somatosensory neocortex of the raccoon. *Brain Res.* 129, 61, 1977.

103. Iwamura, Y., Tanaka, M., Sakamoto, M., and Hikosaka, O., Functional subdivisions representing different finger regions in area 3 of the first somatosensory cortex of the conscious monkey. *Exp. Brain Res.* 51, 315, 1983.

104. Warren, S., Hämäläinen, H.A., and Gardner, E.P., Coding of the spatial period of gratings rolled across the receptive fields of somatosensory cortical neurons in awake monkeys. *J. Neurophysiol.* 56, 623, 1986.

105. Hyvärinen, J. and Poranen, A., Receptive field integration and submodality convergence in the hand area of the post-central gyrus of the alert monkey. *J. Physiol. (Lond.)* 283, 539, 1978.

106. Phillips, J.R., Johnson, K.O., and Hsiao, S.S., Spatial pattern representation and transformation in monkey somatosensory cortex. *Proc. Natl. Acad. Sci. (U.S.A.)* 85, 1317, 1988.

107. Hsiao, S.S., Johnson, K.O., Twombly, I.A., and DiCarlo, J.J., Form processing and attention effects in the somatosensory system. In: *Somesthesis and the Neurobiology of the Somatosensory Cortex,* O. Franzén, R.S. Johansson, and L. Terenius, Eds., 229, Birkhäuser Verlag, Basel, 1996.

108. DiCarlo, J.J., Johnson, K.O., and Hsiao, S.S., Structure of receptive fields in area 3b of primary somatosensory cortex in the alert monkey. *J. Neurosci.* 18, 2626-2645, 1998.

109. Barlow, H.B. and Levick, W.R., The mechanism of directionally selective units in rabbit's retina. *J. Physiol. (Lond.)* 178, 477, 1965.

110. Warren, S., Hämäläinen, H.A., and Gardner, E.P., Objective classification of motion- and direction-sensitive neurons in primary somatosensory cortex of awake monkeys. *J. Neurophysiol.* 56, 598, 1986.

4 Feeling Surfaces and Objects Remotely

Susan J. Lederman and Roberta L. Klatzky

CONTENTS

4.1 INTRODUCTION

Experimental psychology offers a perspective on the workings of the somatosensory system that is complementary to various "wet-lab" approaches considered by most chapters in this book. The latter address the neural mechanisms and neural representations that underlie how the somatosensory system processes information about the external world and about our bodies. Such approaches examine the relationship between the environment (external, internal) and the underlying **neural** responses. Experimental psychologists study how the intact organism processes those same environmental events in terms of the **behavioral** responses of the whole organism. Ultimately, our joint goal is to understand how organisms sense, perceive, think, represent, and act on the environment.

Much of our previous work has concerned the nature of temporally constrained and unconstrained manual exploration of multi-property objects, and its contribution to the haptic processing and representation of objects and surfaces. Further details of this research program may be obtained in References 1 and 2. In those experiments, observers interacted with the objects and surfaces "directly" with their bare fingers.

In this chapter, we focus on how people process those same external objects and events when they obtain somatosensory inputs by means of an intermediate tool held in the hand, e.g., a stylus or pointer. As living systems frequently use tools to explore, perceive, and act upon their environments, it is therefore important to know more about how, and how well, people perform when the contact is "indirect." What somatosensory information is available, and how is it used to form internal representations? How may this information be used in different application domains?

4.2 RESEARCH BACKGROUND AND METHODOLOGIES: PERCEIVING WITH THE BARE FINGER

In this chapter, we consider two somatosensory tasks that have previously received considerable attention to date: the perception of surface texture and the identification of common objects.

4.2.1 ROUGHNESS PERCEPTION

Studies investigating the perception of surface texture with the bare finger have previously focused on roughness, one of the most prominent attributes of surfaces. Most research has adopted one of two complementary methodologies. Behavioral psychophysical procedures have been used to determine the nature of the functions relating various physical parameters of both the surface geometry and the manual exploration process to perceived roughness magnitude. Single-unit recording techniques have been used to determine the underlying neural code, primarily of peripheral tactile units to date. This work has been described in detail in References 3 and 4.

4.2.1.1 Psychophysics

Lederman[5] showed that the primary surface-geometry determinant of the perceived roughness of gratings was the spacing between the inner edges of the elements ("interelement spacing"). This held as well for abrasive surfaces[6] and raised two-dimensional (2-D) dot surfaces.[4] Perceived roughness magnitude tended to rise steeply and monotonically as a function of interelement spacing (but see Reference 7). Other factors investigated — ridge width, spatial period, and interelement spacing/ridge width ratio — influenced the roughness percept only modestly, if at all.[5,8]

With respect to the exploratory process, increasing finger force increased perceived roughness;[9] however, this effect was relatively less substantial than that of interelement spacing. Finally, speed of relative motion between skin and surface produced only a relatively small effect, compared to the effect of groove width (e.g., Reference 8).

Taylor and Lederman[10] subsequently developed a quasi-static model of perceived roughness. They performed a mechanical analysis of the changes in the proximal signals resulting from skin deformation, as affected by variation in the two most significant parameters for roughness perception, i.e., interelement spacing and finger force. The model most successfully predicted changes in perceived roughness magnitude as a function of the mean deviation of the skin from its initial resting position, summed over the total area of skin contact. Taylor and Lederman described their representation of roughness as "intensive," since the most likely skin-deformation parameter varied along a single intensive dimension. At the time this was proposed, technological constraints prevented the investigators from evaluating the contribution of possible spatial features of the skin deformation.

A number of studies have suggested that although vibratory information is available during relative motion between the bare skin and a textured surface, the vibrotactile inputs are not typically used to judge roughness (see References 5, 8, 11, 12, and 13). However, more recently, Hollins, Bensmaia, and Risner[14] and Hollins and Risner[15] have offered a duplex model of human roughness perception, in which vibration-based signals are used for judging fine textures (with element sizes below about 100 microns), but not for more coarse surfaces (e.g., those used by either Lederman or Johnson, and their colleagues).

4.2.1.2 Neurophysiology

Johnson and Connor[7,16,17] have argued for a "spatial-intensive" neural code for representing the neural events that underlie the perception of roughness magnitude. Their work suggests a spatial code at the peripheral level in terms of the relative activity across spatially distributed Slowly Adapting Type I (SAI) mechanoreceptors. This spatial code is maintained in SI cortex (areas 3b and 1), using the neural differences in activity of adjacent SAI units separated by about 1 mm. The neural difference operations must be performed by cortical neurons whose receptive fields possess neighboring excitatory and inhibitory areas. The researchers further suggest that the spatially distributed neural differences are passed on to neurons in area SII, which integrate the information received from areas 3b and 1 into a single intensive code.

Most recently, Dorsch, Yoshioka, Hsiao, and Johnson[18] have provided additional evidence that suggests that roughness perception below the limit of cutaneous spatial resolution (i.e., for gratings with interelement spacings as low as .1 mm) is also mediated by spatial variation in the SAI population response. These data converge nicely with the psychophysical findings reported by Hollins and Risner.[15]

4.2.2 COMMON-OBJECT IDENTIFICATION

Behavioral experiments have also been conducted in the study of the haptic identification of common objects. The haptic system uses sensory inputs from mechanoreceptors in the skin, as well as from those embedded in muscles, tendons, and joints.

In 1985, Klatzky, Lederman, and Metzger[19] noted that touch alone can be quite successful in identifying common objects. Participants typically achieved near-perfect accuracy within just a couple of seconds.

Subsequent research by Lederman and Klatzky[20] suggested that the key to such excellent performance lay in the consistent selection and execution of stereotypical patterns of hand movements ("exploratory procedures"). Each exploratory procedure was shown to be optimally associated with the extraction of one or more specific object properties.

For example, to obtain the most precise information about qualities pertaining to surface texture, the perceiver would execute a "Lateral Motion" exploratory procedure, which involved quick, repetitive, back-and-forth shearing motions across a surface. To obtain fine details about object compliance, the perceiver would apply a "Pressure" exploratory procedure, a force normal to the surface or a torque about some axis of the object. Participants executed a "Static Contact" procedure — static contact between skin and object surface — to extract the thermal properties, and "Unsupported Holding" — lifting an object away from a supporting surface — to obtain information about weight. These four exploratory procedures contribute information about the material properties of objects both quickly and accurately. In contrast, participants were relatively inefficient when extracting geometric information (e.g., shape, size). Both "Contour Following" (i.e., edge following) and "Enclosure" (i.e., grasp) exploratory procedures were used to extract information about shape and size; however, the resulting information was relatively coarse and particularly slow for Contour Following.

The performance characteristics that we originally documented with respect to manual exploratory procedures have led us to highlight an important distinction between how the haptic system processes material as opposed to geometric information. We propose that the skill with which people are able to identify three-dimensional (3-D) common objects haptically may be explained, in part, because multiple properties (especially material cues) are available for processing. In addition, people are able to access the coarse 3-D geometric information that further differentiates objects relatively quickly.

Additional research on haptic object identification[21] supports our proposal. In this experiment, the primary sources of information about common object identity were reduced to residual geometric properties. That is, the objects had no highly compliant parts; they were fixed in place, with all moving parts glued to make them rigid; and finally, the fingertips were gloved. Performance varied with the manner of exploration: a) with unrestrained 5-finger manual search, accuracy was still quite high, compared to Klatzky et al.,[19] i.e., about 94% accuracy and 15s response latency; b) when the subject's five fingers were splinted to prevent finger flexion, accuracy remained about the same, but response latency increased to about 24s; c) when subjects were only allowed to explore with a single splinted finger, accuracy dropped to about 75% and response latencies were also considerably slower — about 45s. When the subjects used their bare fingers, performance in the corresponding conditions was as follows: a) for unrestrained 5-finger exploration, about 94% accuracy and 10s response latency; b) for the 5-splinted finger condition, about 92% accuracy and 16s response latency; c) with a single splinted finger, performance was about 87% and response latencies were about 22s. We concluded that the absence of material properties impaired identification performance, particularly when only a single, splinted finger was used.

In addition, our research[22,23] has shown that because of the relative accessibility of material — as opposed to geometric — information to haptic processing, people tended to weight material information very strongly in their representations of multi-property objects when they used touch alone. In the Klatzky et al. experiment, subjects were asked to sort multi-property objects into piles according to their judged "similarity." Every object varied along each of four property dimensions: texture and compliance (both material), and shape and size (both geometric). The objects were equally differentiable, regardless of the dimension(s) used to judge similarity. Thus, choosing to sort primarily by material properties when touch was used alone presumably occurred because haptically derived inputs about the material properties can be processed more efficiently (speed, accuracy) than can the geometric properties. Conversely, when vision was also permitted, people tended to weight their object representations more strongly in terms of geometric information. Presumably, subjects now emphasized geometric properties in their object representations because they have learned over a lifetime that such information can be extracted considerably more efficiently by the visual system than can any kind of information — geometric or material — via haptic exploration.

4.3 PERCEIVING SURFACE TEXTURES VIA REMOTE TOUCH

People typically produce spatial-intensive representations of the magnitude of perceived roughness when using the bare finger.[10,17] However, they can still use vibration-based cues to evaluate surface roughness remotely with a tool, such as a stylus, grasped in the hand.[11] Recently, we have begun a comprehensive investigation into the remote haptic sensing of surface textures.

Once again our psychophysical questions have been dictated by our search for primary determinants of roughness perception via remote touch. As with Lederman's earlier work with the bare finger, our current investigation has focused on the contributions of both a) the geometric features of the textured surfaces, and b) the manner in which the surfaces are explored (e.g., relative speed, force, and mode of touch — active or passive). In addition, we must now consider a third set of physical parameters, which pertain to c) the geometric characteristics of the intermediate tool (e.g., probe diameter).

4.3.1 EFFECTS OF SURFACE GEOMETRY: INTERELEMENT SPACING

In our initial study, Klatzky and Lederman[4] asked subjects to estimate roughness magnitude (Experiment 1) and to perform a roughness-comparison task (Experiment 2), each with three types of end effectors. The textured stimuli used in the experiments consisted of raised 2-D dot matrices, which varied in interelement spacing from .5 to 3.5 mm. For a given matrix, each element was spatially jittered within a circularly defined area, both radially and angularly, so as to maintain the mean interelement spacing within each plate constant, while creating a "random" appearance to the dot patterns. Of the many possible geometric parameters that could be varied, we selected interelement spacing, inasmuch as this variable had reliably proved to be the single most critical determinant of perceived roughness with bare-finger contact. The end effectors consisted of the bare finger, and two rigid stick probes that varied in the size of the contact area. For various reasons, the tips in this initial study were not identical in shape. In this first study, we wished to determine the extent to which people are able to differentiate the roughness of textured surfaces, which varied in interelement spacing, via direct (bare finger) or remote touch (probes).

With respect to this issue, we obtained several informative results in the magnitude-estimation task used in Experiment 1 (Figure 4.1). First, the psychophysical roughness function (magnitude of perceived roughness as a function of interelement spacing on log scales) for direct touch was excellently fit by a linear equation within the range of interelement spacings over which subjects were comfortable judging surface roughness: perceived roughness increased in magnitude as a function of increasing interelement spacing. In contrast, the corresponding functions for the two probe end effectors were better described by quadratic equations: perceived roughness initially rose to a peak, then subsequently declined, as a function of increasing interelement spacing. Overall, the surfaces varying in interelement spacing were differentiated better with the bare finger than with the probes. Despite this difference, subjects were clearly able to perform the task via remote touch.

In Experiment 2, subjects were required to choose the "rougher" of two textured surfaces. The proportion correct is shown in Figure 4.2 for each end effector as a function of discrimination difficulty, which was measured as the number of 0.125-mm steps between the stimuli with respect to interelement spacing. Each subject took part in one bare-finger and one probe condition. Overall, accuracy was highest with the bare finger and lowest with the small probe (particularly for the narrow interelement spacings).

In summary, our initial probe study confirmed that the haptic system is indeed capable of judging surface roughness remotely via a rigid probe; however, performance with the intermediate tool was not quite as good as with the bare finger when relatively fine discriminations were required. This conclusion was further confirmed by parallel experiments using a rigid sheath to cover the entire volar portion of the fingertip in place of the stick probes (for a magnitude-estimation task: [Reference 4, Expt. 3]; also [Reference 24, Expt. 2]; for a roughness-comparison task: [Reference 4, Expt. 4]). Clearly, interelement spacing had a strong effect on judgments of surface roughness across the range of values over which subjects indicated they were comfortable estimating roughness magnitude. In addition, we noted that for the relatively wide contact areas (i.e., bare finger, sheath), the shape of the psychophysical roughness function (roughness magnitude as a function of interelement spacing, log scales) was well fit by a linear (as opposed to a quadratic) equation.

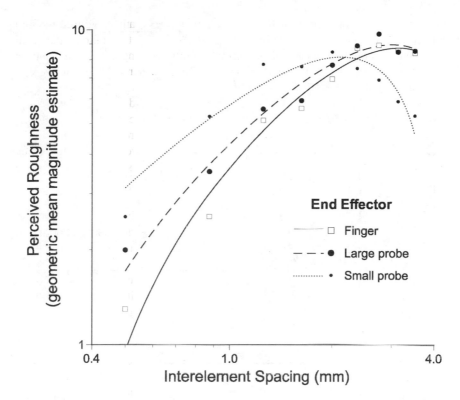

FIGURE 4.1 Perceived roughness magnitude (geometric means of the magnitude estimates) as a function of interelement spacing of the stimulus on log scales, for the three end effectors (Experiment 1). Quadratic functions fit to the log–log data were as follows. Small probe, y = –1.79 * x^2 + 0.84 * x + 0.82; large probe, y = –0.70 * x^2 + 0.98 * x + 0.63; finger, y = –0.93 * x^2 + 1.26 * x + 0.58. The *r* value in all cases was .99. Reprinted from Reference 4, with permission from *Perception & Psychophysics*.

4.3.2 EFFECTS OF GEOMETRIC CHARACTERISTICS OF THE PROBE: PROBE-TIP DIAMETER

Results of the Klatzky and Lederman study[4] were also used to consider the contribution to perceived roughness of probe-tip diameter, a geometric attribute of the probe that we expected would prove highly salient. We are just beginning to evaluate the effects of probe diameter and other physical parameters by fitting second-order polynomials to the empirical data. Typically, a second-order polynomial function is expressed as follows:

$$y = ax^2 + bx + c \qquad (4.1)$$

where *a* is the coefficient for the quadratic term, *b* is the coefficient for the linear component, and *c* is a constant that raises or lowers the function. Alternatively, it is possible to express the same function as follows:

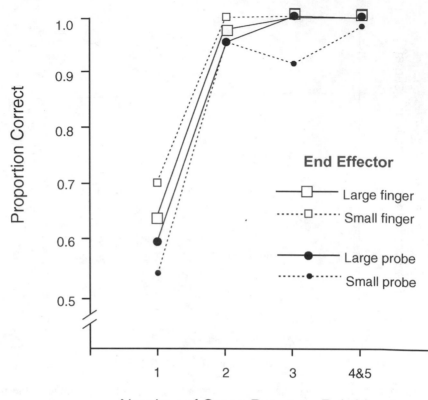

FIGURE 4.2 Roughness-comparison accuracy (proportion correct) as a function of number of 0.125-mm steps between the stimuli with respect to interelement spacing (Experiment 2). Different groups of subjects explored with a large or a small probe or with the bare finger. The legend subscript refers to the size of the probe used in the group. Reprinted from Reference 4, with permission from *Perception & Psychophysics*.

$$y = a(x - x_0)^2 + h \qquad\qquad (4.2)$$

where a now represents the curvature of the quadratic component (i.e., the sharpness with which the function peaks); x_0 is the location of the peak along the x-axis, and h is the height of the peak on the y-axis. In terms of the coefficients from Equation 1, x_0 is equal to $-b/2a$ and h is equal to c – (b²/4a).

Using Equation 2, it is now possible to evaluate the effects of candidate physical variables on perceived roughness more systematically in terms of the curvature of the quadratic component (a), peak position on the x axis (x_0), and peak height on the y-axis (h), which are all meaningful features. We plan to present a comprehensive analysis of the three quadratic parameters with respect to the results of our probe experiments in a future paper. Here, we will report only the most consistent findings. To date, we have noted a consistent effect of probe-tip diameter on perceived

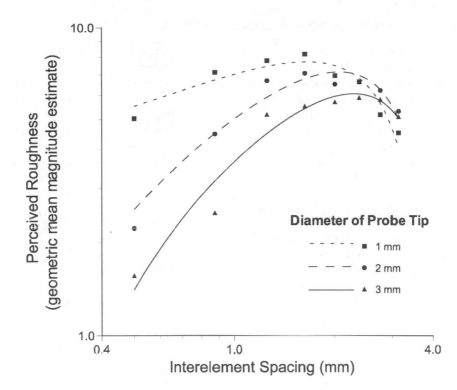

FIGURE 4.3 Perceived roughness magnitude (geometric means of the magnitude estimates) as a function of interelement spacing (log scales) for 1-, 2-, and 3-mm probe diameters. Quadratic functions are fit to the data for each probe-diameter condition. Reprinted from Reference 25, with permission from *Proceedings of ASME International Mechanical Engineering Congress: Dynamic Systems and Control Division.*

roughness: the peak of the quadratic function shifted toward the wider end of the interelement-spacing continuum as probe diameter increased. Presumably, this occurred as the probe switched from riding along the tops of the raised elements to penetrating the interelement spaces. A wider probe tip would penetrate only the wider spacings, leading to an increasingly smooth percept as the probe rode along the base substrate. Another result of the displacement of the peak along the x-axis was a concomitant reversal in the roughness estimates that occurred toward the higher end of the spacing continuum. More specifically, surfaces with interelement spacings below the peak value were perceived to be rougher with the small probe than with the large probe; the order was reversed for interelement spacings beyond the peak value — such surfaces felt rougher with the large probe than with the small probe. The reversal in roughness may have been due to the greater mechanical impact that occurred when the narrower probe penetrated the interelement spaces, subsequently catching on the leading edge of the raised elements.

A subsequent study by Lederman, Klatzky, Hamilton, and Grindley [Reference 25, Expt. 2] has recently confirmed our earlier results, this time using identical spherical shapes for the probe tip and diameters of 1, 2, and 3 mm.

4.3.3 Effects of the Manual Exploration Process

It is important to consider the consequences of varying the manner in which surfaces are explored to determine the extent to which subjects demonstrate roughness constancy. We wished to determine whether, and if so, the extent to which people are capable of adjusting their estimates of roughness to compensate for changes in the proximal stimulation (cutaneous and/or kinesthetic) that results from alterations in the manner in which the surfaces are touched. Accordingly, Lederman, Klatzky, and our colleagues also explored the effect of three potentially important parameters of the manual exploration process that had previously been considered with respect to direct touch (bare finger).

4.3.3.1 Relative Speed of Motion

Lederman, Klatzky, Hamilton, and Ramsay[3] examined the consequences for roughness perception of changing the relative speed of motion between the textured surfaces and the stick probe used to explore them. Both active and passive modes of touch were employed. Active touch required that the subject move the probe over a stationary surface, whereas passive touch required that an external agent move the surface under a stationary probe.

In one experiment, speed was altered by a factor of 10: 20.5 ("slow"), 73.2 ("medium"), and 207.3 ("fast") mm/s. Like the probe studies just mentioned, the psychophysical function for each speed was best fit by a quadratic (as opposed to a linear) equation: perceived roughness initially increased with increases in interelement spacing, then progressively decreased with further increases in interelement spacing. Using Equation 2 to convert to the alternate form of the second-order polynomial, we determined that the peak, x_0, of the quadratic function shifted progressively further toward the wide end of the interelement-spacing continuum with increasing speed. An associated reversal in perceived roughness magnitude was also observed: roughness magnitude diminished with increasing speed for the narrower interelement spacings (i.e., over the monotonically increasing portion of the function); beyond the peak value, the effect of increasing speed was reversed (i.e., perceived roughness increased with decreasing speed for the progressively wider interelement spacings). Increasing speed influenced perceived roughness in ways that were similar to those due to our increasing probe size. The results are shown in Figure 4.4. A second experiment that used only a four-fold change in speed produced similar results in terms of the maximum speed effects obtained.*

We attribute not only the quadratic shape of the function, but also the peak shifts and the reversals in roughness magnitude due to speed and probe size to mechanical changes in the proximal vibratory signal: for the relatively narrow spacings, each probe will ride along the tops of the raised elements. At some spacing, the probe will begin to penetrate the gap between the raised elements, ride along the smooth base, and catch on the leading edges of the raised elements encountered in its

---

* Reconsideration of the mathematical validity of the "doubling ratio," a statistic originally devised to compare the relative magnitude of various factors to that of interelement spacing, led us to conclude subsequently that the range of speeds in fact had no effect on perceived roughness estimates.

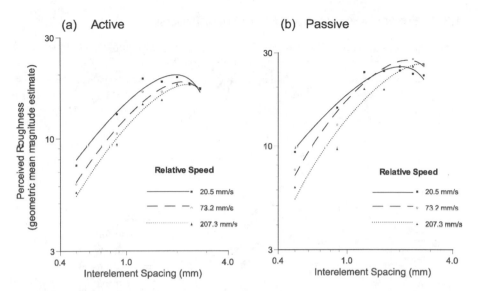

FIGURE 4.4 Perceived roughness magnitude (geometric means of the magnitude estimates) as a function of interelement spacing presented for low, medium and high relative speed of motion. The data are plotted on log scales. The results for active and passive modes of touch are plotted separately in panels a and b, respectively. The probe-tip diameter was 3 mm. Reprinted from Reference 3, with permission from *Haptics-e*.

movement path. The value of the interelement spacing at which this change occurs will likely be determined by both speed of motion and probe size — as well by element height, which we are currently investigating.

4.3.3.2 Mode of Touch

In the experiment above, subjects used either active or passive modes of touch to explore the textured surfaces. The maximum effect of speed on perceived roughness was statistically significant for both modes of touch (see Figure 4.4). These results indicate that subjects failed to show complete roughness constancy; that is, they did not alter their roughness estimates to fully compensate for the effects of changing speed on the proximal cutaneous and kinesthetic inputs. However, the maximum effect of speed was greater for passive than for active touch. This result suggests greater compensation in the active condition, as expected because of the additional kinesthetic cues to velocity.

4.3.3.3 Force

Lederman et al. [Reference 25, Expt. 1] also examined the effects of varying the force applied to the surface while exploring it with a probe. A passive-touch mode was selected, as this provided better control over the constant speed. Once again, subjects provided numeric estimates of the roughness magnitude of a set of raised-dot surfaces. All functions were well fit by quadratic equations. As shown earlier with the bare finger, increasing the applied force increased subjects' estimates of

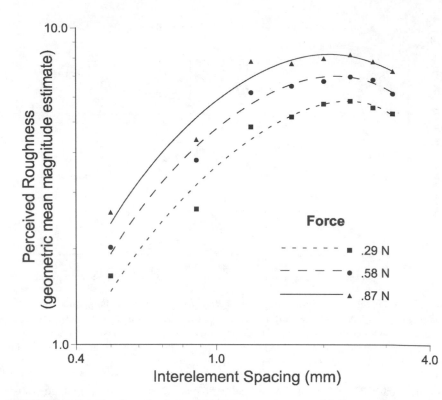

FIGURE 4.5 Perceived roughness magnitude (geometric means of the magnitude estimates) as a function of interelement spacing for 0.29, 0.58, and 0.87 N forces. The data are plotted on log scales, and quadratic equations have been fit to each function. The probe-tip diameter was nominally 3 mm. Reprinted from Reference 25, with permission from *Proceedings of ASME International Mechanical Engineering Congress: Dynamic Systems and Control Division*.

the perceived roughness magnitude across all interelement spacings. Furthermore, *h,* the height of the peak, was shifted upward with increasing force. The results are shown in Figure 4.5.

4.4 IDENTIFYING COMMON OBJECTS HAPTICALLY VIA REMOTE TOUCH

Recall the study by Klatzky et al.,[21] in which subjects identified common objects using different tactual fields of view (1, 5 fingers) and different conditions of finger flexion (unconstrained, splinted). With the consequences of such constraints in mind, Lederman and Klatzky [Reference 26, Expt. 3] performed an experiment in which subjects haptically identified the same set of common objects, this time remotely, using the same small and large probes from the study by Klatzky and Lederman [Reference 4, Expts..1 and 2]. Using a probe may be viewed much like using the single, splinted finger in the 1993 study.

TABLE 4.1
Mean Accuracy and Response Time
(max = 120 s) for the Two Probe Conditions

	Small Probe	Large Probe
Mean accuracy (%)	41.8	38.9
Mean response time (s)	85.6	85.2

Blindfolded subjects explored objects with either the small or large probe, choosing to hold it much as they would a pencil. Each subject was fitted with a fabric knit glove. A total of 2 cm was cut away from each fingertip, thereby restricting contact information to the skin areas in contact with the probe. Objects were placed before the observer in a prototypical orientation. The mean accuracy and mean response times from initial contact to naming are shown in Table 4.1 for the small and large probes.

The observed performance values are considerably poorer than corresponding levels in the Klatzky et al. study (cf. bare index finger: mean accuracy = 84%, mean response time = 22.8 s; gloved index finger: mean accuracy = 74% and mean response time = 45.2 s). T tests indicated that the differences between the two probes did not even approach significance. Relatively poor performance scores with a rigid sheath, relative to unconstrained exploration with multiple bare fingers, were also obtained in this study of object identification via intermediary links. A more comprehensive paper that presents details of the complete set of experiments is now under preparation.

4.5 REMOTE TOUCH IN MEDICINE

The results of our ongoing research program on remote touch also have potentially significant implications for the practice of medicine in the future.

Physical contact offers powerful evidence that an object exists. Only haptics reliably and unambiguously convey information about the forces and vibrations arising from direct mechanical contact between the perceiver/actor and objects in the environment. We describe such immediate contact between skin and surface as non-remote. Because contact sensing offers one of the most primitive and powerful confirmations of object reality, it constitutes the very essence of physical "presence." We propose that touch can provide the missing physical "presence" for remote communication by enhancing telepresence via "haptic interfaces" for teleoperation and virtual-environment systems. A haptic interface measures forces applied to a remote environment; it may also display those forces in the form of directional force feedback, distributed spatial patterns of local forces, and vibrotactile information to the hand of a human operator as s/he explores, perceives, and manipulates that remote environment.

First, let us consider the potential use of haptic interfaces in medical applications for teleoperation. In such cases, the remote environment is real. "Touch is one of the most important cues for a physician; in fact, surgery is often conducted more by touch than by vision."[27] And yet, currently, owing to the length and rigidity of their tools, surgeons who perform minimally invasive surgery (MIS) are denied a sense of touch during the operation. Researchers (see e.g., References 28, 29, and 30) have developed a variety of MIS tools, which have force sensors (contact and array) attached to the distal ends of the surgical tools. The forces sensed remotely by a tool positioned within the body cavity are in turn displayed in real time to the surgeons' hands while they grasp the instruments. The goal of such teleoperator systems is to permit the surgeon to feel as if s/he is touching the tissues directly. Haptic interfaces may also ultimately be used by general practitioners and specialists to examine patients at considerably greater distances, using remote palpation for purposes of diagnosis. Ultimately, haptic interfaces will likely be combined with other sensory interfaces (e.g., visual, auditory) into a multimodal interface, in order to render the sensory information about the distant environment as rich as possible.

A haptic interface can also be used as part of a virtual environment system, where the environments are simulated rather than real. For example, virtual training systems may be used to help teach novice surgeons how to perform complex operations. Trainees would don 3-D goggles that deliver high precision graphics displays, while manipulating virtual organs via surgical tools, attached to haptic interfaces that provide appropriate force feedback to the hands.

As biological scientists, we have argued that in order to build highly efficient haptic interfaces, it is necessary that designers match the design characteristics of the hardware and software systems to the capabilities and limitations of the human operator. The results of our research speak directly to this issue in a number of ways, several of which we outline here.

Probes represent one of the simplest forms of teleoperator and virtual-environment systems that can operate on a remote environment. Indeed, most haptic interfaces that are currently available (commercially or in research labs) involve such single-point contact. Therefore, the results of our fundamental research on perceiving surfaces and objects via a probe are highly relevant to the design of haptic interfaces. We will close this chapter with a consideration of some of the implications of this work.

When discussing remote texture perception via a probe (above), we alluded to supporting research by Lederman and Klatzky.[24] It is appropriate now to describe this study in further detail. Observers were required to perform a battery of cutaneous and haptic tasks, both sensory and perceptual. During half of the trials, observers wore a rigid sheath on the volar surface of the same finger; the sheath was held in place by a surgical glove, into which the sheathed finger was inserted. The sheath served to remove the spatially distributed forces normally available. During the remaining trials, observers used the bare finger, again covered by a surgical glove, to equate for friction. Performance was markedly impaired on the following tasks: 2-point touch threshold, point localization thresholds, perception of the orientation of a bar impressed on the fingertip, and detection via palpation of simulated 3-D lumps (of various sizes) embedded in simulated tissue. These results led us to argue strongly for multiple (array), as opposed to single-element, force sensors and stimulators in the next generation of haptic interfaces. In contrast, we found that performance was only moderately impaired for texture perception tasks, and not at all for detecting vibration. The latter set of findings suggested that our observers could — and did — use information about texture in the form of temporal vibration-based cues, as opposed to the spatial-intensive cues normally used when contact between skin and surface is direct.

These encouraging results with texture perception through a rigid sheath led us to our ongoing research program dealing with texture perception via a probe. Underlying this program is the hypothesis that the vibrations produced when surfaces are explored with a stylus, rather than the bare finger, are sufficient to produce a rich representation of a textured world. One component involves conducting a comprehensive set of human psychophysical experiments to determine a small number of critical parameters that describe how vibration, induced by the interaction between surface and rigid stylus, and passed to the skin, leads to an internal representation of surface roughness (see e.g., Section 4.3). We are also currently assessing the consequences of adding redundant visual and/or auditory texture information. The second component involves modeling the transfer from surface to percept, by developing a theoretically rigorous model of vibratory texture perception, which describes the functional linkage between mechanical interactions and human textural percepts. The last component involves using the vibration-based model to create virtual textures by imposing vibratory forces on a stylus linked to a haptic interface. It also includes a formal evaluation of the virtual textures using a set of scientific, behavioral experiments; to date, this element has been noticeably missing from the design of haptic interfaces.

The studies pertaining to haptic object identification via a probe[26] or single bare finger[21] also speak to design issues with respect to haptic interfaces, inasmuch as they highlight potential constraints in the use of current systems. Recall that when restricted to exploration via a single probe or bare finger, performance was impoverished both in terms of speed and accuracy. Therefore, a haptic interface limited to a single end effector is likely to restrict the amount of information that will be available concerning the geometric features of many remote environments. Thus, we suggest systematically investigating whether tasks that require such information may be assisted by the simultaneous use of multiple haptic interfaces, to provide a wider field of view and to afford stable grasping.

In summary, in this chapter we have presented the results of a research program on the haptic perception of surfaces and objects via a probe. The use of behavioral, psychophysical methods has allowed us the opportunity to address fundamental research questions concerning how people process and represent surface and object properties remotely via an intermediate tool, as opposed to with the bare finger. In addition, we have explored the ramifications of these results for designing haptic interfaces for teleoperator and virtual environment systems in several medical applications.

ACKNOWLEDGMENT

Preparation of this chapter was supported by the Natural Sciences and Engineering Research Council of Canada.

REFERENCES

1. Lederman, S.J. and Klatzky, R.L., The hand as a perceptual system. In: *The Psychobiology of the Hand*, Connolly, K., Ed., London: MacKeith Press, Chapter 2, 16, 1998.
2. Klatzky, R.L. and Lederman, S.J., The haptic glance: A route to rapid object identification and manipulation. In: *Attention and Performance XVII: Cognitive Regulation of Performance: Interaction of Theory and Application*, Gopher, E. and Koriat, A., Eds., Mahwah, NJ: Erlbaum, Chapter 6, 165, 1999.
3. Lederman, S.J., Klatzky, R.L., Hamilton, C.L., and Ramsay, G.I., Perceiving roughness via a rigid probe: Psychophysical effects of exploration speed and mode of touch, *Haptics-e (Electronic Journal of Haptics Research),* 1(1), 1, 1999. Available at *http://www.haptics-e.org.*
4. Klatzky, R.L. and Lederman, S.J., Tactile roughness perception with a rigid link interposed between skin and surface, *Perception and Psychophysics,* 61(4), 591, 1999.
5. Lederman, S.J., Tactile roughness of grooved surfaces: The touching process and effects of macro- and micro-surface structure, *Perception and Psychophysics,* 16(2), 385, 1974.
6. Stevens, S.S. and Harris, J.R., The scaling of subjective roughness and smoothness, *Journal of Experimental Psychology: General,* 64(5), 489, 1962.
7. Connor, C.E., Hsiao, S.S., Phillips, J.R., and Johnson, K.O., Tactile roughness: Neural codes that account for psychophysical magnitude estimates. *Journal of Neuroscience,* 10, 3823, 1990.
8. Lederman, S.J. Tactual roughness perception: Spatial and temporal determinants, *Canadian Journal of Psychology,* 37, 498, 1983.
9. Lederman, S.J. and Taylor, M.M., Fingertip force, surface geometry, and the perception of roughness by active touch, *Perception and Psychophysics,* 12(5), 401, 1972.
10. Taylor, M.M. and Lederman, S.J., Tactile roughness of grooved surfaces: A model and the effect of friction, *Perception and Psychophysics,* 17(1), 23, 1975.
11. Katz, D., *The World of Touch,* L. Krueger, Trans., Hillsdale, NJ: Erlbaum, 1989 (original work published 1925).
12. Johnson, K.O. and Lamb, G.D., Neural mechanisms of spatial tactile discrimination: Neural patterns evoked by braille-like dot patterns in the monkey, *Journal of Physiology,* 310, 117, 1981.
13. Lederman, S.J., Loomis, J.M., and Williams, D., The role of vibration in the tactual perception of roughness, *Perception and Psychophysics,* 32(2), 109, 1982.
14. Hollins, M., Bensmaia, S.J., and Risner, S.R., The duplex theory of tactile texture perception, *Proceedings of the Fourteenth Annual Meeting of the International Society for Psychophysics,* 115, 1998.
15. Hollins, M. and Risner, S.R., Evidence for the duplex theory of tactile texture perception, *Perception and Psychophysics,* 62(4), 695, 2000.
16. Connor, C.E. and Johnson, K.O., Neural coding of tactile texture: Comparison of spatial and temporal mechanisms for roughness perception, *Journal of Neuroscience,* 12, 3414, 1992.

17. Johnson, K.O. and Hsiao, S.S., Evaluation of the relative roles of slowly and rapidly adapting afferent fibers in roughness perception, *Canadian Journal of Physiology and Pharmacology*, 72, 488, 1994.

18. Dorsch, A.K., Yoshioka, T., Hsiao, S.S., and Johnson, K.O., Peripheral neural mechanisms underlying roughness perception of fine stimulus patterns, *Society for Neuroscience Abstracts*, 26, 2000.

19. Klatzky, R.L., Lederman, S.J., and Metzger, V., Identifying objects by touch: An "expert system," *Perception and Psychophysics*, 37(4), 299, 1985.

20. Lederman, S.J. and Klatzky, R.L., Hand movements: A window into haptic object recognition, *Cognitive Psychology*, 19(3), 342, 1987.

21. Klatzky, R.L., Loomis, J.M., Lederman, S.J., Wake, H., and Fujita, N., Haptic identification of objects and their depictions, *Perception and Psychophysics*, 54(2), 170, 1993.

22. Klatzky, R., Lederman, S.J., and Reed, C., There's more to touch than meets the eye: The salience of object attributes for haptics with and without vision, *Journal of Experimental Psychology: General*, 116(4), 356, 1987.

23. Lederman, S., Summers, C., and Klatzky, R., Cognitive salience of haptic object properties: Role of modality-encoding bias, *Perception*, 25(8), 983, 1996.

24. Lederman, S.J. and Klatzky, R.L., Sensing and displaying spatially distributed fingertip forces in haptic interfaces for teleoperator and virtual environment systems, *Presence: Teleoperators and Virtual Environments*, 8(1), 86, 1999.

25. Lederman, S.J., Klatzky, R.L., Hamilton, C.L., and Grindley, M., Perceiving surfaces roughness through a probe: Effects of applied force and probe diameter, *Proceedings of the ASME International Mechanical Engineering Congress: Dynamic Systems and Control Division (Haptic Interfaces for Virtual Environments and Teleoperator Systems,* 2000.

26. Lederman, S.J. and Klatzky, R.L., Feeling through a probe, *Proceedings of the ASME International Mechanical Engineering Congress: Dynamic Systems and Control Division (Haptic Interfaces for Virtual Environments and Teleoperator Systems)*, DSC-Vol. 64, 127, 1998.

27. Laine, L., Is virtual medicine becoming, literally, a reality?, *Medical and Diagnostic Industry*, 40, 1997.

28. Fischer, H., Neisius, B., and Trapp, R., Tactile feedback for endoscopic surgery. In: *Interactive Technology and the New Paradigm for Healthcare*, Morgan, K., Satava, R.M., Seiburg, H.G., Mattheus, R. and Christensen, J.P., Eds., IOS Press and Ohmsha, 114, 1995.

29. Hannaford, B., Trujillo, J., Sinanan, M., Moreyra, M., Rosen, J., Brown, J., Leuschke, R., and MacFarlane, M., Computerized endoscopic surgical grasper, *Medicine Meets Virtual Reality*, 50, 265, 1998.

30. Howe, R.H., Peine, W.J., Kontarinis, D.A., and Son, J.S., Remote palpation technology, *IEEE Engineering in Medicine and Biology*, 14(3), 318, 1995.

5 Somatosensory and Proprioceptive Contributions to Body Orientation, Sensory Localization, and Self-Calibration

James R. Lackner and Paul DiZio

CONTENTS

We take contact of our body with the ground for granted. Except for brief moments during running or jumping, or falling from a tree or ladder, those of us who are not skydivers are rarely without physical contact between our body and the support surface of the earth. The acceleration of earth gravity pulls us downward and the reaction force provided by the surface of the earth or the floors of structures arrests our downward motion and provides a physical base on which stance and locomotion can be carried out. What a scale measures as weight is actually the contact force it provides in supporting the body against the pull of gravity. In the absence of an effective gravitational pull on the body, walking and running would not be possible unless one were stepping in place on a treadmill while being mechanically pulled toward its surface by elastic tethers. Locomotion on the moon, for example, is greatly constrained by the acceleration of moon gravity which is only one-sixth of that on earth. This has consequences for both walking and running. Walking is basically pendular in character and is limited by gravitational force. The maximum possible frequency of walking is proportional to the square root of the gravitational force. Running depends on being able to apply a backwards force with the foot without the foot slipping. The maximum backward thrust that can be exerted without the foot slipping is limited by the coefficient of friction and the normal force provided by gravity. The latter will be only one-sixth on the moon of what is on the earth.[1,2]

Animals and people are bilaterally symmetric and their center of mass lies in or close to the median plane of their body. This symmetry considerably simplifies control of balance and locomotion. The upright body will not fall over during quiet stance so long as the center of mass is maintained above the support area defined by the feet (see Figure 5.1).[3] In this situation, the patterning of shear and normal forces on the soles of the two feet and their time history, when interpreted in relation to command signals to the legs and feet, provides information about the direction of body sway and the relative position of the center of mass of the body (see Note 1). Individuals who lose vestibular function as adults initially have great difficulties with balance and locomotion. During their recovery period, they learn to use the tactile cues on the soles of their feet to control their balance. This becomes easier with continued practice; however, they continue to experience difficulties when they walk on uneven or compliant surfaces under conditions of reduced illumination.[4]

5.1 CONTACT FORCES AND BODY ORIENTATION

An important feature of the contact forces that in a given situation support the body against gravity is that they also indicate the direction of "down." That is, they provide information about the body's orientation to gravity and the location of the body center of mass with respect to the body surfaces in contact with the ground. The otolith organs of the vestibular system also provide signals that relate to the direction of gravity; however, these signals relate to head orientation and, unlike the soma-tosensory contact cues, do not convey information about the relative position of the center-of-mass. Individuals lacking vestibular function still have a firm sense of the direction of down owing to the contact support forces on their body. The importance of contact cues for orientation can be seen in labyrinthectomized and blinded dogfish swimming in an aquarium. When the creature bumps into a wall, it reorients its body to that surface and swims horizontally.[5] That surface is "taken" as the direction of down because the creature's nervous system is interpreting the contact as hitting "bottom."

This is an important observation because aquatic creatures rarely make contact with the shoreline. Contact is more likely to occur with submerged objects or the ocean or lake bottom. The vestibular system appeared phylogenetically in aquatic animals for which contact cues on the body surface dependent on body orientation in relation to gravity would be absent. In later, land-dwelling creatures with an

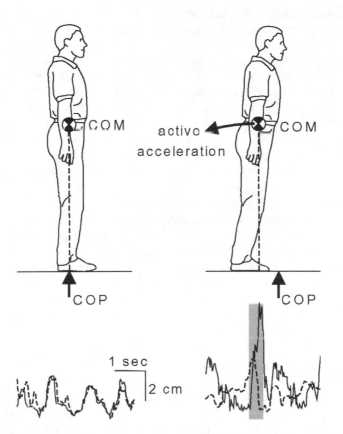

FIGURE 5.1 Left, top: During quiet stance with pendular sway, the vertical projection (dashed line) of the the body center of mass (COM) and the center of foot pressure (COP) remain within the area of foot contact on the ground. **Left, bottom**: The COP (solid trace) and COM (dashed trace) fluctuations coincide. **Right, top**: Leg, ankle, and foot muscle activation can accomplish a restorative acceleration of the COM by displacing COP in the opposite direction, momentarily outside the base of support if necessary. **Right, bottom**: The shaded area indicates where muscle activation occurs. Information about body position and motion can be extracted from pressure on the feet interpreted in relation to postural commands.

independently articulated head and neck, the vestibular system could still be used as a primary source of information about head orientation and, coupled with head-on trunk information, be used to determine body orientation to gravity. Touch and pressure cues on the body surface, however, provide additional information directly representing the direction of down and the location of the body center of mass relative to the parts of the body supporting it against the downward acceleration of gravity. These body contact cues are very important in the control of balance and locomotion. Patients with dorsal column lesions or degeneration (e.g., tabes dorsalis, Friedrich's ataxia) are unstable and adopt a broad-based shuffling gait. Diabetic patients with peripheral neuropathies show similar impairments of function.

5.2 ORIENTATION IN WEIGHTLESS AND HIGH FORCE CONDITIONS

The importance of surface contact stimuli is easy to underestimate because we rarely are exposed to situations where they are absent. The weightless phases of parabolic flight maneuvers represent one such condition. When an aircraft is flown in a parabolic trajectory, periods of high and low force levels alternate (Figure 5.2A). The latter, free-fall conditions represent a state of nearly complete weightlessness. In this situation, the aircraft and its occupants are all being freely accelerated downward by the gravitational attraction of the earth. This means that the aircraft does not support the inhabitants' bodies against the action of gravity. Consequently, it is possible to free float unrestrained and without contact with any surfaces whatsoever (see Figure 5.2B). If an individual free-floating in this fashion closes his or her eyes, then he or she may lose all sense of spatial anchoring to the surroundings and retain awareness only of ongoing body configuration, e.g., whether the legs are flexed. The person does not feel in any orientation relative to the aircraft — a firm sense of spatial orientation and location is gone, albeit cognitive awareness of the spatial layout of the aircraft and body location in relation to it is retained.[6,7] Opening the eyes immediately restores a sense of spatial anchoring to the surroundings. The importance of contact cues for orientation is further highlighted when test subjects are free floating with their eyes closed and pressure is applied to the tops of their heads. They will again feel "oriented;" however, they may feel upside down resting on their heads (Figure 5.2C). The contact cues restore a sense of orientation and are interpreted as the direction of down just as they are on earth.

Exposure to unusual patterns of body motion also reveals the importance of contact support forces for perceived body orientation. Figure 5.3A illustrates an apparatus, the ZARR or z-axis recumbent rotator, that rotates individuals about their long body axis in barbecue spit fashion — although the mode of attachment to the device, straps and a foam-lined contoured body mold, is more humane. When subjects are rotated in this device at a constant velocity, the activity of the angular acceleration-sensitive semicircular canals of their inner ear will decay to baseline levels in less than a minute after completion of the acceleration phase. The otolith organs of the inner ear are activated by linear acceleration and are being continually reoriented in relation to the direction of gravity and thus differentially stimulated throughout the rotation period. If the subjects' eyes are open, visual information about how body orientation is continuously changing relative to the test chamber will also be available, complementing the otolith cues signaling a continuous 360° rotation about the long axis of the body. In this circumstance at rotation rates below about 10 rpm, subjects feel continuous rotation about their body axis; at higher velocities of rotation, they feel continuous rotation, but also feel displaced in front of the rotation axis.[8,9]

Startling changes occur when subjects close their eyes when rotating at velocities above 10 rpm. They no longer feel rotation about the long axis of their body but instead experience vertical plane orbital motion while maintaining either a face-up or face-down direction. The diameter of the apparent orbit can be many feet. Interestingly, most

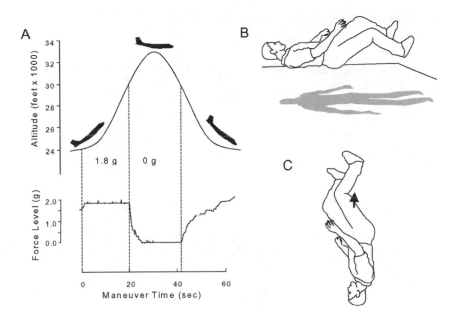

FIGURE 5.2 A. Schematic illustration of a parabolic flight maneuver (above) and a typical trace of the gravitoinertial force level (below). Each parabola contains approximately 20 s of steady 0 g and 1.8 g. **B**. In the free-fall phase of parabolic flight, the aircraft and passengers are accelerating toward earth at the same rate so it is possible to float. Lack of contact leads to loss of orientation relative to the aircraft if the eyes are closed. **C**. A sense of orientation is restored and subjects feel upside down if pressure is applied to their head (arrow) while they are free floating, eyes closed.

individuals experience face-up or face-down orbital motion.* The direction of experienced orbital motion is opposite the direction of actual body rotation. In addition, for the face-up and face-down experienced orientations there is a 180° phase shift in relation to actual body position for the two experienced orientations. Figure 5.3B shows the relationship between experienced orbital motion and the actual rotary position. For the face-up orbital motion, the person feels at the top of the orbit when actually face down; in the left quadrant, when turning onto the left side; at the bottom, when physically face-up, and in the right quadrant, when actually right-side down. By contrast, for face-up motion, the subject is physically face up when at the top of the orbital path. Most subjects can shift their apparent orientation between face-up and face-down by straining in the apparatus and applying pressure to the front or back of their head. Pressure cues can also affect the apparent diameter of the orbital motion. By increasing the actual pressure on the front of their head as they rotate, subjects can make their apparent orbits increase in diameter by several feet.

Putting pressure on the soles of their feet or the top of the head can cause changes in apparent orientation in relation to the test chamber. Foot pressure makes the

---

* Occasionally, a subject will feel as if he/she is undergoing orbital motion in a horizontal plane while always face-up or face-down.

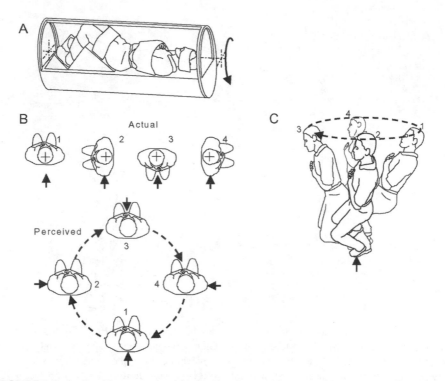

FIGURE 5.3 **A**. A subject restrained in the Z-axis recumbent rotation apparatus. **B**. During actual counterclockwise rotation in the apparatus, subjects experience clockwise orbital rotation (broken arrows) with a constant heading, face-up is illustrated. The actual and perceived patterns are both consistent with the rotation of contact cues around the body (solid arrows). **C**. Increasing plantar pressure by straining against the restraints makes subjects feel upright, pivoting in a cone about their feet while maintaining a constant heading.

subjects feel upright while describing their orbital path (see Figure 5.3C), and pressure on the top of the head makes them feel upside down. Thus, the overall pattern of touch and pressure cues impinging on the body dramatically influences experienced orientation. The transition from horizontal to vertical orbital motion is not experienced as a physical rotation from horizontal to vertical, but rather as a gradual feeling of being less and less in one orientation and then progressively more compelling in the new orientation, until it is completely vivid. The absence of a sense of spatial displacement from horizontal to vertical is likely related to the absence of the semicircular canal and otolith activity that would accompany such a change. A similar fading-out, fading-in of experienced orientation occurs in the orientation illusions experienced in parabolic flight described above.

The ZARR studies emphasize the importance of touch and pressure cues for orientation and indicate that dynamic patterns of somatosensory stimulation can supplement and even supplant vestibular contributions to perceived orientation. This is especially obvious in parabolic flight experiments involving the ZARR where background force level varies between 0 and 1.8 g. When blindfolded subjects ride in the ZARR at 30 rpm during straight and level flight, they experience the patterns illustrated in Figure 5.3 with comparable orbital diameters. However, as the aircraft pulls "g's" and increases the subject's effective weight, the forces exerted on the subject's body by the apparatus increase and the diameter of the apparent orbit rapidly becomes larger and larger. It peaks and remains constant as the g force levels off at its peak and can seem as much as 20 feet in diameter. As the g force begins to decline, the orbit diameter shrinks more and more until in weightlessness no orbital motion whatsoever is experienced. Even though the subject is rotating at 30 rpm, no motion is experienced whatsoever; the subject feels totally stationary.[6] In this circumstance, at constant velocity rotation, there is no differential touch and pressure stimulation of the body surface because the subject is weightless. If the subject is allowed unrestricted vision during rotation, then in the high force periods he or she will experience continuous body rotation while simultaneously undergoing orbital motion in the opposite direction. The orbit diameter is more than 50% smaller when the eyes are open. In 0 g, some subjects experience constant velocity rotation about their long body axis, while others feel completely stationary and perceive the aircraft to be rotating about them.

The importance of tactile cues for orientation can be demonstrated under normal terrestrial conditions as well. If the soles of a blindfolded seated subject's feet are brushed by a smooth rotating surface while the feet are prevented from moving by a yoke, illusory self-rotation will be elicited (see Figure 5.4). The subject, within 10–20 seconds, will perceive the rotating platform under his feet as stationary and experience his body as rotating in the direction opposite that of the platform. Simultaneously, the subject's eyes will exhibit a nystagmoid pattern of eye movements with slow phase direction compensatory for the direction of apparent body displacement and with velocity scaled to that of experienced velocity [10] Even though subjects know the platform can rotate under their feet, they still come to perceive it to be stationary and themselves as turning.

5.3 "TOP-DOWN" COGNITIVE INFLUENCES ON ORIENTATION

Contact cues from the feet can also override vestibular signals to orientation. It has been known for hundreds of years that if a seated person is passively rotated in the dark at constant velocity, then he or she will soon come to feel stationary. The adequate stimulus for the semicircular canals is angular acceleration and canal afferent discharge returns to resting levels during constant velocity turning, thus the canals cannot discriminate rest from constant velocity. If an individual is rapidly decelerated to rest from constant velocity rotation, body rotation in the opposite direction will be perceived. This occurs because during the rapid deceleration the cupula in each horizontal semicircular canal becomes deviated in the direction opposite that during initial acceleration to constant velocity. During constant velocity rotation, the canals gradually re-equilibrate and during deceleration, opposite direction rotation is sensed. Importantly, if immediately post-rotation the subject lowers his feet from the footrest of the chair to the floor of the experimental chamber, then there will be a truncation or a reversal of the post-rotation motion aftereffect. The subject may perceive his direction of motion reverse so that it is in the same direction

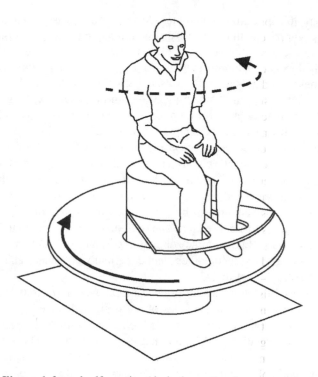

FIGURE 5.4 Illusory leftward self-rotation (dashed arrow) and compensatory eye movements (left beating nystagmus) are elicited when the "floor" moves rightward (solid arrows) under the soles of the yoked feet of the stationary, blindfolded observer. The observer perceives the moving surface to be stationary under his feet.

as per-rotation. Lifting and lowering the feet will cause the aftereffect to alternately change direction and to dissipate faster than the usual 20–30 seconds. With stops from higher velocities of rotation, the duration of the aftereffect is increased. It is important to note that when the subject's feet remain on the footrest of the rotating chair, he experiences his feet as stationary in relation to the footrest and himself and the entire chair as turning. On lowering his feet to touch the floor, he experiences it to be stationary and his body and the chair to be turning in relation to it. This pattern means that contact of the feet with different reference frames — chair vs. floor — greatly influences the apparent motion of the body. Such an influence points to "top-down" cognitive effects on orientation rather than a "bottom-up" integration of sensory signals.

Similar top-down influences can be demonstrated for hand and arm movement control. When a subject, in darkness, is accelerated in a rotating chair nystagmus is elicited. The nystagmus has a slow phase that is opposite the direction of rotation. If a target light is attached to a boom on the chair and fixed relative to the observer, during angular acceleration, the target will be seen to change its position relative to the subject. The subject will see it displace in the direction of acceleration. After reaching a peak subject-relative displacement, the target will still seem to be moving relative to the observer but no longer increasing its subject-relative displacement.

Simultaneously, the observer will perceive rotation of self and target together relative to the unseen experimental chamber. This phenomenon is known as the oculogyral illusion.[11]

If the subject is allowed to maintain hand contact with the target light, then when the chair is accelerated, the apparent displacement and motion of the target relative to the subject will be suppressed.[12] If the target light is extinguished and the subject maintains hand contact with the target light mount and attempts to fixate it, the vestibular nystagmus normally evoked by chair acceleration will be greatly attenuated.[12] In both the target light and the darkness conditions, suppression is contingent on the subject's hand making physical contact with the target or target holder. Having the hand spatially even a few millimeters away has no suppressing effect. These results demonstrate that somatosensory plus proprioceptive information about target location influences visual localization and oculomotor control, and can suppress visual mislocalizations induced by vestibular stimulation.

Hand contact can also suppress other forms of illusory visual motion. The autokinetic effect — the apparent visual motion of a stationary target light in an otherwise dark room — has been one of the classic illusions studied by psychologists and physiologists. This illusion was found to be greatly suppressed in magnitude and total time of occurrence if the subject grasped the target light holder with his or her index finger next to the light.[13] In addition, the amplitude of involuntary losses of fixation of the target light was diminished as well. These results point to a proprioceptive and somatosensory contribution to stabilization of visual direction and oculomotor control. A subsequent study using a scleral search coil technique to evaluate the quantitative reduction of fixation instability provided by holding vs. not holding the target light failed to show any differences.[14] It turns out, however, that having a contact lens in the eye or having a piece of surgical tape against the lower eyelid provides somatosensory cues about eye position that are as effective as holding the target light in suppressing autokinesis.[15] Together these observations emphasize the importance of somatosensory and proprioceptive cues — distributed across the body — in influencing sensory localization and oculomotor control and the broad range of cues the body can use to enhance accuracy.

5.4 PROPRIOCEPTIVE AND SOMATOSENSORY INFLUENCES ON VISUAL AND AUDITORY LOCALIZATION

Proprioception and somatosensory cues can also be used to create errors in sensory localization. When the hand is in contact with a target light in an otherwise dark room, if apparent motion of the arm is induced by muscle vibration, then apparent visual motion of the physically stationary target will be perceived. The subject will see the target change its apparent position even though recordings of eye position will indicate that stable fixation is maintained.[16] Subjects often report that it feels as if their eyes are moving in their orbits, tracking the visual target's "motion." Vibration of an arm muscle activates its muscle-spindle receptors, eliciting a reflexive contraction. It has been known since the classic observations of Matthews and his colleagues,[17] that if the motion of a limb moving under the action of a tonic vibration

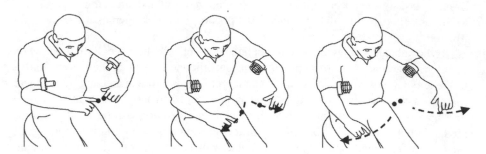

FIGURE 5.5 Perceptual remapping of visual angle elicited by vibratory myesthetic illusions.
Left: Small target lights (filled circles) are attached to the index fingers of both arms, which
are immobilized (restraints not shown, for clarity). Vibrators are positioned over both biceps
brachii muscles, and the subject stably fixates one of the targets in the otherwise dark room.
Middle: Activation of the vibrators entrains spindle afferent activity, artificially signaling
biceps elongation and eliciting illusory extension of both unseen forearms (arrows). The
perceived distance between the two lights increases during the felt arm displacement. **Right**:
Cutaneous contact is denied by attaching the target lights to the immobile restraint (not shown)
1 mm away from the fingers, increasing dramatically the vibration-induced felt displacements
of the fingers and abolishing the illusory visual separation of the targets.

reflex is resisted, apparent motion of that limb will be experienced in the direction
that would be associated with lengthening of the vibrated muscle. Thus, biceps
brachii activation leads to apparent forearm extension and triceps brachii activation
to apparent flexion. The arm movement illusion has two components: a spatial
displacement and a velocity one. The limb will seem to move physically to a new
spatial location and then to still be moving but no longer displacing. The induced
motion of a visual target attached to the hand of the vibrated arm has the same
features and shows a peaking of apparent spatial displacement while still being
perceived to be in continuous motion.

If a target light is attached to the index finger of each hand and the fingers are
separated by a few centimeters, as illustrated in Figure 5.5, then when illusory
extension of both arms is induced by vibration of both bicep muscles, the targets
will be seen to increase in separation. Bilateral triceps vibration induces apparent
flexion of the two arms and the targets will be seen to move closer together. This
visual direction remapping occurs for both monocular and binocular viewing and
eye position recordings indicate that steady fixation is maintained.[18] The illusory
visual effect is eliminated if the fingers are not in direct contact with the targets.
These results demonstrate that proprioceptive-tactile information about visual target
location can influence the apparent direction of gaze, visual direction, and the
apparent physical separation of two targets (the latter, without affecting apparent
distance, which means that there must be a central remapping of retinal local signs).
These findings indicate that somatosensory and proprioceptive information provides,
through hand contact, a mechanism for calibrating the direction of gaze and visual
spatial localization.

FIGURE 5.6 Illusions experienced in darkness during biceps brachii vibration with a speaker emitting noise bursts attached to the hand. **Left**: The forearm and the sound source are actually immobilized (restraint not shown for clarity) in a sagittal plane but are experienced as moving down; apparent movement of the sound is consistent with the physical pattern of non-changing auditory arrival time and intensity cues at the ears. **Right**: The forearm and speaker are actually restrained in a horizontal plane but are experienced as moving rightward relative to the head which feels stationary, a pattern inconsistent with non-changing binaural cues.

Auditory localization can be similarly remapped.[19] Figure 5.6 shows a subject seated with an auditory speaker emitting noise bursts attached to his hand. Two arm orientations are illustrated — a horizontal and a vertical one. When the biceps brachii is vibrated, illusory extension of the forearm will be experienced horizontally in one case and vertically in the other. Simultaneously, the sound source will be heard to change in position in keeping with the hand's apparent displacement, reaching a peak displacement while still seeming to be moving. The illusions can be 20 or 30 degrees or more, in magnitude. Vertical movement of a sound source in the subject's sagittal plane is consistent with non-changing arrival time and intensity cues at the ears. However, lateral physical movement of a sound source would necessarily change these cues. Since the head is movable, one might expect if lateral displacement of the hand and auditory target was experienced that a change in head orientation would also be experienced. For example, in the dark with a single target light attached to the hand, a change in eye position with respect to the head is experienced during illusory motion of the target. But this does not happen with the sound source. A remapping of auditory localization cues occurs instead, analogous with the two visual targets case, in which the retinal loci associated with particular visual directions are remapped.

5.5 SOMATOSENSORY AND PROPRIOCEPTIVE CONTRIBUTIONS TO SELF-CALIBRATION

The remappings of arm position, eye position, and auditory and visual localization, associated with muscle vibration, are transient. After the vibration is terminated or contact of the hand with the target is broken, the original localizations are re-achieved. Nevertheless, the influence of hand position on sensory and postural localization provides a mechanism by which calibration of auditory and visual maps may be achieved normally. It is clear from recent physiological work that the cortical maps devoted to vision, somatosensation, and audition are highly plastic and subject to reorganization.[20,21,22,23,24] The mobility of the hand and the possibility of gaining veridical information about the external environment and objects within it by exploration with the hand underscore its flexible use as a tool for self-recalibration.

The hand can also be used for calibration of the body schema. This concept refers to the knowledge the CNS has of the dimensions and spatial characteristics of the body that allows an individual to touch accurately without visual guidance different parts of the body and that allows appropriate guidance of the body relative to the surroundings. The "Pinocchio illusion" illustrates how the hand may be involved. Figure 5.7 shows an individual, with eyes closed, grasping his nose while the biceps brachii of his arm is vibrated to create an illusion of arm extension. The subject perceives his nose to grow longer and longer in keeping with the apparent position of his hand which seems to move farther and farther from his face. With appropriate positioning of the hands on the body, broad-ranging perceptual remappings of body configuration, body dimensions, and body orientation can be generated.[25] Figure 5.8 shows an individual with arms akimbo. Bilateral vibration of the biceps brachii leads to apparent expansion of the waist as the hands seem to move farther apart. With bilateral triceps vibration, the pattern most often experienced by subjects is the waist diminishing in size. Although these effects are transient and the correct body size is experienced with cessation of vibration, they nevertheless demonstrate how veridical mappings could be generated and refreshed by hand contact under conditions of accurate hand movement control. In other words, these studies provide a fast-forward view of how the normal calibration process may be achieved.

5.6 SOMATOSENSORY CONTRIBUTIONS TO ARM MOVEMENT CONTROL

The use of the hand in calibration of the body schema raises the issue of how the calibration of the arm itself is achieved. The hand and arm change size during development and arm length affects the physical amplitude of hand displacement through space. For example, if the arm grows an additional 10 cm in length, then a 1° change in joint angle at the shoulder will produce a larger linear displacement of the hand than prior to the growth change. Normally, when the hand is moved over a stationary surface, that surface is perceived to be stationary. This means that the changing tactile input at the fingers is attributed to the motion of the hand. Katz[26] recognized the significance of this fact in his classic studies of touch. Gibson[27] later

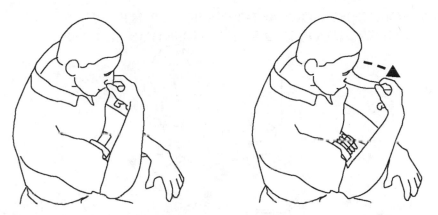

FIGURE 5.7 The Pinocchio illusion. **Left:** The person being tested holds his nose in a pincer grip while a vibrator is positioned over the biceps brachii muscle of his right arm. **Right:** When the vibrator is turned on, illusory extension of the forearm is felt and the spatially fixed point of contact is also experienced as moving, resulting in apparent extension of the nose (right).

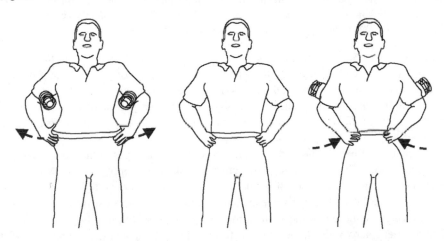

FIGURE 5.8 Perceptual remapping of body dimensions in the vicinity of contact with an appendage in which a vibratory myesthetic illusion is induced. In the test situation, the subject holds his arms akimbo in darkness (center) during bilateral biceps (left) or triceps (right) brachii muscle vibration. Biceps vibration elicits illusory forearm extension and body expansion at the point of hand contact. Triceps vibration induces illusory forearm flexion and pinching of the waist.

measured the threshold for detecting the motion of a surface moving under the fingertips. He found that the stationary, restrained fingertips resting on a surface could detect motion of 1 mm/s or even less. He also had subjects move their hands back and forth laterally with their fingertips contacting a yardstick that also could be in motion. He found that the direction and approximate speed of yardstick motion could be detected although he did not provide details about the frequencies, velocities, and amplitudes of arm and yardstick movements studied. The apparatus illus-

FIGURE 5.9 Apparatus for manipulating the coupling between voluntary movements and tactile slip cues. In the situation illustrated, the surface on which the finger tips are dragging moves rightward an equal amount to the hand movement thereby eliminating slip cues. Exposure to novel couplings of slip and voluntary movement (without vision) causes immediate alterations in detection of surface motion and estimation of voluntary arm movement magnitude. Longer term exposure leads to adaptive remappings.

trated in Figure 5.9 permits related questions to be asked. It couples motion of the contact surface with motion of the hand in the same or opposite direction by a fraction thereof. Consequently, the tactile feedback associated with voluntary movements of the hand over a surface can be systematically remapped. This allows measurement of the detection thresholds for identifying motion of the surface during conjoint hand movements.

Subjects are better able to detect surface motion when it is in the same direction as arm motion. The displacement detection threshold is about 7% for "with" displacement of the surface but 40% for "against" displacements. Thus, subjects are much more able to detect displacements of the contact surface during voluntary motion of the hand when the surface moves so as to reduce the magnitude of slip displacement across the fingertips. Surface displacement against the direction of hand motion is considerably more difficult to detect reliably. An interesting additional feature is that subjects also tend to overestimate the magnitude of their arm movements in experimental conditions involving counter-displacements of the contact surface and to underestimate them when the surface moves in the same direction as their arm movements. This pattern suggests that the slip cues at the fingertips as the hand sweeps over a surface contribute to the appreciation of the magnitude of the arm movement.

A contribution of tactile input at the fingertip to the appreciation of arm position is consistent with recent observations demonstrating that when pointing movements are made to targets on a surface, the pattern of shear force stimulation of the fingertip provides a spatial coding of hand position with respect to the body.[1] The region of the fingertip stimulated also indicates far vs. near and left vs. right. The shear reaction force vectors generated on the fingertip during the first 30 msec after impact are oriented to a locus near the shoulder of the reaching arm and code the spatial position of the hand with respect to the body. Thus, when the finger touches the surface, the pattern of contact cues on the fingertip literally specifies where the finger is in relation to the body. They code hand location in an egocentric reference frame. The shear forces abate within 100 msec of the finger contacting the surface after which only a small normal force remains as the arm is actively supported with the finger in contact with the surface.

The importance of the finger impact forces for calibration of reaching movements is emphasized by two additional experimental paradigms. When a head-mounted visual display is used to present virtual targets, a subject can point to them but even if his or her finger is in spatial register with the virtual target there will be no somatosensory feedback related to the target position because there is no physical contact. The head-mounted display obstructs sight of the hand so visual feedback about hand position is also absent. In this circumstance, if the subject reaches repeatedly to the virtual target, a remarkable thing will happen. The subject's reaching movements to the target will become more and more inaccurate, showing greater and greater dispersion.[28] Importantly, if a solid surface is placed at the spatial plane of the virtual targets so they coincide with its surface, then, even after one single contact with the surface, subsequent reaches will be much more accurate. If the surface is removed, movement accuracy again will rapidly degrade. This pattern means that absence of contact at the end of movements coupled with absence of sight of the hand leads to degradation of limb position sense. Contact is necessary to preserve calibration; proprioceptive information alone is not adequate.

The importance of terminal contact cues for calibration is also apparent in experiments involving Coriolis perturbations of reaching movements.[29] If a reaching movement is made in a room rotating at constant velocity, an inertial force known as a Coriolis force will be generated on the arm. Its magnitude is proportional to the velocity of room rotation and the velocity of the moving arm relative to the room; consequently, it is only present during a reaching movement itself. The transient Coriolis force deflects the path of the arm in the direction opposite rotation of the room. When the room is turning at 60°/s, subjects will miss visual targets by many centimeters (see Figure 5.10 on left).

Subjects rapidly adapt to the Coriolis perturbations when allowed visual feedback of their movements. Adaptation takes less than 10 reaches, when repeated reaches are made to a single target. If subjects point to the location of a visual target that is extinguished at the onset of their reach (in the otherwise dark rotating room), they still are able to adapt within 10–15 reaches, if their hand makes contact with the surface in which the target is embedded. Terminal contact allows them to adapt fully to the Coriolis perturbations even though the surface provides no textural cues to the target's location (see Figure 5.10 on right, top). By contrast, if subjects point in the air just above the location of the just-extinguished visual target, they do not regain accurate movement endpoints and will continue to show endpoint deviations in the direction of the Coriolis forces generated by their movements (see Figure 5.10 on right, bottom). The absence of terminal contact cues prevents the sensorimotor control mechanism from registering or detecting the discrepancy between desired

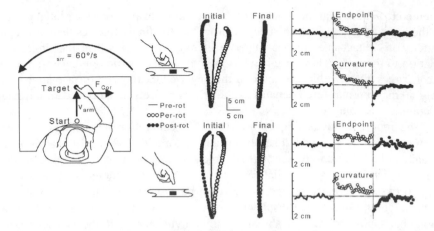

FIGURE 5.10 Role of fingertip contact in adaptation of reaching movements to Coriolis force perturbations in a rotating room. **Left**: Forward reaching movements (v_{arm}) during counter-clockwise rotation (w_{srr}) generate rightward Coriolis forces on the arm (F_{Cor}). **Right, top**: When reaches end on a smooth surface encasing the target, they are initially deviated by Coriolis forces but return to baseline endpoint and curvature patterns within 40 movements. The initial post-rotation reaches show mirror image deviations to the initial per-rotation reaches. **Right, bottom**: Reaches ending in the air above the target per-rotation show no adaptation to Coriolis deviations of their endpoint but return to normal curvature. Post-rotation, there is little endpoint deviation but curvature is mirror image to initial per-rotation reaches.

hand position and actual hand position. Thus, full endpoint adaptation cannot be achieved.

In this experimental situation, the adaptation takes place "automatically" if terminal contact is allowed. With repeated reaches, the subject becomes progressively more accurate without awareness of what he or she is doing. Moreover, after adaptation is complete subjects no longer perceive the presence of the Coriolis forces generated by their reaching movements. Their movements seem completely natural and normal, and the Coriolis forces, although still generated during movements, seem absent. Space does not permit pursuit of this important fact and its implications for perception of the forces encountered during self-generated movements but this issue is discussed further elsewhere by Lackner and DiZio.[1]

These studies demonstrate the importance of somatosensory cues from the hand in the calibration and maintenance of limb position sense and in the adaptive recalibration of arm movement control. A final series of studies will be described to illustrate the contribution of haptic cues from the hand to balance.

5.7 HAPTIC CONTRIBUTIONS TO POSTURAL CONTROL

Standing subjects greatly stabilize their balance if they touch a stationary surface with the index finger of their outstretched arm. Subjects are at least as stable in the dark with finger contact as they are with full sight of their surroundings without

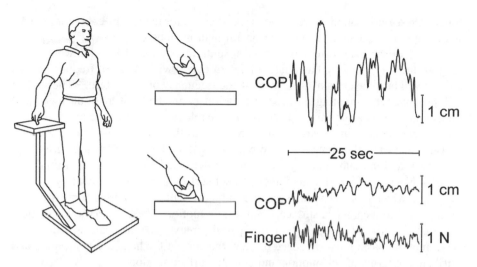

FIGURE 5.11 Left: Experimental setup used to investigate the influence of light touch of the finger on posture. **Right, top**: Without finger contact in darkness, there is substantial sway of the center of pressure (COP). **Right, bottom**: Fingertip contact at <1 N attenuates sway. (1 N ≅100 g.) Surreptitiously, oscillating the touch surface entrains posture (not shown).

finger contact.[30,31] If they are allowed both vision and touch, they are most stable of all. Touch stabilizes posture even though the levels of applied force at the fingertip are far below those adequate to provide mechanical stabilization. Nearly all subjects spontaneously adopt about a 40 g level of applied force at their fingertip. This value is interesting because it corresponds to the region of maximum dynamic sensitivity of the somatosensory receptors in the fingertip as measured by Westling and Johansson.[32] Variations about this level give the greatest modulation of somatosensory afferent activity. Analyses of the relationship among the force changes at the fingertip, EMG activity in the leg muscles controlling sway, and the displacement of the body center-of-pressure reveal that the fingertip is not being used as a passive probe to detect sway, but rather is being actively controlled. Changes in the leg EMG signals follow changes at the fingertip by about 150 msec, and changes in body position occur another 150 msec later (see Figure 5.11).[33]

Individuals without labyrinthine function cannot stand more than a few seconds in the dark without falling if they are in a heel-to-toe stance. Nevertheless, if allowed fingertip contact, they can balance as stably as normal subjects who are standing without finger touch but with sight of their surroundings.[34] This observation demonstrates the great functional significance for spatial orientation of haptic inputs and the possibility of substituting them for vestibular input in some circumstances. Touch cues can also override proprioceptive misinformation about body orientation. If the leg muscles of standing subjects are vibrated to elicit a reflexive contraction through a tonic vibration reflex, they will be greatly destabilized and may fall over. However, light touch of the index finger completely overrides the destabilizing influence of leg muscle vibration.[35] In fact, subjects are more stable during leg muscle vibration in the dark if they are allowed touch than they are in the absence of vibration without touch. These observations have obvious significance for the rehabilitation of patients with labyrinthine deficits or with peripheral neuropathies.

Touch of the hand can also be used to entrain posture. If unbeknownst to the subject, the surface is oscillated laterally, then over a wide range of frequencies the subject will become entrained and sway at the frequency of the surface.[36] The subject will perceive the oscillating surface to be stationary and be unaware of his own sway being entrained. By minimizing his own motion relative to the finger contact surface, the subject actually increases his sway relative to the ground. If subjects are told that in some trials the contact surface will be oscillated and that in others it will be stationary and are subsequently exposed to intermixed trials, then in virtually all trials, subjects will perceive the finger contact surface as not being spatially stationary even when it is. Nevertheless, the subjects will still exhibit some postural entrainment to the contact surface when it is in motion. These observations emphasize that cognitive knowledge about the spatial context, e.g., the contact surface may be in motion, has a great influence on how the cues at the hand are interpreted and attributed in terms of self-motion and contact surface motion.

The force levels and finger contact areas that are adequate to enhance postural stability are surprisingly small. Studies demonstrating this have made use of von Frey filaments that bend at different buckling force levels and a rigid steel filament.[37] The filaments were mounted vertically so that subjects could touch their ends with their index fingers. Von Frey filaments of different diameters have different compressional bending strengths. When a filament bends, it exerts a constant force on the fingertip so long as the fingertip does not move laterally in relation to it. If the finger moves laterally, the force generated by the filament on the finger will be considerably less.[38] Finger contact with a steel rigid filament 1 mm in diameter was fully as effective in attenuating sway as contact with a flat surface. With the von Frey hairs force levels as small as 7 grams significantly stabilized sway relative to conditions in which subjects held out their fingertip and imagined contact with a stable surface. In these experiments, the subjects standing with arm outstretched and index finger contacting a flexible von Frey filament could control and modulate the force at their finger more precisely than is possible in "fine force resolution" studies. In these latter studies, subjects are seated with their forearm supported and the minimal change in force that they can exert against an object is measured.[39] Thus, subjects may actually be able to perform with finer fingertip force resolution in a context that mobilizes their entire body.

5.8 SUMMARY

Mechanical contact on the body surface is an omnipresent feature of terrestrial life. We have shown here that such contact provides spatial information about the orientation of the body with respect to the environment. During reaching movements, the pattern of forces generated on the reaching hand when it contacts a surface provides a spatial mapping of hand location relative to the body. This spatial coding of terminal hand position is also extremely important for allowing adaptive changes in motor control to be implemented. In fact, in the absence of vision and contact cues, reaching accuracy rapidly degrades. Contact of the hand with the body itself provides a means by which the central nervous system representations of the body surface and its dimensions can be calibrated and updated over time. Hand contact with external auditory and visual objects localizes them spatially relative to the body, allowing directional calibration of vision and audition. The direction of gaze can be similarly calibrated. Light hand contact with stationary environmental objects greatly enhances the stability of stance. Finally, the control of the hands and feet themselves can be updated through physical interaction with objects in the external environment.

REFERENCES

1. Lackner, J. R. and DiZio, P. Aspects of body self-calibration. *Trends Cog. Sci.*, 4, 279, 2000.
2. Alexander, R. M. *Locomotion of Animals*. Blackwell, Oxford, 1982.
3. Winter, D. A. *Biomechanics and Motor Control of Human Movement*, 2nd Ed. John Wiley & Sons, New York, 1990.
4. Bigby, J. V. Some problems of postural sway. In *Ciba Foundation Symposium on Myotactic Kinethetic and Vestibular Mechanism*, A. V. S. de Reuck and J. Knight, Eds., Churchill, London, 1967.
5. Maxwell, S. S. *Labyrinth and Equilibrium*. Lippencott, Philadelphia, 1923.
6. Lackner, J. R. and Graybiel, A. Parabolic flight: Loss of sense of orientation. *Science*, 206, 1105, 1979.
7. Lackner, J. R. and Graybiel, A. Perceived orientation in free fall depends on visual, postural, and architectural factors. *Aviat, Space Environ. Med.*, 54, 47, 1983.
8. Lackner, J. R. and Graybiel, A. Postural illusions experienced during Z-axis recumbent rotation and their dependence on somatosensory stimulation of the body surface. *Aviat. Space Environ. Med.*, 49, 484, 1978.
9. Lackner, J. R. and Graybiel A. Some influences of touch and pressure cues on human spatial orientation. *Aviat. Space Environ. Med.*, 49, 798, 1978.
10. Lackner, J. R. and DiZio, P. Some efferent and somatosensory influences on body orientation and oculomotor control. In *Sensory Experience, Adaptation, and Perception*, L. Spillman and B. R. Wooten, Eds., Erlbaum Associates, Clifton, NJ, 281, 1984.
11. Graybiel, A. and Hupp, D. I. The oculogyral illusion: a form of apparent motion which may be observed following stimulation of the semicircular canals. *J. Aviat. Med.*, 17, 3, 1946.
12. Evanoff, J. N. and Lackner, J. R. Some proprioceptive influences on the spatial displacement component of the oculogyral illusion. *Perception and Psychophysics*, 43, 526, 1988.
13. Lackner, J. R. and Zabkar, J. Proprioceptive information about target location suppresses autokinesis. *Vision Res.*, 17, 1225, 1977.
14. Winterson, B. J., Steinman, R. M. Proprioceptive information neither improves fixation stability nor reduces autokinesis. *Vision Res.* 19, 1289, 1979.
15. Lackner, J. R., Fox, C., and DiZio, P. Autokinesis is suppressed by eye-lid contact cues related to eye position. Submitted, 2001.
16. Levine, M. S. and Lackner, J. R. Some sensory and motor factors influencing the control and appreciation of eye and limb position. *Exp. Brain Res.*, 36, 275, 1979.
17. Goodwin, G. M., McCloskey, D. I., and Matthews, P. B. C. The contribution of muscle afferents to kinaesthesia shown by vibration induced illusions of movement and by the effects of paralysing joint afferents. *Brain*, 95, 705, 1972.

18. DiZio, P., Lackner, J. R., and Lathan, C. E. The role of brachial muscle-spindle signals in assignment of visual direction. *J. Neurophysiol.*, 70(4), 1578, 1993.

19. Lackner, J. R. and Shenker, B. Proprioceptive influences on auditory and visual spatial localization. *J. Neurosci.*, 5, 579, 1985.

20. Yuste, R. and Sur, M. Development and plasticity of the cerebral cortex: from molecules to maps. *J. Neurobiol.*, 41, 1, 1999.

21. More, C. I., Nelson, S. B., and Sur, M. Dynamics of neuronal processing in rat somatosensory cortex. *Trends Neurosci.* 22, 513, 1999.

22. Kaas, J. H. The transformation of association cortex into sensory cortex. *Brain Res. Bull.*, 50, 425, 1999.

23. Kaas, J. H. Organizing principles of sensory representations. *Novartiz Foundation Symp.*, 228, 188, 2000.

24. Kaas, J. H. The reorganization of somatosensory and motor cortex after peripheral nerve or spinal cord injury in primates. *Prog. in Brain Res.*, 128, 173, 2000.

25. Lackner, J. R. Some proprioceptive influences on the perceptual representation of body shape and orientation. *Brain*, 111, 281, 1988.

26. Katz, D. *The World of Touch*, Lawrence Erlbaum Associates, Mahwah, NJ, 1989.

27. Gibson, J. J. Observations on active touch. *Psychol. Rev.* 69, 477, 1962.

28. DiZio, P. and Lackner, J. R. Proprioceptive adaptation and aftereffects. In *Handbook of Virtual Environment Technology*, K. Stanney, Ed., Lawrence Erlbaum Associates, New Jersey, in press.

29. Lackner, J. R. and DiZio, P. Rapid adaptation to Coriolis force perturbations of arm trajectory. *J. Neurophysiol.*, 72(1), 299, 1994.

30. Holden, M., Ventura, J., and Lackner, J. R. Stabilization of posture by precision contact of the index finger. *J. Vest. Res.*, 4(4), 285, 1994.

31. Jeka, J. J. and Lackner, J. R. Fingertip contact influences human postural control. *Exp. Brain Res.*, 100(3), 495, 1994.

32. Westling, G. and Johansson, R. S. Responses in glaborous skin mechanoreceptors during precision grip in humans. *Exp. Brain Res.*, 66, 128, 1987.

33. Jeka, J. and Lackner, J. R. The role of haptic cues from rough and slippery surfaces on human postural control. *Exp. Brain Res.*, 103, 267, 1995.

34. Lackner, J. R., DiZio, P., Jeka, J. J., Horak, F., Krebs, D., and Rabin, E. Precision contact of the fingertip reduces postural sway of individuals with bilateral vestibular loss. *Exp. Brain Res.*, 126, 459, 1999.

35. Lackner, J. R., Rabin, E., and DiZio, P. Fingertip contact suppresses the destabilizing influence of leg muscle vibration. *J. Neurophysiol.*, 84(5), 2217, 2000.

36. Jeka, J. J., Schoner, G., Dijkstra, T., Ribeiro, P., and Lackner, J. R. Coupling of fingertip somatosensory information to head and body sway. *Exp. Brain Res.*, 113, 475, 1997.

37. Lackner, J. R., Rabin, E., and DiZio, P. Stabilization of posture by precision touch of the index finger with rigid and flexible filaments. *Exp. Brain Res.*, 139, 454, 2001.

38. Gordon, J. E. *Structures/or Why Things Don't Fall Down*. Plenum Press, New York, 1978.

39. Henningson, H., Knechts, N., and Henningson, B. Influence of afferent feedback on isometric fine force resolution in humans. *Exp. Brain Res.*, 113, 207, 1997.

6 Locomotor Control: From Spring-like Reactions of Muscles to Neural Prediction

Arthur Prochazka and Sergiy Yakovenko

CONTENTS

0-8493-2336-3/02/$0.00+$1.50
© 2002 by CRC Press LLC

6.1   INTRODUCTION

Dr. P.R. Burgess organized a Society for Neuroscience symposium in 1992 to discuss his contention that "You can only control what you sense." The question is, what is being controlled in locomotion and which of the many sensory inputs to the CNS are the main players? There are numerous ways of answering this, each implying a different level of control and different neural systems. For example, at one level, that which is controlled is support and movement of the body with respect to uneven terrain. The control problem at this level is to cope with support surfaces of variable orientation, consistency, stability, friction, and compliance. At another level, that which is controlled is movement of the body with respect to a moving target (e.g., as in the hunting of prey). At this level, the problem is to anticipate future positions of the target and to control and adapt one's own trajectory accordingly, taking into account obstacles and hazards in the way.

6.2   CELLS VS. SYSTEMS

Stuart et al. (2001) recently pointed out that sensorimotor control has been studied either "inside-out" from cellular and molecular mechanisms within small neuronal networks (the cellular level) or "outside-in" from complex behaviors to reflexes (the systems level). Although these two approaches often remain far apart, more and more laboratories are tackling specific problems from each end (Rossignol 1996; Jordan 1998; Kiehn and Kjaerulff 1998; O'Donovan et al. 1998; Grillner et al. 2001).

6.2.1   HISTORICAL DEVELOPMENT

The outside-in approach started centuries ago, when it was suggested that complex behaviors including locomotion comprised chains or assemblies of simple behaviors or reflexes (Descartes 1664; Mettrie 1745; Spencer 1855; Sechenov 1863). These ideas gained credibility with early experimental work that showed that after removal of the cerebrum in birds, frogs, and quadruped mammals, the brainstem and spinal cord could still generate complex movements such as righting reflexes and locomotion (Flourens 1823; Goltz 1869; Freusberg 1874; Goltz 1892; Brown 1911).

The inside-out approach gained momentum with the neuroanatomical studies of Ramon y Cajal (Cajal 1894) and the technical breakthrough of electronic recordings of single-neuron activity (Adrian and Zotterman 1926). Within years the glass microelectrode had allowed intracellular potentials to be measured and the scene was set for the detailed study of the ionic mechanisms of the action potential, the synaptic actions of sensory axons on motoneurons and interneurons (Eccles et al.

1957a; Eccles et al. 1957b), the role of neurotransmitters in simple reflex behavior (Eccles et al. 1954; Jankowska et al. 1967; Jankowska et al. 2000) and the neuronal analysis of reflexes elicited in anesthetized or mid-collicular decerebrated animals (Chen and Poppele 1978; Terzuolo et al. 1982). In the 1970s, patch-clamping and molecular techniques allowed the functioning of membrane channels and their associated intracellular mechanisms to be studied in detail (Neher and Sakmann 1976).

In the 1960s, it was shown by the Moscow group of Shik, Orlovsky, and their colleagues that locomotion in decerebrate and spinal animals provided an excellent basis for electrophysiological studies at both cellular and systems levels (Shik et al. 1969). This in-between approach has provided much useful insight and, combined with pharmacological and molecular techniques, has begun to allow the first comprehensive analyses of locomotor control ranging from ions to behavior (Grillner et al. 2000).

To return to the main theme of this chapter, deafferentation studies, modelling, and lessons that have been learned from designing walking robots all show that it is important to have sensory information throughout the step cycle about the terrain and obstacles ahead; ground reaction forces, and displacements; internal forces and displacements; and relative velocities of the body segments. Numerous reviews have been written in the last few years on one or more of these topics (Pearson 1995; Horak and MacPherson 1996; Prochazka 1996b; Rossignol 1996; Büschges and Manira 1998; Marder and Pearson 1998; Pearson et al. 1998; Duysens et al. 2000). We will focus mainly on control mediated by mechanoreceptors, but key aspects of the visual control of locomotion will be included toward the end of the chapter.

6.3 THE EFFECT ON LOCOMOTION OF THE LOSS OF SENSATION

It has long been assumed that one's sense of movement derives from internal sensory signals generated in the moving tissues. Bell (1834) discussed "muscle sense" in detail both in relation to movement control and as a conscious sensation. He spoke of the pleasure of muscle sense during vigorous exercise. In a particularly striking passage, he described a woman afflicted with the loss of muscle sense, who could not hold her child safely without constant conscious effort. Since then, literally hundreds of deafferentation studies in animals and humans have underscored this dramatic loss of motor control in the absence of input from muscle receptors. One of us recently reviewed the deafferentation literature (Prochazka 1996b) and summarized the main points as follows:

1. The basic ability to produce voluntary force and move limbs is preserved after deafferentation. However, movements are generally uncoordinated and inaccurate, especially when visual guidance is absent.
2. Coordination of the different segments of the primate hand in precision tasks is particularly impaired. The accuracy of spatial orientation, fractionated movements, and anticipatory pre-shaping of the hand is reduced, and writing may be severely affected.

3. Gait is possible after deafferentation, but again it tends to be irregular and uncoordinated. This holds true in vertebrates and invertebrates alike. In humans who have lost limb proprioception, gait is severely impaired and requires conscious attention. If neck proprioception is also lost, gait becomes virtually impossible.
4. Control of tasks involving simultaneous changes in several variables, coordination of several limb segments, or adaptation to changes in the external environment is impaired. Thus fastening buttons or holding a cup is difficult and sometimes impossible without visual guidance.

Regarding locomotion in particular, the effects of deafferentation have not always been clear-cut, because under certain restricted circumstances the nerve or muscle activation patterns or the locomotor movements themselves can appear to be reasonably normal (Grillner and Zangger 1975). For example, though deafferented animals and human subjects with somatosensory loss have great difficulty in walking at first, in time they can learn to use residual sensory cues from the limbs and trunk and other sensory inputs such as vision to control their locomotion quite successfully. Hulliger has identified some of the pitfalls in interpreting deafferentation studies in a recent review (Allum and Honegger 1998). He concludes that complete deafferentation of the limbs and trunk results in a massive deficit in the coordination of locomotor movements, though some of the deficits may be overcome through adaptation and intensive training.

6.4 RECEPTORS INVOLVED IN THE CONTROL OF LOCOMOTION

6.4.1 MECHANORECEPTORS

Sherrington (1906) elaborated on muscle sense, and defined proprioceptors as receptors mediating the conscious sensation of the body's own movements. He assumed proprioception to be mediated by receptors in muscles, joints, and ligaments (Sherrington 1906; Sherrington 1947). From the 1930s to the 1960s, physiologists (but not clinicians) came to believe that muscle receptor activity and therefore proprioception did *not* involve conscious sensation, but this view had to be reversed again when it was shown that tendon vibration, which fairly selectively excites muscle spindles, produced clear illusions of movement (Goodwin et al. 1972).

Skin receptors were originally assumed to respond mainly to external forces impinging on the body and so were classified by Sherrington as exteroceptors, along with visual, auditory, and olfactory receptors. However, skin receptors also respond to the stretching of skin that occurs during most limb movements and under these circumstances they contribute to kinaesthesia (Moberg 1983; Collins and Prochazka 1996), so this would seem to qualify skin receptors as proprioceptors too. By the same token, it is now clear that muscle receptors can respond very sensitively to stimuli applied to the body. The classification of mechanoreceptors as "proprioceptors" or "exteroceptors" has thus become rather confusing and of dubious usefulness.

In 1926, the first electrical recordings from single sensory axons were obtained (Adrian and Zotterman 1926) and for the next 50 or more years, the response properties of different types of sensory mechanoreceptors have been studied in great detail in many species. In mammals, muscle spindle and tendon organ endings have been characterized and modelled mathematically. Various types of skin receptor have been identified and their responses to skin indentation of different amplitudes and frequencies have been elucidated in detail (Birder and Perl 1994).

6.4.2 VISION, VESTIBULAR INPUT, AND HEARING

The other senses that have a profound influence on the control of locomotion are vision, vestibular input, and, to a lesser extent, hearing. Though locomotion can be generated by all animals in the absence of vision, in most cases if the terrain is unpredictable, or if there are obstacles in the way, the control of locomotion is degraded. As we shall see, vision is crucial for the higher levels of locomotor control, in which the external context of locomotion must be taken into account in the predictive and adaptive aspects of control (Schubert et al. 1999). Sensory input from the vestibular apparatus also has a powerful effect on the control of posture, balance, and locomotion (Horak and MacPherson 1996; Zelenin et al. 2000). The contribution of hearing to locomotor control has rarely been studied and is assumed to be of lesser importance. However, under certain circumstances, auditory input produces rapid orienting responses and can affect postural adjustments (Valls-Sole et al. 1999).

6.5 STRUCTURE AND RESPONSE PROPERTIES OF PROPRIOCEPTORS

There are several reviews in the literature on the morphology and response properties of mechanoreceptors that signal locomotor movements (vertebrates: Granit 1970; Matthews 1972; Hulliger 1984; Prochazka 1989; Johansson et al. 1991; Jami 1992; Prochazka 1996b; Prochazka and Gorassini 1998a; Prochazka 1999; invertebrates: Bassler 1983; French 1988; Burrows 1992; Bassler and Buschges 1998). In the following, we will concentrate mainly on mammalian proprioceptors, muscle spindles, and tendon organs.

6.5.1 MUSCLE SPINDLES: STRUCTURE

There are 25,000–30,000 muscle spindles in the human body, including about 4000 in each arm and 7000 in each leg (Voss 1971; Hulliger 1984). The average number of spindles in a mammalian muscle is roughly 38*(cube root of mass in grams) (Banks and Stacey 1988; Prochazka 1996b). Thus a 64 g muscle contains about 152 spindles. On average, a single limb muscle in the cat contains 50 to 200 spindles, each ranging from 2 mm to 6 mm in length (Voss 1971; Boyd and Gladden 1985). A muscle spindle consists of a dozen or so intrafusal muscle fibers (Latin "fusus" = spindle) attached at each end to the surrounding extrafusal muscle fiber, with a central region innervated by sensory endings encased in a capsule (Boyd and Gladden 1985). The spindle lengthens and shortens along with the extrafusal muscle fibers to which it is attached. The typical spindle in cats, monkeys, and humans contains three types of intrafusal muscle fiber: a dynamic bag$_1$ (DB1 or b$_1$), a static bag$_2$ (SB2 or b$_2$), and 2–11 chain (c) fibers. These intrafusal fibers receive motor input from 10–12 fusimotor axons and sometimes from a β-skeletofusimotor axon, which also innervates neighboring extrafusal muscle fibers (Emonet-Denand et al. 1975; Hulliger 1984; Banks 1994). The central encapsulated region of the spindle contains 1 primary and up to 5 secondary sensory endings spiraled around the non-contractile portions of the intrafusal fibers (Boyd and Gladden 1985). The primary endings are those of group Ia afferents (conduction velocity 72–120 m/s in cats) and the secondary endings are those of group II afferents (20–72 m/s) (Matthews 1972).

6.5.2 PASSIVE RESPONSE PROPERTIES OF SPINDLE AFFERENTS

Spindle primary and secondary endings respond to muscle length variations similarly in cats, monkeys, and humans (for detailed comparisons and a discussion of scaling for different muscle lengths see Prochazka 1981; Prochazka and Hulliger 1983). In the absence of fusimotor action, group Ia and II afferents respond to muscle length changes dynamically, Ia afferents being more sensitive to muscle velocity and acceleration (e.g., they respond to ramp stretches with larger jumps in firing rate and they show more phase advance in response to sinusoidal inputs). There is a continuum from the smallest diameter group II afferents with low velocity sensitivity, to the largest diameter Ia afferents with high velocity- and acceleration-sensitivity. Group Ia and II afferents both have non-linear aspects of response: e.g., stretch sensitivity that depends on amplitude and offset, after-effects of muscle and fusimotor contraction and non-linear velocity scaling (Hulliger 1984; Prochazka 1996b).

6.5.3 FUSIMOTOR ACTION

The b$_1$ fiber and its associated Ia sensory spirals are selectively activated by dynamic fusimotor (α_d) or β_d skeletofusimotor axons (Boyd and Gladden 1985; Boyd et al. 1985; Banks 1994). The b$_2$ and chain fibers are activated by static fusimotor or skeletofusimotor (α_s or β_s) axons and rarely by α_d or β_d axons (Banks et al. 1998). Up to 30% of hindlimb spindles lack b$_1$ fibers and their group Ia afferents are then called b$_2$c afferents (Boyd and Gladden 1985; Taylor et al. 1992; Taylor et al. 1998; Taylor et al. 1999).

The main fusimotor actions can be summarized as follows. When muscle length changes are small (< 0.5% rest length), pure α_d action, mediated by b$_1$ fibers, increases the background firing rate (bias), decreases the stretch-sensitivity (gain) and reduces the phase advance of group Ia afferents (Emonet-Denand et al. 1977). For larger-amplitude length changes, α_d action increases group Ia stretch-sensitivity up to five-fold (Boyd et al. 1985) and either increases or decreases Ia phase advance slightly (Hulliger et al. 1977; Chen and Poppele 1978). Type b$_2$c Ia afferents do not exhibit dynamic fusimotor effects, the b$_1$ intrafusal fibers being absent. α_s action strongly increases the bias of both group Ia and II sensory endings and reduces group Ia stretch sensitivity (gain) by 50% or more for all amplitudes of length change (Cussons et al. 1977; Chen and Poppele 1978). Paradoxically, *weak* α_s action can *increase* Ia gain (Hulliger et al. 1985). In either case, phase is little changed. Of the 6 to 9 α_s axons acting on a group

II ending, each produces some bias, most attenuate its sensitivity to small stretches (< 1% rest length) but one or two of the α_s axons substantially *increase* group II sensitivity to stretches, presumably by activating b_2 intrafusal fibers (Jami and Petit 1978). The sensitizing action of these α_s fibers on group II endings is similar to the action of α_d fibers on Ia endings. The terms *dynamic* and *static* fusimotor action are in fact misleading, in that both types alter mainly the *gain* and *offset* rather than the *dynamics* of group Ia and II responses to stretch (Prochazka 1996b).

6.5.4 SPINDLE MODELS

Mathematical models of spindle response characteristics were originally developed from results obtained in acute experiments (Matthews and Stein 1969; Poppele and Bowman 1970; Poppele and Kennedy 1974; Chen and Poppele 1978; Poppele 1981). These models have recently been tested and compared in ensembles of group Ia and II afferent activity recorded in freely moving cats (Prochazka and Gorassini 1998a; Prochazka and Gorassini 1998b). Because all of the Ia models have a velocity-sensitive term and this tends to dominate the response to the relatively fast changes in muscle length that occur in locomotion, most of the models were reasonably successful in predicting the group Ia responses from the muscle length and EMG activity profiles, as exemplified in the ensemble firing profile of hamstrings Ia afferents shown in Figure 6.1. In this example, the small EMG term was added to provide a small amount of alpha-linked biasing of the Ia firing rate, to represent alpha-gamma linkage. In both the group Ia and II models, fusimotor action is usually represented as a single gain parameter though a more comprehensive model has been developed that incorporates some of the non-linear features of fusimotor action (Schaafsma et al. 1991; Otten et al. 1995). Modelling has provided some crucial insights into locomotor control in recent years and the mathematical models of spindle and tendon organ responses have played an important role in this.

6.5.5 TENDON ORGANS: STRUCTURE AND RESPONSE PROPERTIES

Tendon organs are encapsulated structures 0.2–1 mm long, usually located at musculo–tendinous junctions (Barker 1974). Generally speaking, there are about 80% as many tendon organs in a typical limb muscle as spindles. Their sensory endings, which become group Ib afferent axons, are entwined among the tendinous strands of 10-20 motor units, a given motor unit affecting 1–6 tendon organs (Proske 1981; Jami 1992). The sensory endings are compressed when the tendon is put under tensile stress. When non-contracting muscles are stretched, the force transmitted by the tendon is low except at the extreme end of the physiological range. Because many tendon organs only fire significantly at such extremes, it was initially posited that they were overload protectors. It was then found that their sensitivity to force changes produced by activating selected motor units was much higher than to force changes produced by passive stretch of the whole muscle (Houk and Henneman, 1967). This led to the idea that they were more sensitive to active force than to passive force. However when their force-sensitivity was compared in active contractions and passive stretching *of the whole muscle,* no significant difference was found (Stuart et al. 1970, Stephens et al. 1975).

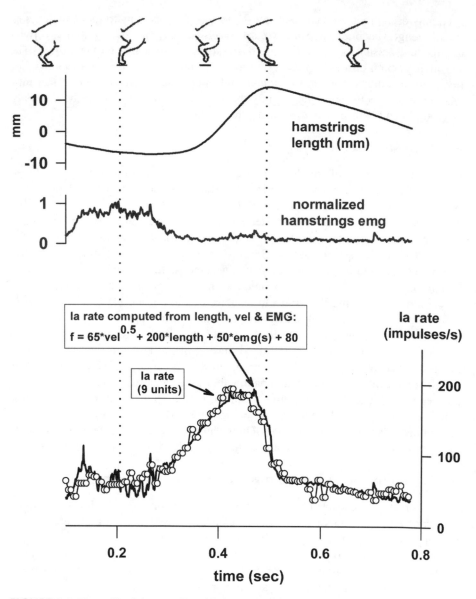

FIGURE 6.1 Ensemble firing profile of 9 hamstrings spindle primary (group Ia) afferents recorded during overground locomotion in normal cats. **Top**: muscle length, **middle**: electromyogram (EMG), **bottom**: firing rate profile with superimposed predicted rate derived from the length and EMG signals. Reprinted from Prochazka, A., *Prog. Brain Res.*, ©1999. pp. 133–142, with permission from Elsevier Science.

6.5.6 TENDON ORGAN MODELS

Because it is impossible to monitor the net force produced by the particular group of motor units "sampled" by a tendon organ, it is difficult to determine precise input/output characteristics for the receptor. Nonetheless, frequency analyses have been performed by applying feedback-controlled force signals to the whole muscle (Houk and Simon 1967; Anderson 1974; Stephens et al. 1975). These showed that tendon organs had transfer functions comparable to those of spindle group II endings (Alnaes 1967). Like group II afferents, tendon organ group Ib afferents fire fairly regularly, except at very low levels of active force, when unfused twitch contractions of newly recruited motor units may cause bursts of group Ib firing (Jami et al. 1985). Various types of non-linearity in tendon organ transduction have been described. For example, a given Ib ending may be unloaded by contractions of muscle fibers not inserting into the receptor capsule (Houk and Henneman 1967; Stuart et al. 1972). As group Ib afferents sample from a restricted subset of motor units, they do not necessarily signal whole muscle force linearly (Jami et al. 1985). The transfer function models describing the relationship between whole muscle force and Ib firing rate (Houk and Simon 1967; Anderson 1974) have recently been tested on firing profiles of small ensembles of Ib afferents recorded during locomotion in normal cats. The EMG activity from the receptor-bearing muscles was used in lieu of muscle force. In spite of this substitution, and in spite of concomitant length changes which would have modulated force somewhat due to inherent force-length properties of the muscles, the tendon organ firing rate profiles were surprisingly well predicted (Figure 6.2).

6.5.7 Mechanoreceptors in Joints, Ligaments, and Skin

Mechanoreceptors in joint capsules, joint ligaments, and skin are strategically placed to provide proprioceptive feedback, but proving this role has been difficult. Until the late 1960s, joint receptors were assumed to signal joint position over the full range of motion (Boyd and Roberts 1953). However, it was then reported that most joint receptors in the cat knee and wrist only fired at the extremes of the range (Burgess and Clark 1969; Tracey 1979). Subsequently, several groups reported full-range signaling (Godwin-Austen 1969; Zalkind 1971; Carli et al. 1979; Ferrell 1980; Lund and Matthews 1981) though some of the full-range afferents in the cat knee joint may have been either muscle spindle or tendon organ afferents (McIntyre et al. 1978). Loading of the joint capsule by muscle contraction sensitizes joint receptors, in some cases enough to confer mid-range responsiveness on them (Grigg and Greenspan 1977).

On balance, it seems that joint capsular and ligamentous afferents are capable of signaling limb position and movement at the extremes of motion and in some joints over the full range of motion. Single-unit discharges are detectable in recordings from whole joint nerves (Ferrell 1980), so the total number of joint receptors signaling mid-range movement is probably low compared to the number of muscle and skin receptors responding to the same movement. Joint afferents have conduction velocities mainly in the group II range (Burgess and Clark 1969) and their segmental reflex connections with α-motoneurons are less direct than those of muscle spindles (Johansson et al. 1991; Jankowska 1992). It has been suggested that they have a special role in reflexly inhibiting motoneurons of muscles near joints that are damaged (Iles et al. 1990).

Cutaneous receptors overlying joints and muscle respond phasically as well as tonically to movement (Edin and Abbs 1991; Edin 1992; Edin and Johansson 1995;

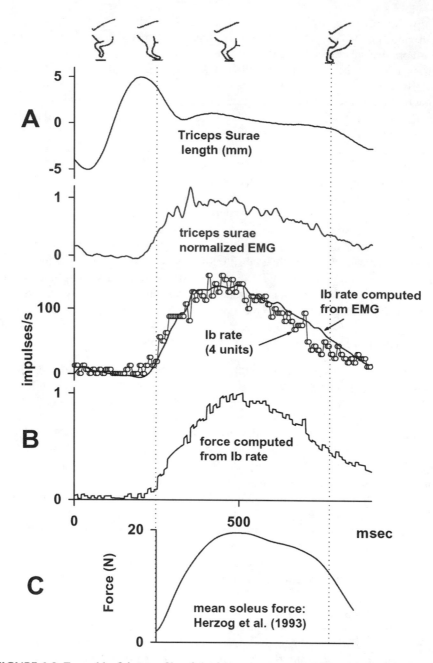

FIGURE 6.2 Ensemble firing profile of 4 triceps surae tendon organ (group Ib) afferents recorded during overground locomotion in normal cats. **A. Top**: muscle length, **second panel**: electromyogram (EMG), **third panel**: ensemble Ib firing rate profile with superimposed predicted rate derived from the EMG signal, **B**: muscle force as predicted from the firing rate, **C**: mean force in soleus obtained in separate experiments in another laboratory (Herzog et al. 1993). Reprinted from Prochazka, A., *Prog. Brain Res.*, ©1999. pp. 133–142, with permission from Elsevier Science.

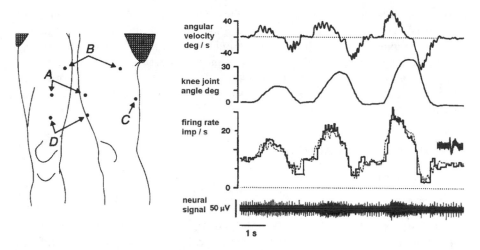

FIGURE 6.3 Firing rate profile of a slowly adapting type III cutaneous receptor located at "D" in the left panel recorded in a human subject with microneurography (letters A, B, and C refer to other afferents recorded in this subject). The afferent responded to knee joint displacement. The third trace on the right superimposes the instantaneous firing rate (bold) and the predicted rate derived from the joint angular velocity and joint angle signals in the two top traces. The bottom trace shows the raw action potentials of this unit. Reproduced with permission from Edin, B.B., *J. Physiol.*, 531, 289, 2001.

Edin 2001) and it has been argued that they probably contribute to the sense of position and motion of the extremities. There are massive numbers of skin receptors in the limbs. For example, it has been estimated that there are about 17,000 skin mechanoreceptors with myelinated afferent fibers on the surface of the human hand (Johansson and Vallbo 1979), compared to about 4000 muscle spindles, 2500 tendon organs, and a few hundred mid-range joint receptors in the whole arm (Voss 1971; Hulliger 1984). Type I skin receptors and hair follicle receptors are responsive to rapidly varying skin stimuli. Slowly adapting type II and III receptors respond sensitively to stretching of the skin and continue to signal maintained stretch (Horch et al. 1977). In two recent studies, it was shown that stretching of the skin overlying finger joints produced illusions of movement of the fingers, reinforcing the idea of a proprioceptive role for skin receptors (Edin and Johansson 1995; Collins and Prochazka 1996). Figure 6.3 shows the firing rate of a slowly adapting type III cutaneous receptor located at "D" in the left panel recorded in a human subject with microneurography. The afferent clearly provided information about knee joint displacement and velocity. Its firing profile was extremely well fitted with a first order transfer function similar to that used in the simpler models of group Ia transduction.

6.5.8 OVERVIEW OF PROPRIOCEPTIVE FIRING DURING LOCOMOTION

Figure 6.4 summarizes the current knowledge regarding the firing rate profiles of ensembles of group Ia, Ib, and II muscle afferents during medium-speed stepping in cats. The data was quantified or estimated from numerous single-unit recordings obtained with microwire electrodes implanted in dorsal root ganglia of free-to-move cats (Prochazka et al. 1976; Loeb and Duysens 1979; Loeb 1981; Loeb 1984; Prochazka and Gorassini 1998a). If we assume a mean firing rate per receptor of about 75 impulses/s during locomotion (Figure 6.4), at any given moment the net input to the spinal cord from the 10,000 or so muscle afferents in each leg is between 0.5 and 1 million impulses/sec. Figure 6.4 also shows that the firing rates of muscle afferents are deeply modulated during stepping, and so in principle this would provide highly detailed information to the CNS for locomotor control.

6.6 SIMPLE LOCOMOTOR REFLEXES

The neurophysiological significance of the age-old observation that decapitated animals can display coordinated locomotion had been recognized and documented by the mid-18th Century (Mettrie 1745). Clearly, the neuronal machinery in the spinal cord was capable of controlling quite complex activities without descending input from the brain. The notion that all movements were simply chains of reflexes developed a century later (Spencer 1855). In the late 19th and early 20th centuries, the elementary movements involved in locomotion began to be studied in detail, particularly in spinalized or decerebrated dogs and cats (Freusberg 1874; Sherrington 1910; Brown 1911). Sherrington's (1910) study was a tour de force of careful experimentation and highly detailed description and remains a definitive reference work to this day. In it, he described the biomechanical actions of most of the muscles of the cat hindlimb, how these muscles were activated or fell silent in flexion and extension reflexes, and how they participated in what he termed "reflex stepping" (now referred to as air stepping) in spinal cats and "reflex walking," i.e., weight-bearing gait, in decerebrate cats.

6.6.1 THE STRETCH REFLEX

The "stretch reflex" differs in decerebrate and intact animals and since Sherrington's time it has come to be realized that several CNS mechanisms may contribute components of different latency to the stretch reflex response. At the segmental level, muscle spindle Ia afferents activated by muscle lengthening monosynaptically excite homonymous alpha motoneurons which in turn cause the muscle to resist the stretch. In static postures, Ib input generally results in homonymous inhibition, but recently it has been shown that this switches to longer-latency homonymous excitation during locomotion (Conway et al. 1987), at least in cat extensor muscles (the evidence for Ib homonymous excitation in human locomotion is a matter of controversy: see below). Group II input from muscle spindles has also been implicated in long-latency components of stretch reflexes (Matthews 1991; Sinkjaer et al. 2000; Grey et al. 2001). Ia homonymous excitation represents negative *displacement* feedback, which augments the intrinsic stiffness of active muscles in the face of length perturbations. Ib homonymous feedback, on the other hand, represents positive *force* feedback. Positive feedback is synonymous with instability and oscillation in engineering systems, but when muscles are the actuators, their non-linear length-tension properties turn out to stabilize the positive feedback loop (Prochazka et al. 1997a; Prochazka et al. 1997b).

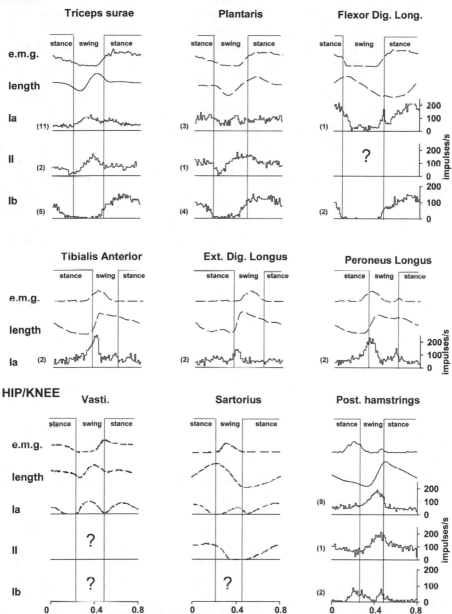

FIGURE 6.4 Summary figure showing firing rate profiles of ensembles of group Ia, Ib, and II muscle afferents during medium-speed stepping in normal cats. The data were compiled from numerous single-unit recordings obtained with implanted dorsal root electrodes. Reproduced with permission from Prochazka A. and Gorassini, M., *J. Physiol.* 507, 293, 1998.

Segmental stretch reflexes and their electrically-elicited counterparts, H-reflexes, have been studied intensively for many years, partly because they are modulated in interesting ways, but mainly because it is technically relatively easy to elicit and measure them. Studies of this type have been reviewed many times in recent years (Dietz 1996; Prochazka 1996b; Brooke et al. 1997; Dietz 1998; Duysens et al. 2000; Schneider et al. 2000). Human H-reflexes are smaller during locomotion than during static postures and they show phase-dependent fluctuations (Garrett et al. 1981; Garrett and Luckwill 1983; Garrett et al. 1984; Capaday and Stein 1986). It is often assumed that H-reflexes represent transmission in the short-latency pathway from Ia afferents to homonymous motoneurons, though an oligosynaptic contribution cannot be ruled out (Burke 1983). There is evidence that the faster the locomotion and the more difficult the terrain, the greater the suppression of H-reflexes (Capaday and Stein 1987; Llewellyn et al. 1990). It is vigorously debated whether this modulation is centrally generated (Schneider et al. 2000) or due to reafferent signals (Misiaszek et al. 1998). Stretch reflexes elicited during locomotion by rapidly stretching individual muscles (Akazawa et al. 1982; Hiebert et al. 1996) or imposing sudden rotations about joints (Orlovsky and Shik 1965; Sinkjaer 1997; Gritsenko et al. 2001) also show phase-dependent modulations, but these are less predictable than H-reflex modulations (Sinkjaer 1997; Christensen et al. 2001). This is not too surprising, as stretch reflexes comprise not only the Ia-mediated short latency responses but also longer-latency responses involving group II reflexes, group Ib reflexes, and higher-level processing in the CNS (Sinkjaer et al. 1999).

Before one delves too deeply into the mechanisms and modulation of stretch reflexes, however, one should ask how important they are in the control of locomotion and movement anyway. Would an animal be simply unable to support its weight if it had no stretch reflexes? The answer is no, on both experimental and theoretical grounds. Provided that the alpha motoneurons of load-bearing muscles are activated from *some* source in the nervous system, the muscles develop an intrinsic stiffness that resists stretching and that can in fact be represented as displacement feedback (Partridge 1966). This is easily seen in the stretch reflex model of Appendix Figure 1, whereby in the inner feedback loops muscle stretch results in reactive forces due to the force-length and force-velocity properties of muscle. Muscle afferent feedback mediated by Ia and Ib pathways in the outer loops of the model provides a source of input to alpha motoneurons but it is not the only source. There are many descending and propriospinal pathways that also activate motoneurons, represented by just one input in the model of Appendix Figure 1.

Analysis of responses to loading with this model showed that simple stretch reflexes mediated by spindle Ia afferents for example can at most triple the prevailing muscle stiffness without causing instability (Prochazka et al. 1997a). This is illustrated in Figure 6.5, where Ia gains of 1 and 2 (negative feedback) educed the muscle stretch caused by applied force by factors of about 2 and 3, respectively; the loop became unstable when the gain in the Ia-mediated pathway was set to 3. Though there are several simplifying assumptions in models of this type, their predictions have been shown to be fairly accurate in hybrid experiments in which actual muscles were stretched and "reflexly" activated by electrical stimulation modulated by a signal obtained by filtering the displacement signal with a spindle Ia transfer function (Bennett et al. 1994).

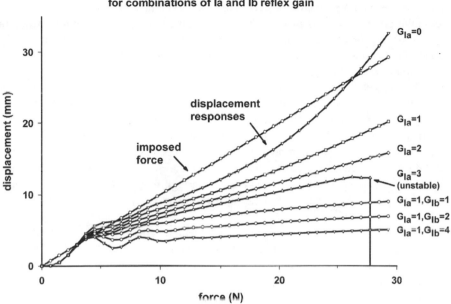

FIGURE 6.5 Responses of stretch reflex model (Appendix Figure 1, page 170) to ramps of force, with 7 different combinations of Ia gain (G_{Ia}) and Ib gain (G_{Ib}), as indicated. The slope of the displacement curve marked $G_{Ia} = 0$ corresponds to the intrinsic stiffness of the active muscle alone. As negative feedback Ia gain increases, the stiffness also increases, but for $G_{Ia} = 3$, the loop goes unstable at the point indicated by the arrow on the right. The addition of positive Ib feedback to negative Ia feedback greatly increases the stiffness of the system.

Figure 6.5 also illustrates another aspect of proprioceptive reflexes that was quite unexpected from a control theoretical point of view, and that has only recently been understood analytically: positive force feedback. To the surprise of many in the field, it was discovered in 1987 that signals from Ib afferents, which normally inhibit homonymous and synergistic alpha motoneurons (MNs) in the absence of locomotion, switched to *exciting* these motoneurons *during* fictive locomotion in the acute spinal cat (Conway et al. 1987). This has been confirmed in the cat by several other groups since (Pearson and Collins 1993; Guertin et al. 1995; Angel et al. 1996), though the evidence for this switch in humans is equivocal (Dietz 1998; Stephens and Yang 1999; Capaday 2000; Misiaszek et al. 2000; Pang and Yang 2000). MN excitation causes muscles to generate more force, further increasing Ib firing, which in turn would produce yet more excitation, i.e., positive feedback. Would this not result in an explosive increase to maximal force? The surprising answer is that because the force response of muscles to increments in activation declines as the muscle gets shorter, an equilibrium length is reached at which the force-related excitation does not cause further muscle shortening. Thus in Figure 6.5, for three different levels of positive force-feedback gain, the "muscle" resisted stretching with a much higher stiffness than could be achieved with spindle Ia feedback. Not only is this scheme remarkably stable, but as the large increase in stiffness in this example shows, positive force feedback is potentially more effective in load compensation than spindle-mediated feedback. Evolution was evidently not constrained by conventional control systems theory! Since the publication of the analytical work in 1997, positive force feedback has been implemented in a cockroach robot with muscle-like actuators, and has been found to be effective and useful in load compensation (Nelson and Quinn 1999).

Thus stretch reflexes should be seen as adding a controlled level of stiffness to the underlying intrinsic muscle stiffness. Mammals, including humans, with large-fiber sensory loss and absent tendon jerks, though severely incapacitated and uncoordinated, can regain full weight-support and can walk if supported laterally. It has been confirmed in such experiments that there has been no recovery in sensory input (Allum et al. 1998). This shows that the intrinsic stiffness of muscles activated through non-reflexive pathways can generate enough force to bear the weight of the body and propel it forward. We will support this position with further modelling below.

If stretch reflexes are not the primary mechanism of load compensation in locomotion, does this mean that sensory input from the limbs only plays a minor role in locomotor control? Again, we will argue that in general the answer is no: sensory input does play a vital role, but through mechanisms other than the stretch reflex.

6.6.2 FLEXION AND EXTENSION RESPONSES, "REFLEX STEPPING," AND "REFLEX WALKING"

As mentioned above, the neural basis of flexion and extension reflexes was first studied in spinal and decerebrate dogs and cats in the late 19th Century. These reflexes were elicited by aversive stimuli applied to the skin or cutaneous nerves of the ipsilateral or contralateral limb. The movement synergies thus evoked were similar, though not identical, to those in the flexion and extension phases of locomotion. "Reflex stepping" was studied in decapitated dogs by Freusberg (Freusberg 1874) and then in decerebrate and spinal cats by Sherrington and Brown (Sherrington 1910; Brown 1911). An hour or so after spinalization, locomotor movements could be triggered by lifting the animal and dropping one hindlimb from a semi-flexed starting position. An important observation of both Sherrington and Brown was that position and loading of the limbs were crucial in triggering and maintaining reflex stepping. If the downward motion of the hindlimb was again stopped in semi-flexion, stepping movements ceased in this limb, as well as in the other, free limb. These results have been verified in numerous experiments since (Rossignol 1996; Orlovsky et al. 1999).

Reflex stepping could also be elicited by kneading, squeezing, or continuous electrical stimulation of the skin of the tail, perineum, back, neck, or pinna. It could also be elicited by continuous electrical stimulation applied to the transected spinal cord at cervical level. Sherrington struggled to explain how these remote, continuous stimuli applied to "exteroceptive" sensory receptors interacted with phasic sensory signals from the limb, which he argued were mainly proprioceptive, to produce cyclical movement. He suggested a scheme whereby the remote stimuli evoked a "primary reflex" (causing either extension or flexion, depending on the site of stimulation), which evoked proprioceptive stimuli in the moving limb. These elicited a secondary reflex, which alternated with or "interrupted" the primary reflex to produce cyclical activity. One flaw in this argument was preliminary evidence that deafferentation did not necessarily abolish the rhythm. Sherrington wondered whether the central rebound effect described a few years earlier by Magnus and von Uexkuell ("Umkehr") could be contributing. A year later, Sherrington presented a paper to the Royal Society by his student Graham Brown (Brown 1911), confirming the deafferentation results in decerebrated and spinalized cats. After a low thoracic transection, the deafferented lumbosacral spinal cord could still generate rhythmical movements in an isolated pair of hindlimb muscles in the absence of descending input or sensory input from either hind limb. Graham Brown dubbed the underlying mechanism the "intrinsic factor," a term that has now been superceded by the more specific term "Central Pattern Generator" (CPG) (Grillner and Zangger 1975).

6.7 CENTRAL PATTERN GENERATORS AND SENSORY FEEDBACK

6.7.1 CYCLICAL MOTOR PATTERNS GENERATED WITHOUT SENSORY INPUT

Literally hundreds of studies in which sensory input has been reduced or abolished by deafferentation have since demonstrated beyond doubt that isolated neuronal networks in the CNS can generate the basic rhythmical motor patterns involved not only in walking but also in the activities of breathing, chewing, swimming, flying, scratching, paw-shaking, and autonomic functions such as micturition and sexual "reflexes" (Grillner 1975; Prochazka 1996b; Kiehn et al. 1998; Orlovsky et al. 1999).

There has been much discussion about the ability of CPGs, in the absence of sensory input, to generate complex coordinated patterns of muscle activity such as those required for overground locomotion. After deafferentation in the cat, many subtle features of normal activation sequences can still be seen (e.g., the small burst of activity in knee flexor muscles at the end of the swing phase in cat locomotion [Grillner and Zangger 1975]). However, the locomotor rhythm is generally more labile (Grillner and Zangger 1975; Wetzel et al. 1976; Goldberger and Murray 1980; Grillner and Zangger 1984; Giuliani and Smith 1987; Koshland and Smith 1989) and above all, there is an inability to compensate for the changes in loading or terrain (Allum et al. 1998).

This latter defect is more serious than it may sound. If information on joint angles, body posture, loading, and displacement of the extremities are all unavailable to the central controller, the amplitude and timing of the cyclical motor output can only be set to some default level and cannot be matched to the varying requirements. To quote Brown (1911):

> A purely central mechanism of progression ungraded by proprioceptive stimuli would clearly be inefficient in determining the passage of an animal through an uneven environment. Across a plain of perfect evenness the central mechanism of itself might

drive an animal with precision. Or it might be efficient for instance in the case of an elephant charging over ground of moderate unevenness. But it alone would make impossible the fine stalking of a cat over rough ground. In such a case each step may be somewhat different to all others, and each must be graded to its conditions if the whole progression of the animal is to be efficient. The hind limb which at one time is somewhat more extended in its posture as it is in contact with the ground, in another step may be more flexed. But the forward thrust it gives as its contribution to the passage of the animal must be of a comparatively uniform degree in each consecutive step. It may only be so if it is graded by the posture of the limb when in contact with the ground, and by the duration of its contact with the ground. This grading can only be brought about by peripheral stimuli. Of these we must regard the proprioceptive stimuli from the muscles themselves as the most important, and the part which they play is essentially the regulative — not the causative.

The loss of proprioceptive input in humans is more devastating than in quadru-peds: without vision and/or external supports, people with large-fiber sensory loss find it difficult to take more than a step or two without stumbling and falling, even on a flat floor (Lajoie et al. 1996). Though evidence has been adduced for the existence of a locomotor CPG in the human lumbosacral spinal cord in people with spinal cord injury, the types of rhythmical movement observed were weak and inadequate for weight-bearing locomotion (Calancie et al. 1994; Dimitrijevic et al. 1998) and because the legs were suspended and sensory input was intact, it is possible that they could have resulted from reciprocating stretch reflexes. In a recent study in normal subjects, steady vibration of the legs resulted in cyclical air stepping (Gurfinkel et al. 1998). It was suggested that the steady sensory input activated "the central structures responsible for stepping generation." However, vibration of pha-sically contracting muscles is known to produce phasic firing of muscle spindles (Matthews and Watson 1981; Prochazka and Trend 1988), so the development of pendular reflexive motion in the suspended legs cannot be ruled out here either.

In certain cases, it has been claimed that accurate, goal-directed movements in a normal animal can only be attributed to central programming, because the moment-to-moment participation of sensory input can be ruled out. For example, in cock-roaches, it has been shown that sensory feedback in fast walking is too delayed to have a reflex effect within a given cycle, the duration of which is 40 ms or less (Zill 1985; Delcomyn 1991a; Delcomyn 1991b). In the most rapid ballistic movements in humans, sensory feedback is also too slow to modify the movements once they are underway (Desmedt and Godaux 1979).

Does this mean that in very rapid movements sensory input is unimportant? If a very rapid movement is considered in isolation it is true that the sensory input related *to that movement* may come too late to contribute to its control. But that movement, like all others, was preceded by sensory input that provided the CNS with information on the overall biomechanical state of the limbs and the rest of the body in the immediately preceding period. In locusts, it has been shown that sensory input influences the wing-beat cycle *following* the one in which it was elicited (Wolf and Pearson 1988). Similarly, it has been suggested that in the example of cockroach locomotion given above, the sensory input from one step cycle provides postural information for the control of *ensuing* step cycles (Prochazka 1985). This could either be viewed as delayed feedback or as prediction, depending on the way the signals are handled. We would argue that in the generation of ballistic movements that accurately reach their target, the CNS has taken into account the biomechanical initial conditions and the relative position of the target, which is only possible with prior sensory input. The only movements that might be controlled entirely open-loop (without any significant involvement of sensory input) would be escape behaviors in which the animal propels itself forward as rapidly as possible without regard for stability or the likelihood of a fall.

6.7.2 Interaction between CPGs and Sensory Feedback

In the mid-1960s, it was found that cats, decerebrated rostral to the superior colliculus, were more likely to walk than those with a mid-collicular decerebration, particularly with steady electrical stimulation of a region that has since become known as the midbrain locomotor region (MLR) (Shik et al. 1966; Shik et al. 1969). This allowed the spinal mechanisms of locomotion to be studied with microelectrodes, pharmacological interventions, and lesioning experiments. Some of the early observations of Sherrington, Brown, and their predecessors on the effect of limb position on locomotor-like reactions were soon confirmed and extended to stable treadmill locomotion. In one such experiment, both hindlimbs were de-efferented, leaving sensory input to the spinal cord intact (Orlovsky and Feldman 1972). MLR-evoked locomotor rhythms recorded electrically in S1 ventral root filaments were then found to be entrained by cyclical imposed movements of one of the limbs. The rhythm could be halted by moving the limb into extreme flexion or extension. After partial deafferentation, the rhythm could also be halted by arresting locomotor movements of one of the limbs in mid-cycle, just as Sherrington had described. In a more sophisticated version of this experiment a few years later, it was found that the hindlimbs of chronic spinal kittens walking on a split-belt treadmill could adapt their cadence to each belt separately, i.e., the sensory input to each limb was entraining that limb individually (Forssberg et al. 1980).

Entrainment and override of the locomotor rhythm by sensory input from a limb has been confirmed and studied in detail in dozens of experiments since the early 1970s in the high decerebrate MLR cat, chronic spinal cats, and acute spinal cats treated with clonidine (rev: Rossignol 1996). By the mid-1980s, two separate sensory variables had been identified as being capable of entraining or overriding the locomotor rhythm: hip position (Andersson and Grillner 1983; Kriellaars et al. 1994) and extensor force (Duysens and Pearson 1980). In a recent review of these findings, it was realized that the same two variables had been implicated in triggering the switch from stance to swing in species as widely separated as cats, crayfish, and locusts (Prochazka 1996b). In simple terms, the sensory rule that seemed to prevail in all these animals could be stated as follows: IF the leg has become very extended *and* extensor force has become very low, THEN initiate swing. This same rule had been quite independently "discovered" in the technological control of above-knee prostheses as well as in the control of functional electrical stimulation in human gait.

It is clear from all of the above that the basic locomotor pattern can be generated by the CNS without sensory input, but sensory input can promote, delay, or even block the switching between stance and swing phases and thereby completely determine step cycle frequency. The question of the relative importance that should be placed on central vs. sensory control has been quite bothersome to neurophysiologists over the years, some taking the view that sensory input is only important when the CPG-generated pattern fails to produce the required movements (the centralist view), while others suggest that the CPG rhythm is a default pattern that is only manifested when sensory input is withdrawn (the peripheralist view).

The paradox can be partly understood by considering the family of electronic circuits called multivibrators (also known as flip-flops). These consist of a pair of switchable elements such as transistors, interconnected such that when one is active, it suppresses the other by applying an "off" signal to its gate. In a free-running ("astable") flip-flop, the "off" signal discharges through a capacitor from the moment it is applied. Consequently, halfway through the cycle, the suppressed partner is "released," turns "on," and becomes the "oppressor" for the next half-cycle. The frequency can be modulated by controlling the rate of discharge or decay of the "off" signals by varying passive circuit components or by applying external inputs to the gates. These can override the oscillation completely and hold the circuit in one or the other half-cycle indefinitely. Thus, although the core circuit can generate a default rhythm, the additional circuit components promote, delay, or block phase switching and are therefore integral parts of the system as a whole. The flip-flop analogy was recognized in the late 1970s (Miller and Scott 1977) and forms the basis of some CPG models to this day (Orlovsky et al. 1999).

The usual compromise position between centralists and peripheralists is that the CPG generates the basic locomotor pattern, but this is "sculpted" or "fine-tuned" by sensory input. However, given the fact that sensory input can entrain and halt the rhythm, and given the extreme disability in human bipedal gait caused by even partial deafferentation (Lajoie et al. 1996), "sculpting" and "fine-tuning" seem to understate the case.

6.7.3 HUMAN LOCOMOTION

A puzzle remains about the role of local reflexes mediated by muscle receptors in *human* locomotion. As already mentioned, large-fiber deafferentation (which can eliminate input from muscle spindle primary and tendon organ afferents from the legs and trunk) can have a devastating effect on human locomotion (Lajoie et al. 1996). Yet, when these receptors are excited by test mechanical stimuli applied to muscles during gait, the effects are disappointingly small. For example, bursts of powerful vibration of the lower leg muscles, which most likely entrain the firing of many spindle group Ia spindle afferents and probably numerous group Ib afferents too, have virtually no effect on the trajectory of locomotor movements (Ivanenko et al. 2000a; Ivanenko et al. 2000b). Interestingly, tonic vibration of the upper leg muscles or the subject's neck did have a generalized effect on body tilt and speed of locomotion.

Short-latency EMG responses can certainly be elicited during gait by electrical stimulation of the large afferents (Garrett et al. 1984; Capaday and Stein 1986), by tapping on tendons (Llewellyn et al. 1987) or by applying rapid joint rotations via pneumatic orthoses (Sinkjaer et al. 1996). But the muscle stretches have to be faster than those occurring in unimpeded locomotion for the EMG responses to be significantly larger than the prevailing levels. Ischaemic block of large afferents had very little effect on short-latency soleus EMG unloading responses elicited by the pneumatic orthosis, though longer-latency responses persisted (Sinkjaer et al. 2000). This led to the suggestion that presynaptic inhibition effectively eliminates any significant contribution of Ia signals to homonymous muscle activation during normal locomotion, and it is only in very rapid perturbations that Ia-mediated activation of MNs "breaks through." A significant contribution to EMG activation (up to 50%) was claimed for the longer-latency pathways, though these reactions were delayed enough and of long enough duration to involve more complex central processing rather than simple segmental reflexes.

6.7.4 ROBOTS

Recently there have been two interesting developments in technology that may provide insight into these issues. The first is the design of walking robots. Some of the most advanced work in this area is of a corporate nature (e.g., the humanoid robots developed by the Honda company), and the information available on the control strategies used tends to be sketchy. Nonetheless, it is quite clear that the most versatile robots, including the extraordinary Honda Asimo P4, rely heavily on sensory input to generate locomotion over uneven terrain. The robot designers carefully reviewed the essential aspects of biological locomotor control in insects, quadruped mammals, and humans and then implemented the most promising aspects in their machines. The instructive thing here is to consider which control strategies turned out to be effective. Two hexapod robots, a "stick insect" and a "cockroach," developed in Cleveland, provide useful information in this regard (Espenschied et al. 1996; Nelson and Quinn 1999). In both cases, the robots have six two- or three-segment limbs. Each joint has a passive spring for compliance, an actuator, and a position sensor. The actuator is under proportional position feedback control of variable gain, providing a controllable stiffness that adds to the passive "inherent" spring stiffness. The stiffness properties of each joint therefore mimic the intrinsic properties of biological muscles under stretch reflex control. Locomotion is achieved by a mixture of processes local to joints and legs, and two governing ("global") algorithms. The local processes include the active stiffness control just mentioned, as well as If–Then control rules based on end-point position for stance-swing and swing-stance transitions, and special rules for adaptive responses to tripping (stumble reaction) and "foot-in-hole" (Gorassini et al. 1994). Figure 6.6 shows the reactions of a leg of the hexapod robot in three situations, perturbed stance, placing reaction, and "foot-in-hole." For comparison, we have included a panel showing the searching movements of a locust in which the leading limb enters a hole (at 2), searches, and is eventually placed (at 3) (from a movie kindly provided by K. G. Pearson).

Regarding the two "global" algorithms, one adjusts leg trajectories to distribute force equally among weight-bearing legs, and to match leg "lengths" according to the terrain (i.e., to keep the body horizontal in the face of slopes and other unevenness of ground support). Interestingly, leg length (the distance from the end of the paw

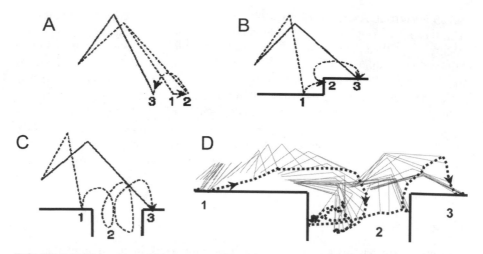

FIGURE 6.6 Kinematics of leg movement in a hexapod robot (A, B, and C) and a locust (D). **A.** perturbed stance: imposed movement of the foot from 1 to 2 evokes a corrective reaction with placement at 3. **B.** tripping reaction: foot contacts obstacle at 2, is lifted and placed beyond the obstacle at 3. **C.** "foot-in-hole:" foot enters hole, which triggers searching movements at 2 and eventually a placing reaction at 3. **D.** locust "foot-in-hole." Stick figures traced from frames of a movie. Leading limb enters a hole at 2, searches, and is eventually placed at 3. A, B, and C reproduced with permission from Espenschied et al., *Robotics and Autonomous Systems*, 18, 59, 1996.

to the hip joint) has recently been identified as an emergent variable that accounts best for the firing behavior of dorsal spinocerebellar tract neurons in the cat, even when individual joints are constrained (Bosco and Poppele 2001). These authors suggest that the spinocerebellar system may be viewed as the end-point of processing of proprioceptive sensory information in the spinal cord. The other global algorithm quoted "encourages stance legs to lift into their swing phases in a coordinated manner, swing forward, and transition to stance" (Quinn and Ritzmann 1998).

Like the hexapod robots, the Honda robot also utilizes local active and passive joint compliances, IF-THEN phase transitions, special adaptive reactions similar to the ones just described and a novel global strategy of setting a moving target of ground reaction force as the command for forward or backward locomotion. The actual ground reaction force is continually computed from the sensor signals. The difference between this vector and the target vector is referred to as the "falling moment." This falling moment is minimized by "reflexes" to the joint actuators, which are presumably synergistically coupled.

Notice that in neither of the above robots is it easy to extract the notion of an autonomous CPG from the various global algorithms. Rather, the central controllers respond to external requirements by issuing general commands to move in particular directions (e.g., by proposing a virtual trajectory for the ground reaction force), selecting sensory rule bases appropriate to the task and context, and evaluating performance for predictive adjustments. Hazard rules are also computed (e.g., the Honda robot resists lateral imposed forces by stiffness control, but when sway exceeds the imbalance point, it yields and takes a step in the direction of the imposed force).

6.7.5 Virtual Animals

The other useful advance in technology is the development of powerful biomechanical modelling tools suitable for personal computers. These programs allow one to design quite complex models of "virtual animals" with limbs, joints, and muscles whose intrinsic properties can be approximated to those of real animals. Muscle activation profiles based on EMG patterns recorded during locomotion can be used to activate the muscles. The ensuing locomotor performance, which is displayed as a slow-motion movie as the computations proceed, provides similar types of insight to the mechanical robots above, but because changes can be made easily and tested quickly, many different sensorimotor rules and parametric variations can be explored.

Figure 6.7 shows a biomechanical locomotor model we have been working on for some time. The model is based on the cat hindquarters but it is intentionally not an accurate replica. Many muscles are absent. The origins and insertions of the muscles that are represented do not correspond exactly to those in real cats. The model uses a Hill-based force–velocity relationship and monotonic passive and active force-length curves (Figure 6.7B), as our recent results indicated that the static isometric force-length curve with its descending limb is invalid in continuous movements (Gillard et al. 2000). Short-range stiffness properties are neglected. The purpose of the model is to test some general hypotheses, not to provide a definitive analysis of gait in any given species.

Once we had "fine-tuned" the EMG activation patterns of its various actuators (Figure 6.7C), which were based on EMG profiles of cat locomotion (Prochazka et al. 1989), the model produced stable locomotion on a flat surface indefinitely, in spite of being "deafferented" (Figure 6.8A). Each step was slightly different from the last, which showed that the intrinsic stiffnesses of the muscles provided enough flexibility to make continuous adjustments to compensate for small variations in body speed, height, and the relative positions of the limb segments. The deafferented virtual cat could also adapt to modest uphill slopes (Figure 6.8A). As mentioned above, intrinsic muscle stiffness is equivalent to a length feedback system which resists deviations from some set equilibrium length. Thus, locomotion in a deafferented animal is not entirely open-loop.

The stability of the deafferented model was unexpected, because we were aware of the difficulty Gerritsen at al. (Gerritsen et al. 1998) had experienced in generating more than three or four steps in a similar model of bipedal human locomotion, though other groups have been able to overcome this by optimizing the EMG patterns with inverse dynamics or neural network learning techniques (Taga et al. 1991; Taga 1995a; Taga 1995b; Yamazaki et al. 1996; Taga 1998; Neptune et al. 2001; Ogihara and Yamazaki 2001). Of course, the cart that supports the front of our virtual cat greatly simplifies the problem of maintaining a stable upright posture. It is also a simplification of quadrupedal gait, which requires forelimb–hindlimb coupling for stability. It is very interesting that Gerritsen and Nagano (1999) recently obtained a far more stable performance when they incorporated some sensory feedback into their bipedal model.

The deafferented virtual cat immediately gets into trouble when the read-out rate of its EMG patterns is increased or decreased, producing a higher or lower gait

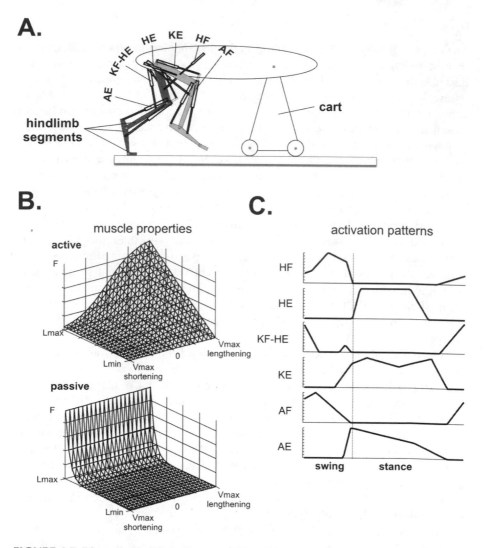

FIGURE 6.7 Biomechanical locomotor model based loosely on the cat hindlimb. **A.** Muscle groups are represented by actuators HF (hip flexors), HE (hip extensors), KF (knee flexors), KE (knee extensors), AF (ankle flexors), and AE (ankle extensors). **B.** Hill-based force–velocity relationship and passive and active force-length curves used in each actuator. **C.** Step-cycle activation profiles of the actuators.

velocity (Figure 6.8B). It does not help just to add stretch reflexes, because these merely augment the muscle intrinsic stiffnesses without affecting step cycle phase-switching. On the other hand, If–Then rules are useful in coping with variations of this type. In Figure 6.8B, the If–Then rules governing the transitions from stance to swing and back (see below) allowed the model to adapt to the increased EMG read-out rate and hence increased velocity, though it did not fare so well with the lowered read-out rate.

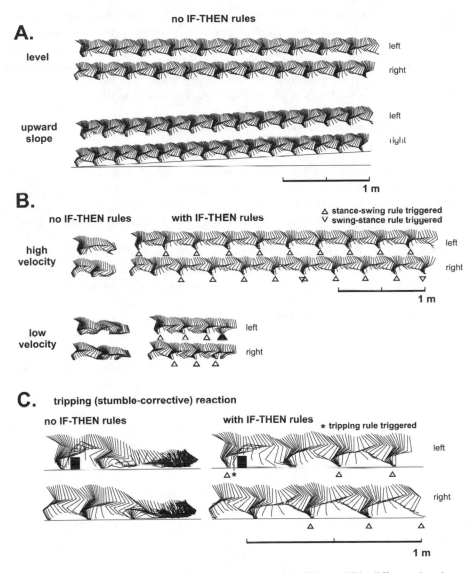

FIGURE 6.8 Kinematic analysis of behavior of the model of Figure 6.7 in different situations. A. locomotion of the model without feedback rules. The intrinsic stiffnesses of the muscles sufficed to compensate for kinematic and kinetic variations in locomotion over flat surface and up a small slope. B. Gait velocity increased (top) or decreased (bottom) by ~20%. Locomotion failed in the "deafferented" case, but improved with If–Then rules (triggering of stance-swing transitions marked with upward triangles; note that the contralateral leg was phase-locked to ipsilateral leg, so it did not "fire" its rules except in one case of a swing-to-stance transition marked by downward triangle). C. Obstacle impeded forward swing of one leg. The deafferented model (left) failed to compensate and dragged its foot over obstacle, causing a subsequent fall. Inclusion of a "tripping rule" (triggered at *) resulted in good reactive compensation with ensuing stable steps.

For the challenging situation of an obstacle impeding the forward swing of one leg, these simple rules were not enough to prevent a fall either. In this case, a separate "hazard" rule was needed, which was invoked when the front of the foot contacted the obstacle (see "tripping reaction" rule below). This initiated an EMG sequence similar to that previously recorded in the tripping reaction of normal cats (Wand et al. 1980). The result was a response that looked very true to life (Figure 6.8C).

6.7.6 If–Then Rules Governing Phase Switching and the Selection of "Hazard" Responses

The control of forces in muscles within the flexion or extension phases of locomotion is a smoothly graded process involving continuous proportional feedback, whereas the switching between these phases is usually discontinuous and abrupt. Control systems that switch between well-defined states are known as finite-state systems and switching is triggered when certain sensory conditions are met (e.g., IF swing AND hip angle small (flexed) AND extensor force low THEN terminate flexion, initiate extension) (Tomovic and McGhee 1966; Tomovic et al. 1990). Note that "small" and "low" are imprecise terms. This is intentional: the nervous system most likely uses sensory inputs in a probabilistic way rather than setting precise threshold values and requiring each threshold to be met before firing the rule.

A control systems analogy for probabilistic finite-state control is "fuzzy logic," in which sensory variables are accorded weighting ("membership") functions, the sum of the weighted sensory signals determining the motor outcomes. If–Then rules are crafted to suit specific behavioral states. More than one rule may be "active" in a behavior or a part of a behavior (e.g., in the stance phase of gait, the stance-swing rule and the "foot-in-hole" rule are both active: see below). The system constantly monitors the "firing strength" of the rules and retires behaviors and/or recruits new behaviors according to predetermined thresholds of the firing strengths. A higher level "arbitration mechanism" may decide which behaviors are appropriate for the current global state (Prochazka 1996a).

In such systems, each input effectively "votes" for a range of possible motor outcomes according to the currently active If–Then rules, the sum of the votes determining what actually happens (Chen et al. 1997; Jacobs 1997; Davoodi and Andrews 1999; Jonic et al. 1999). Bässler has suggested a similar process in the control of locomotion in stick insects and used the term the "Parliamentary Principle" as a metaphor for the voting mechanism (Bässler 1993). Fuzzy logic has also been discussed in relation to the changes that occur in forward, sideways, and backward treadmill walking in infants (Pang and Yang 2000).

On the understanding that If–Then rules may be used in this probabilistic way, let us now consider some of the rules that might underlie phase switching in biological locomotor control.

Stance phase (forward step)

Rule 1 (stance-swing transition): *IF* stance *AND* extensor force low *AND* hip angle large (extended) *AND* contralateral limb supported, *THEN* switch to swing.

Rule 2 ("foot-in-hole" reaction): *IF* mid-stance (hip angle medium) *AND* no ground contact *AND* contralateral limb supported, *THEN* switch to placing reaction.

Swing phase (forward step)
   Rule 1 (swing-stance transition): *IF* swing *AND* hip angle small AND knee angle large (knee extended) *THEN* switch to stance.
   Rule 2 ("tripping" reaction): *IF* swing *AND* skin stimulus to front of foot *THEN* switch to placing reaction; *IF* stance *AND* skin stimulus to front of foot *THEN* prolong stance.

Stance phase (backward step)
   Rule 1 (stance-swing transition): *IF* backward gait *AND* extensor force low *AND* hip flexed *AND* contralateral limb supported *THEN* initiate backward swing.

Swing phase (backward step)
   Rule 2 (swing-stance transition): *IF* backward gait *AND* hip angle large *THEN* initiate stance.
   The following rule describes the way the Honda robot invokes the "hopping reaction" of Magnus (Magnus 1924) when it is pushed in the face:
   Static Postural Rule 1 (hopping reaction): *IF* falling moment is small *AND* gravity vector is within support surface, activate extensors; *IF* falling moment is large *AND* gravity vector is outside support surface, take a step back.

Notes to the rules:
   a. The above list of rules describes sensorimotor interactions that occur in locomotion in many animals and that are explicitly programmed in robots, prostheses, and bio-mimetic models. Identifying the neural control systems responsible is no trivial matter, just as it remains no trivial matter to identify Brown's "intrinsic factor" (the locomotor CPG) after 100 years of research.
   b. Not only is the list far from complete, but the variables chosen are not necessarily the *only* ones that would "work," nor the ones that might be used in biological systems. For example, a possible alternative to "hip is extended" is "leg length is long" (Bosco and Poppele 2000; Bosco et al. 2000; Bosco and Poppele 2001). Likewise, the variables in the postural rule must clearly be derived from several sets of sensors in different parts of the body.
   c. As gait speed increases, it is important to phase-advance the switching back and forth between stance and swing to compensate for the delays due to inertia and muscle properties. Rather than invoke more rules using angular velocities, if we assume that muscle-spindle signals are used in place of joint angles, their velocity dependence would automatically provide this phase advance.

6.8 PREDICTION AND ADAPTATION

It goes without saying that when animals move they take into account global information on body posture, the environment and the context of the task from visual, auditory, and mechanosensory inputs. The greater the motor requirements to maintain stability (e.g., bipedal vs. quadrupedal gait), the more crucial are these predictive inputs. The controlling processes are clearly extremely complex, but a start has been made in recording the kinematic and neural correlates of predictive and adaptive responses (e.g., Drew et al. 1996; McFadyen et al. 1999; Rho et al. 1999). Some basic concepts regarding prediction and adaptation have been proposed over the years:

1. *"Einstellung" ("set")*. Animals prepare themselves in advance either to initiate movements or to react to impending perturbations (James 1890; Ach 1905; Watt 1905; Gibson 1941). Mental set, orienting, postural set, Bereitschaft (readiness), reflex set, and fusimotor set are just some examples (review: Prochazka 1989).
2. *"Degrees of Freedom."* The control of multi-segmented limbs is simplified when the number of degrees of freedom is reduced, either by co-contracting antagonist muscles, or by coordinating the activation of synergists (Bernstein 1967).
3. *"Efference copy" "Internal models."* In the 1950s, it was posited that the cerebral cortex generates a copy of motor commands from which re-afferent signals are subtracted (von Holst and Mittelstaedt 1950; von Holst 1954). Corrections would only be required if there were differences between the predicted and actual sensory inputs. The same idea was developed independently to overcome delays in industrial control processes (Smith 1959) and it has been rephrased and refined many times since, the most recent example being "Internal Models" (Blakemore et al. 1998; Bhushan and Shadmehr 1999; Vetter and Wolpert 2000; Wolpert and Ghahramani 2000).
4. *"Fixed action patterns" "Motor Programs" "Preprogrammed movements" "Movement Primitives."* Spencer proposed that reflexes were the *atoms of the psyche*, the psyche was an assemblage of reflexes, and "instincts" were reflex assemblies consolidated by repetition and transmitted in an hereditary manner (Spencer 1855). The idea of stored motor programs or subroutines has been reiterated many times since (review: Prochazka et al. 2000). The spinal central pattern generator is essentially an example of this idea. The tripping reaction may be another example.

The cerebellum is thought to be key in nearly all the above operations, in particular generating motor programs, modulating reflex gains, and scaling the size of movement sequences (Bloedel and Bracha 1995; Bloedel and Bracha 1998; Mummel et al. 1998; Serrien and Wiesendanger 1999). The recent conclusions of

Bosco and Poppele on the signaling of derived variables by spinocerebellar tract neurons is therefore of great interest (Bosco and Poppele 2000; Bosco et al. 2000; Bosco and Poppele 2001). It has also been suggested that motor cortical activity may be viewed as a summing junction for feedforward signals providing visuospatial information about the environment and feedback signals providing information on the state of the interneuronal pattern generating networks in the spinal cord (Drew et al. 1996). In human locomotion, subjects plan foot placement one or two steps ahead and avoid obstacles by anticipatory high-stepping (Patla and Vickers 1997; Patla et al. 1999). It is interesting that the reaction of many animals to unpredictable terrain is simply to "high-step" a gait modification that can be elicited by injecting a droplet of Lidocaine into the interpositus nucleus of the cerebellum (Gorassini et al. 1993).

One of the important conclusions drawn from the modelling work presented above is that the locomotor CPG, even when aided by segmental reflexes and the simplest of the If–Then rules, has a very limited ability to respond to variations in speed, slope, and uneven terrain. The detection of "hazard states" and the advance scaling of step cycles to deal with them is clearly a vital predictive and adaptive mechanism (Drew 1991).

6.9 SUMMARY POINTS

   a. The spinal cord of many vertebrates, possibly including humans, can autonomously generate detailed patterns of muscle activity which form the substrate of locomotion.
   b. Deafferentation studies and modelling show that these patterns alone would be of little use in the real world. They are continuously modulated, switched, and overridden by control mechanisms that take into account many sensory inputs.
   c. If-Then rules can be identified, which describe some of the simpler phase-switching operations as well as some of the higher-level responses to "hazards."
   d. Predictive and adaptive control, operating though supraspinal centers, are indispensable for locomotion in daily life.

6.10 APPENDIX

6.10.1 STRETCH REFLEX MODEL

Appendix Figure 1 shows a Matlab Simulink model originally developed to test the effect of excitatory Ib input to homonymous and heteronymous muscles (positive force feedback). In this slightly modified version, we have added a feedback pathway and product element marked by the A and B, to take into account the fact that transmission from group Ia afferents to MNs scales with the number of MNs recruited. The digital version of this model can be downloaded from the authors' website, or obtained upon request.

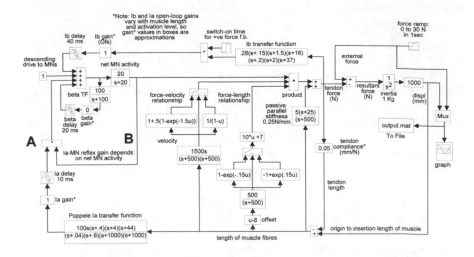

APPENDIX FIGURE 1 Matlab Simulink model of stretch reflex used to generate Figure 6.5. Adapted from Prochazka et al. 1997a. In this slightly modified version of the original model, the feedback pathway and product element marked "A" and "B" were added, to take into account the scaling of transmission from group Ia afferents to MNs with the number of MNs recruited. Reprinted with permission from Prochazka et al., *J. Neurophysiol.*, 77, 3237, 1997.

ACKNOWLEDGMENT

We thank Dr. Vivian Mushahwar for help with the graphics, Dr. K.G. Pearson for providing us with the locust film, and Mr. Allen Denington for digitizing parts of this film. This work was supported by the Canadian Institutes for Health Research and the Alberta Heritage Foundation for Medical Research.

REFERENCES

Ach, N. Über die Willenstätigkeit und das Denken. Göttingen 1905.
Adrian, E.D. and Zotterman, Y. The impulses produced by sensory nerve endings. Part 2. The responses of a single end-organ. *J. Physiol.* 61, 151, 1926.
Akazawa, K., Aldridge, J.W., Steeves, J.D., and Stein, R.B. Modulation of stretch reflexes during locomotion in the mesencephalic cat. *J. Physiol. (Lond.)* 329, 553, 1982.
Allum, J.H., Bloem, B.R., Carpenter, M.G., Hulliger, M., and Hadders-Algra, M. Proprioceptive control of posture: a review of new concepts. *Gait Posture* 8, 214, 1998.
Allum, J.H. and Honegger, F. Interactions between vestibular and proprioceptive inputs triggering and modulating human balance-correcting responses differ across muscles. *Exp. Brain Res.* 121, 478, 1998.
Alnaes, E. Static and dynamic properties of Golgi tendon organs in the anterior tibial and soleus muscles of the cat. *Acta Physiol. Scand.* 70, 176, 1967.
Anderson, J.H. Dynamic characteristics of Golgi tendon organs. *Brain Res.* 67, 531, 1974.

Andersson, O. and Grillner, S. Peripheral control of the cat's step cycle. II. Entrainment of the central pattern generators for locomotion by sinusoidal hip movements during "fictive locomotion." *Acta Physiol. Scand.* 118, 229, 1983.

Angel, M.J., Guertin, P., Jimenez, T., and McCrea, D.A. Group I extensor afferents evoke disynaptic EPSPs in cat hindlimb extensor motorneurones during fictive locomotion. *J. Physiol.* 494, 851, 1996.

Banks, R.W. The motor innervation of mammalian muscle spindles. *Prog. Neurobiol.* 43, 323, 1994.

Banks, R.W., Hulliger, M., and Scheepstra, K.A. Correlated histological and physiological observations on a case of common sensory output and motor input of the bag1 fiber and a chain fiber in a cat tenuissimus spindle. *J. Anat.* 193, 373, 1998.

Banks, R.W. and Stacey, M.J. Quantitative studies on mammalian muscle spindles and their sensory innervation. In: *Mechanoreceptors: Development, Structure, and Function,* edited by: P. Hnik, T. Soukup, R. Vejsada, and J. Zelena. London, Plenum, 1988.

Barker, D. The morphology of muscle receptors. In: *Handbook of Sensory Physiol. (Muscle Receptors),* edited by: C. C. Hunt. Berlin, Springer, 1974.

Bassler, U. *Neural Basis of Elementary Behavior in Stick Insects.* Berlin, Springer, 1983.

Bassler, U. The femur-tibia control system of stick insects — a model system for the study of the neural basis of joint control. *Brain Res. — Brain Res. Reviews* 18, 207, 1993.

Bassler, U. and Buschges, A. Pattern generation for stick insect walking movements — multisensory control of a locomotor program. *Brain Res. — Brain Res. Reviews* 27, 65, 1998.

Bell, C. *The Hand. Its Mechanism and Vital Endowments as Evincing Design.* London, William Pickering, 1834.

Bennett, D.J., Gorassini, M., and Prochazka, A. Catching a ball: contributions of intrinsic muscle stiffness, reflexes, and higher order responses. *Can. J. Physiol. Pharmacol.* 72, 525, 1994.

Bernstein, N.A. Trends and problems in the study of investigation of physiology of activity. In: *The Coordination and Regulation of Movements.* (Originally published in *Questions of Philosophy* (Vopr. Filos., 1961), 6, 77)., edited by: N.A. Bernstein. Oxford Pergamon, 1967.

Bhushan, N. and Shadmehr, R. Computational nature of human adaptive control during learning of reaching movements in force fields. *Biol. Cybern.* 81, 39, 1999.

Birder, L.A. and Perl, E.R. Cutaneous sensory receptors. *J. Clin. Neurophysiol.* 11, 534, 1994.

Blakemore, S.J., Goodbody, S.J., and Wolpert, D.M. Predicting the consequences of our own actions: the role of sensorimotor context estimation. *J. Neurosci.* 18, 7511, 1998.

Bloedel, J.R. and Bracha, V. On the cerebellum, cutaneomuscular reflexes, movement control, and the elusive engrams of memory. *Behavioral Brain Res.* 6 8, 1, 1995.

Bloedel, J.R. and Bracha, V. Current concepts of climbing fiber function. *Anatomical Rec.* 253, 118, 1998.

Bosco, G. and Poppele, R.E. Reference frames for spinal proprioception: kinematics based or kinetics based? *J. Neurophysiol.* 83, 2946, 2000.

Bosco, G. and Poppele, R.E. Proprioception from a spinocerebellar perspective. *Physiol. Rev.* 81, 539, 2001.

Bosco, G., Poppele, R.E., and Eian, J. Reference frames for spinal proprioception: limb endpoint based or joint-level based? *J. Neurophysiol.* 83, 2931, 2000.

Boyd, I.A. and Gladden, M. Morphology of mammalian muscle spindles. Review. In: *The Muscle Spindle,* edited by: I.A. Boyd and M. Gladden. London, Macmillan, 1985.

Boyd, I.A., Murphy, P.R., and Moss, V.A. Analysis of primary and secondary afferent
    responses to stretch during activation of the dynamic bag1 fiber or the static bag2
    fiber in cat muscle spindles. In: *The Muscle Spindle*, edited by: I.A. Boyd and M.
    Gladden. London, Macmillan, 1985.
Boyd, I.A. and Roberts, T.D.M. Proprioceptive discharges from stretch receptors in the knee
    joint of the cat. *J. Physiol.* 122, 38, 1953.
Brooke, J.D., Cheng, J., Collins, D.F., McIlroy, W.E., Misiaszek, J.E., and Staines, W.R.
    Sensori-sensory afferent conditioning with leg movement: gain control in spinal reflex
    and ascending paths. *Prog. Neurobiol.* 51, 393, 1997.
Brown, T.G. The intrinsic factors in the act of progression in the mammal. *Pro. R. Soc. Lond.,
    Series B* 84, 308, 1911.
Burgess, P.R. and Clark, F.J. Characteristics of knee joint receptors in the cat. *J. Physiol.* 203,
    317, 1969.
Burke, D. Critical examination of the case for or against fusimotor involvement in disorders
    of muscle tone. In: *Motor Control Mechanisms in Health and Disease*, edited by:
    J.E. Desmedt. New York, Raven, 1983.
Burrows, M. Local circuits for the control of leg movements in an insect. *Trends Neurosci.*
    15, 226, 1992.
Büschges, A. and Manira, A.E. Sensory pathways and their modulation in the control of
    locomotion. *Curr. Opin. Neurobiol.* 8, 733, 1998.
Cajal, S.R.y. The Croonian Lecture: La fine structure des centres nerveux. *Proc. R. Soc. Lond.
    Biol.* 55, 444, 1894.
Calancie, B., Needham-Shropshire, B., Jacobs, P., Willer, K., Zych, G., and Green, B.A.
    Involuntary stepping after chronic spinal cord injury. Evidence for a central rhythm
    generator for locomotion in man. *Brain* 117, 1143, 1994.
Capaday, C. Control of a 'simple' stretch reflex in humans. *Trends Neurosci.* 23, 528, 2000.
Capaday, C. and Stein, R.B. Amplitude modulation of the soleus H-reflex in the human during
    walking and standing. *J. Neurosci.* 6, 1308, 1986.
Capaday, C. and Stein, R.B. Difference in the amplitude of the human soleus H reflex during
    walking and running. *J. Physiol.* 392, 513, 1987.
Carli, G., Farabollini, F., Fontani, G., and Meucci, M. Slowly adapting receptors in cat hip
    joint. *J. Neurophysiol.* 42, 767, 1979.
Chen, J.J., Yu, N.Y., Huang, D.G., Ann, B.T., and Chang, G.C. Applying fuzzy logic to control
    cycling movement induced by functional electrical stimulation. *IEEE Trans. Rehabil.
    Eng.* 5, 158, 1997.
Chen, W.J. and Poppele, R.E. Small-signal analysis of response of mammalian muscle spindles
    with fusimotor stimulation and a comparison with large-signal responses. *J. Neuro-
    physiol.* 41, 15, 1978.
Christensen, L.O.D., Andersen, J.B., SinkjFr, T., and Nielsen, J. Transcranial magnetic stim-
    ulation and stretch reflexes in the tibialis anterior muscle during human walking. *J.
    Physiol.* 531, 545, 2001.
Collins, D.F. and Prochazka, A. Movement illusions evoked by ensemble cutaneous input
    from the dorsum of the human hand. *J. Physiol. (Lond.)* 496, 857, 1996.
Conway, B.A., Hultborn, H., and Kiehn, O. Proprioceptive input resets central locomotor
    rhythm in the spinal cat. *Exp. Brain Res.* 68, 643, 1987.
Cussons, P.D., Hulliger, M., and Matthews, P.B. Effects of fusimotor stimulation on the
    response of the secondary ending of the muscle spindle to sinusoidal stretching. *J.
    Physiol. (Lond.)* 270, 835, 1977.

Davoodi, R. and Andrews, B.J. Optimal control of FES-assisted standing up in paraplegia using genetic algorithms. *Med. Eng. Phys.* 21, 609, 1999.

Delcomyn, F. Perturbation of the motor system in freely walking cockroaches. I. Rear leg amputation and the timing of motor activity in leg muscles. *J. Exp. Biol.* 156, 483, 1991a.

Delcomyn, F. Perturbation of the motor system in freely walking cockroaches. II. The timing of motor activity in leg muscles after amputation of a middle leg. *J. Exp. Biol.* 156, 503, 1991b.

Descartes, R. *Traité de l'homme (Treatise of Man)*. French text with English translation and commentary. Cambridge, MA, Harvard University Press 1972 (originally by Le Gras, Paris, 1664).

Desmedt, J.E. and Godaux, E. Voluntary motor commands in human ballistic movements. *Ann. Neurol.* 5, 415, 1979.

Dietz, V. Interaction between central programs and afferent input in the control of posture and locomotion. *J. Biomech.* 29, 841, 1996.

Dietz, V. Evidence for a load receptor contribution to the control of posture and locomotion. *Neurosci. Biobehav. Rev.* 22, 495, 1998.

Dimitrijevic, M.R., Gerasimenko, Y., and Pinter, M.M. Evidence for a spinal central pattern generator in humans. *Ann. N Y Acad. Sci.* 860, 360, 1998.

Drew, T. The role of the motor cortex in the control of gait modification in the cat. In: *Neurobiological Basis of Human Locomotion*, edited by: M. Shimamura, S. Grillner, and V.R. Edgerton. Tokyo, Japan, Scientific Societies Press, 1991.

Drew, T., Jiang, W., Kably, B., and Lavoie, S. Role of the motor cortex in the control of visually triggered gait modifications. *Can. J. Physiol. & Pharmacol.* 74, 426, 1996.

Duysens, J., Clarac, F., and Cruse, H. Load-regulating mechanisms in gait and posture: comparative aspects. *Physiol. Rev.* 80, 83, 2000.

Duysens, J. and Pearson, K.G. Inhibition of flexor burst generation by loading ankle extensor muscles in walking cats. *Brain Res.* 187, 321, 1980.

Eccles, J.C., Eccles, R.M., and Lundberg, A. The convergence of monosynaptic excitatory afferents on to many species of alpha motoneurones. *J. Physiol.* 137, 22, 1957a.

Eccles, J.C., Eccles, R.M., and Lundberg, A. Synaptic actions on motoneurones caused by impulses in Golgi tendon organ afferents. *J. Physiol.* 138, 227, 1957b.

Eccles, J.C., Fatt, P., and Koketsu, K. Cholinergic and inhibitory synapses in a pathway from motor-axon collaterals to motoneurones. *J. Physiol.* 126, 524, 1954.

Edin, B.B. Quantitative analysis of static strain sensitivity in human mechanoreceptors from hairy skin. *J. Neurophysiol.* 67, 1105, 1992.

Edin, B.B. Cutaneous afferents provide information about knee joint movements in humans. *J. Physiol.* 531, 289, 2001.

Edin, B.B. and Abbs, J.H. Finger movement responses of cutaneous mechanoreceptors in the dorsal skin of the human hand. *J. Neurophysiol.* 65, 657, 1991.

Edin, B.B. and Johansson, N. Skin strain patterns provide kinaesthetic information to the human central nervous system. *J. Physiol.* 487, 243, 1995.

Emonet-Denand, F., Jami, L., and Laporte, Y. Skeleto-fusimotor axons in the hind-limb muscles of the cat. *J. Physiol.* 249, 153, 1975.

Emonet-Denand, F., Laporte, Y., Matthews, P.B., and Petit, J. On the subdivision of static and dynamic fusimotor actions on the primary ending of the cat muscle spindle. *J. Physiol. (Lond.)* 268, 827, 1977.

Espenschied, K.S., Quinn, R.D., Beer, R.D., and Chiel, H.J. Biologically based distributed control and lcal reflexes improve rough terrain locomotion in a hexapod robot. *Robotics and Autonomous Syst.* 18, 59, 1996.

Ferrell, W.R. The adequacy of stretch receptors in the cat knee joint for signaling joint angle throughout a full range of movement. *J. Physiol.* 299, 85, 1980.

Flourens, P. Recherches experimentales sur les proprietes et les fonctions du systeme nerveux dans les animaux vertebres. *Arch. Gen. Med.* 2, 321, 1823.

Forssberg, H., Grillner, S., Halbertsma, J., and Rossignol, S. The locomotion of the low spinal cat. II. Interlimb coordination. *Acta Physiol. Scand.* 108, 283, 1980.

French, A. Transduction mechanisms of mechanosensilla. *Ann. Rev. Entomol.* 33, 39, 1988.

Freusberg, A. Reflexbewegungen beim Hunde. *Pflueger's Archiv fuer die gesamte Physiologie* 9, 358, 1874.

Garrett, M., Ireland, A., and Luckwill, R.G. Changes in excitability of the Hoffman reflex during walking in man. *J. Physiol.* 355, 23, 1984.

Garrett, M. and Luckwill, R.G. Role of reflex responses of knee musculature during the swing phase of walking in man. *Eur. J. Appl. Physiol.* 52, 36, 1983.

Garrett, M., Luckwill, R.G., and McAleer, J.J.A. The sensitivity of the monosynaptic reflex arc in the leg extensor muscles of the walking man. In: *Biomechanics VII*, edited by: A. Morecki. Baltimore, University Park Press, 1981.

Gerritsen, K.G., van den Bogert, A.J., Hulliger, M., and Zernicke, R.F. Intrinsic muscle properties facilitate locomotor control — a computer simulation study. *Motor Control* 2, 206, 1998.

Gerritsen, K.G.M. and Nagano, A. *The role of sensory feedback during cyclic locomotor activities.* XVII Congress of the International Society of Biomechanics, Calgary, Canada, 1999.

Gibson, J.J. A critical review of the concept of set in contemporary experimental psychology. *Psychol. Bull.* 38, 781, 1941.

Gillard, D.M., Yakovenko, S., Cameron, T., and Prochazka, A. Torque production at the human wrist. *J. Biomech.* In Press 2000.

Giuliani, C.A. and Smith, J.L. Stepping behaviors in chronic spinal cats with one hindlimb deafferented. *J. Neurosci.* 7, 2537, 1987.

Godwin-Austen, R.B. The mechanoreceptors of the costovertebral joints. *J. Physiol.* 202, 737, 1969.

Goldberger, M.E. and Murray, M. Locomotor recovery after deafferentation of one side of the cat's trunk. *Exp. Neurol.* 67, 103, 1980.

Goltz, F. Der Hund ohne Grosshirn. *Pflueger's Archiv gestaltender Physiologie* 51, 570, 1892.

Goltz, F.L. *Beitraege zur Lehre von den Functionen der Nervencentren des Frosches.* Berlin, Hirschwald, 1869.

Goodwin, G.M., McCloskey, D.I., and Matthews, P.B. Proprioceptive illusions induced by muscle vibration: contribution by muscle spindles to perception? *Science* 175, 1382, 1972.

Gorassini, M., Prochazka, A., and Taylor, J.L. Cerebellar ataxia and muscle spindle sensitivity. *J. Neurophysiol.* 70, 1853, 1993.

Gorassini, M.A., Prochazka, A., Hiebert, G.W., and Gauthier, M.J. Corrective responses to loss of ground support during walking. I. Intact cats. *J. Neurophysiol.* 71, 603, 1994.

Granit, R.L.A. *The Basis of Motor Control.* London, Academic Press, 346, 1970.

Grey, M., Ladouceur, M., Andersen, J.B., Nielsen, J.B., and Sinkjaer, T. Contribution of group II muscle afferents to the medium latency soleus stretch reflex during walking in man. *J. Physiol.* In Press 2001.

Grigg, P. and Greenspan, B.J. Response of primate joint afferent neurons to mechanical stimulation of knee joint. *J. Neurophysiol.* 40, 1, 1977.

Grillner, S. Locomotion in vertebrates: central mechanisms and reflex interaction. *Physiol. Rev.* 55, 247, 1975.

Grillner, S., Cangiano, L., Hu, G., Thompson, R., Hill, R., and Wall, P. The intrinsic function of a motor system — from ion channels to networks and behavior. *Brain Res.* 886, 224, 2000.

Grillner, S., Wallén, P., Hill, R., Cangiano, L., and El Manira, A. Ion channels of importance for the locomotor pattern generation in the lamprey brainstem-spinal cord. *J. Physiol.* 533, 23, 2001.

Grillner, S. and Zangger, P. How detailed is the central pattern generation for locomotion? *Brain Res.* 88, 367, 1975.

Grillner, S. and Zangger, P. The effect of dorsal root transection on the efferent motor pattern in the cat's hindlimb during locomotion. *Acta Physiol. Scand.* 120, 393, 1984.

Gritscnko, V., Mushahwar, V.K., and Prochazka, A. Locomotor stretch reflexes increase after partial denervation of cat triceps suare muscles. *J. Physiol.* 533, 299, 2001.

Guertin, P., Angel, M.J., Perreault, M.C., and McCrea, D.A. Ankle extensor group I afferents excite extensors throughout the hindlimb during fictive locomotion in the cat. *J. Physiol.* 487, 197, 1995.

Gurfinkel, V.S., Levik, Y.S., Kazennikov, O.V., and Selionov, V.A. Locomotor-like movements evoked by leg muscle vibration in humans. *Eur. J. Neurosci.* 10, 1608, 1998.

Herzog, W., Leonard, T.R., and Guimaraes, A.C. Forces in gastrocnemius, soleus, and plantaris tendons of the freely moving cat. *J. Biomech.* 26, 945, 1993.

Hiebert, G.W., Whelan, P.J., Prochazka, A., and Pearson, K.G. Contribution of hind limb flexor muscle afferents to the timing of phase transitions in the cat step cycle. *J. Neurophysiol.* 75, 1126, 1996.

Horak, F.B. and MacPherson, J.M. Postural orientation and equilibrium. In: *Exercise: Regulation and Integration of Multiple Systems.*, edited by: L. Rowell and J.T. Sheperd. New York, American Physiological Society, 1996.

Horch, K.W., Tuckett, R.P., and Burgess, P.R. A key to the classification of cutaneous mechanoreceptors. *J. Invest. Dermatol.* 69, 75, 1977.

Houk, J. and Henneman, E. Responses of Golgi tendon organs to active contractions of the soleus muscle of the cat. *J. Neurophysiol.* 30, 466, 1967.

Houk, J.C. and Simon, W. Responses of Golgi tendon organs to forces applied to muscle tendon. *J. Neurophysiol.* 30, 1466, 1967.

Hulliger, M. The mammalian muscle spindle and its central control. [Review]. *Rev. Physiol. Biochem. Pharmacol.* 101, 1, 1984.

Hulliger, M., Emonet-Denand, F., and Baumann, T.K. Enhancement of stretch sensitivity of cat primary spindle afferents by low-rate static gamma action. In: *The Muscle Spindle*, edited by: I.A. Boyd and M. Gladden. London, Macmillan, 1985.

Hulliger, M., Matthews, P.B., and Noth, J. Static and dynamic fusimotor action on the response of Ia fibers to low frequency sinusoidal stretching of widely ranging amplitude. *J. Physiol. (Lond.)* 267, 811, 1977.

Iles, J.F., Stokes, M., and Young, A. Reflex actions of knee joint afferents during contraction of the human quadriceps. *Clin. Physiol.* 10, 489, 1990.

Ivanenko, Y.P., Grasso, R., and Lacquaniti, F. Influence of leg muscle vibration on human walking. *J. Neurophysiol.* 84, 1737, 2000a.

Ivanenko, Y.P., Grasso, R., and Lacquaniti, F. Neck muscle vibration makes walking humans accelerate in the direction of gaze. *J. Physiol.* 525, 803, 2000b.

Jacobs, R. Control model of human stance using fuzzy logic. *Biol. Cybern.* 77, 63, 1997.

James, W. *The Principles of Psychology.* New York, Henry Holt, 1890.

Jami, L. Golgi tendon organs in mammalian skeletal muscle: functional properties and central actions. *Physiol. Rev.* 72, 623, 1992.

Jami, L. and Petit, J. Fusimotor actions on sensitivity of spindle secondary endings to slow muscle stretch in cat peroneus tertius. *J. Neurophysiol.* 41, 860, 1978.

Jami, L., Petit, J., Proske, U., and Zytnicki, D. Responses of tendon organs to unfused contractions of single motor units. *J. Neurophysiol.* 53, 32, 1985.

Jankowska, E. Interneuronal relay in spinal pathways from proprioceptors. *Prog. Neurobiol.* 38, 335, 1992.

Jankowska, E., Hammar, I., Chojnicka, B., and Heden, C.H. Effects of monoamines on interneurons in four spinal reflex pathways from group I and/or group II muscle afferents. *Eur. J. Neurosci.* 12, 701, 2000.

Jankowska, E., Jukes, M.G., Lund, S., and Lundberg, A. The effect of DOPA on the spinal cord. 6. Half-center organization of interneurones transmitting effects from the flexor reflex afferents. *Acta Physiol. Scand.* 70, 389, 1967.

Johansson, H., Sjolander, P., and Sojka, P. Receptors in the knee joint ligaments and their role in the biomechanics of the joint. *CRC Crit. Rev. Biomed. Eng.* 18, 341, 1991.

Johansson, R.S. and Vallbo, A.B. Tactile sensibility in the human hand: relative and absolute densities of four types of mechanoreceptive units in glabrous skin. *J. Physiol. (Lond)* 286, 283, 1979.

Jonic, S., Jankovic, T., Gajic, V., and Popovic, D. Three machine learning techniques for automatic determination of rules to control locomotion. *IEEE Trans. Biomed. Eng.* 46, 300, 1999.

Jordan, L.M. Initiation of locomotion in mammals. *Ann. N. Y. Acad. Sci.* 860, 83, 1998.

Kiehn, O., Harris-Warrick, R.M., Jordan, L.M., Hultborn, H., and Kudo, N., Eds., *Neuronal Mechanisms for Generating Locomotor Activity.* New York, New York Academy of Sciences, 1999.

Kiehn, O. and Kjaerulff, O. Distribution of central pattern generators for rhythmic motor outputs in the spinal cord of limbed vertebrates. *Ann. N. Y. Acad. Sci.* 860, 110, 1998.

Koshland, G.F. and Smith, J.L. Mutable and immutable features of paw-shake responses after hindlimb deafferentation in the cat. *J. Neurophysiol.* 62, 162, 1989.

Kriellaars, D.J., Brownstone, R.M., Noga, B.R., and Jordan, L.M. Mechanical entrainment of fictive locomotion in the decerebrate cat. *J. Neurophysiol.* 71, 2074, 1994.

Lajoie, Y., Teasdale, N., Cole, J.D., Burnett, M., Bard, C., Fleury, M., Forget, R., Paillard, J., and Lamarre, Y. Gait of a deafferented subject without large myelinated sensory fibers below the neck. *Neurology* 47, 109, 1996.

Llewellyn, M., Prochazka, A., and Vincent, S., Transmission of human tendon jerk reflexes during stance and gait. *J. Physiol., (Lond.),* 1987.

Llewellyn, M., Yang, J.F., and Prochazka, A. Human H-reflexes are smaller in difficult beam walking than in normal treadmill walking. *Exp. Brain Res.* 83, 22, 1990.

Loeb, G.E. Somatosensory unit input to the spinal cord during normal walking. *Canadian J. Physiol. & Pharmacol.* 59, 627, 1981.

Loeb, G.E. The control and responses of mammalian muscle spindles during normally executed motor tasks. *Exercise & Sport Sci. Rev.* 12, 157, 1984.

Loeb, G.E. and Duysens, J. Activity patterns in individual hindlimb primary and secondary muscle spindle afferents during normal movements in unrestrained cats. *J. Neurophysiol.* 42, 420, 1979.

Lund, J.P. and Matthews, B. Responses of temporomandibular joint afferents recorded in the Gasserian ganglion of the rabbit to passive movements of the mandible. In: *Oral-facial Sensory and Motor Functions*, edited by: Y. Kawamura. Tokyo, Quintessence, 1981.

Magnus, R. *Koerperstellung (English translation: Body Posture).* New Delhi, Amerind, 1924.

Marder, E. and Pearson, K.G. Editorial overview: motor control from molecules to bedside [editorial]. *Curr. Opin. Neurobiol.* 8, 693, 1998.

Matthews, P.B. The human stretch reflex and the motor cortex. *Trends Neurosci.* 14, 87, 1991.

Matthews, P.B. and Stein, R.B. The sensitivity of muscle spindle afferents to small sinusoidal changes of length. *J. Physiol. (Lond.)* 200, 723, 1969.

Matthews, P.B. and Watson, J.D. Effect of vibrating agonist or antagonist muscle of the reflex response to sinusoidal displacement of the human forearm. *J. Physiol. (Lond.)* 321, 297, 1981.

Matthews, P.B.C. *Mammalian Muscle Receptors and Their Central Actions.* London, Arnold, 1972.

McFadyen, B.J., Lavoie, S., and Drew, T. Kinetic and energetic patterns for hindlimb obstacle avoidance during cat locomotion. *Exp. Brain Res.*125, 502, 1999.

McIntyre, A.K., Proske, U., and Tracey, D.J. Afferent fibers from muscle receptors in the posterior nerve of the cat's knee joint. *Exp. Brain Res.* 33, 415, 1978.

Mettrie, d.L., J.O. *L'Homme-Machine. (English translation: Man and a Machine, Man and a Plant, 1994.)* Indianapolis, Maya Hackett, 1745.

Miller, S. and Scott, P.D. The spinal locomotor generator. *Exp. Brain Res.* 30, 387, 1977.

Misiaszek, J.E., Cheng, J., Brooke, J.D., and Staines, W.R. Movement-induced modulation of soleus H reflexes with altered length of biarticular muscles. *Brain Res.* 795, 25, 1998.

Misiaszek, J.E., Stephens, M.J., Yang, J.F., and Pearson, K.G. Early corrective reactions of the leg to perturbations at the torso during walking in humans. *Exp. Brain Res.*131, 511, 2000.

Moberg, E. The role of cutaneous afferents in position sense, kinaesthesia, and motor function of the hand. *Brain* 106, 1, 1983.

Mummel, P., Timmann, D., Krause, U.W., Boering, D., Thilmann, A.F., Diener, H.C., and Horak, F.B. Postural responses to changing task conditions in patients with cerebellar lesions. *J. Neurol. Neurosurg. Psyc.*, 65, 734, 1998.

Neher, E. and Sakmann, B. Single-channel currents recorded from membrane of denervated frog muscle fibers. *Nature* 260, 799, 1976.

Nelson, G.M. and Quinn, R.D. Posture control of a cockroach-like robot. *IEEE Control Syst.* 19, 9, 1999.

Neptune, R.R., Kautz, S.A., and Zajac, F.E. Contributions of the individual ankle plantar flexors to support forward progression and swing initiation during walking. *J. Biomech.* In Press 2001.

O'Donovan, M.J., Wenner, P., Chub, N., Tabak, J., and Rinzel, J. Mechanisms of spontaneous activity in the developing spinal cord and their relevance to locomotion. *Ann. N.Y. Acad. Sci.* 860, 130, 1998.

Ogihara, N. and Yamazaki, N. Generation of human bipedal locomotion by a bio-mimetic neuro-musculo-skeletal model. *Bio. Cybern.* 84, 1, 2001.

Orlovsky, G.N., Deliagina, T.G., and Grillner, S. *Neuronal control of locomotion.* Oxford, Oxford University Press, 1999.

Orlovsky, G.N. and Feldman, A.G. On the role of afferent activity in generation of stepping movements. *Neurofiziologia (Russian),* English translation in *Neurophysiology* 4, 304-310 4, 401, 1972.

Orlovsky, G.N. and Shik, M.L. Standard elements of cyclic movement. *Biofizika* 5, 847, 1965.

Otten, E., Hulliger, M., and Scheepstra, K.A. A model study on the influence of a slowly activating potassium conductance on repetitive firing patterns of muscle spindle primary endings. *J. Theo. Biol.* 173, 67, 1995.

Pang, M.Y. and Yang, J.F. The initiation of the swing phase in human infant stepping: importance of hip position and leg loading. *J. Physiol.* 528, 389, 2000.

Partridge, L.D. Signal-handling characteristics of load-moving skeletal muscle. *Am. J. Physiol.* 210, 1178, 1966.

Patla, A.E., Prentice, S.D., Rietdyk, S., Allard, F., and Martin, C. What guides the selection of alternate foot placement during locomotion in humans? *Exp. Brain Res.*128, 441, 1999.

Patla, A.E. and Vickers, J.N. Where and when do we look as we approach and step over an obstacle in the travel path? *Neuroreport* 8, 3661, 1997.

Pearson, K.G. Proprioceptive regulation of locomotion. *Curr. Opin. Neurobiol.* 5, 786, 1995.

Pearson, K.G. and Collins, D.F. Reversal of the influence of group Ib afferents from plantaris on activity in medial gastrocnemius muscle during locomotor activity. *J. Neurophysiol.* 70, 1009, 1993.

Pearson, K.G., Misiaszek, J.E., and Fouad, K. Enhancement and resetting of locomotor activity by muscle afferents. *Ann. N. Y. Acad. Sci.* 860, 203, 1998.

Poppele, R.E. An analysis of muscle spindle behavior using randomly applied stretches. *Neuroscience* 6, 1157, 1981.

Poppele, R.E. and Bowman, R.J. Quantitative description of linear behavior of mammalian muscle spindles. *J. Neurophysiol.* 33, 59, 1970.

Poppele, R.E. and Kennedy, W.R. Comparison between behavior of human and cat muscle spindles recorded *in vitro. Brain Res.* 75, 316, 1974.

Prochazka, A. Muscle spindle function during normal movement. *Int. Rev. Physiol.* 25, 47, 1981.

Prochazka, A. Afferent input during normal movements. In: *Feedback and Motor Control in Vertebrates and Invertebrates*, edited by: W. J. P. Barnes and M. Gladden. London, Croon Helm, 1985.

Prochazka, A. Sensorimotor gain control: a basic strategy of motor systems? *Prog. Neurobiol.* 33, 281, 1989.

Prochazka, A. The fuzzy logic of visuomotor control. *Canadian J. Physiol. & Pharmacol.* 74, 456, 1996a.

Prochazka, A. Proprioceptive feedback and movement regulation. In: *Handbook of Physiology. Section 12. Exercise: Regulation and Integration of Multiple Systems*, edited by: L. Rowell and J. T. Sheperd. New York, American Physiological Society, 1996b.

Prochazka, A., Clarac, F., Loeb, G.E., Rothwell, J.C., and Wolpaw, J.R. What do reflex and voluntary mean? Modern views on an ancient debate. *Exp. Brain Res.*130, 417, 2000.

Prochazka, A. Quantifying proprioception. *Prog. Brain Res.* 123, 133, 1999.

Prochazka, A., Gillard, D., and Bennett, D.J. Implications of positive feedback in the control of movement. *J. Neurophysiol.* 77, 3237, 1997a.

Prochazka, A., Gillard, D., and Bennett, D.J. Positive force feedback control of muscles. *J. Neurophysiol.* 77, 3226, 1997b.

Prochazka, A. and Gorassini, M. Ensemble firing of muscle afferents recorded during normal locomotion in cats. *J. Physiol.* 507, 293, 1998a.

Prochazka, A. and Gorassini, M. Models of ensemble firing of muscle spindle afferents recorded during normal locomotion in cats. *J. Physiol.* 507, 277, 1998b.

Prochazka, A. and Hulliger, M. Muscle afferent function and its significance for motor control mechanisms during voluntary movements in cat, monkey, and man. *Adv. Neurol.* 39, 93, 1983.

Prochazka, A., Trend, P., Hulliger, M., and Vincent, S. Ensemble proprioceptive activity in the cat step cycle, towards a representative look-up chart. *Prog. Brain Res.* 80, 61-74; discussion 57, 1989.

Prochazka, A. and Trend, P.S. Instability in human forearm movements studied with feedback-controlled muscle vibration. *J. Physiol. (Lond)* 402, 421, 1988.

Prochazka, A., Westerman, R.A., and Ziccone, S.P. Discharges of single hindlimb afferents in the freely moving cat. *J. Neurophysiol.* 39, 1090, 1976.

Proske, U. The Golgi tendon organ. Properties of the receptor and reflex action of impulses arising from tendon organs. In: *MTP Int. Rev. Physiol., Neurophysiology IV*, edited by: R. Porter. Baltimore, MTP University Park Press, 1981.

Quinn, R.D. and Ritzmann, R.E. Biologically based distributed control and local reflexes improve rough terrain locomotion in a hexapod robot. *Connection Science* 10, 239, 1998.

Rho, M.J., Lavoie, S., and Drew, T. Effects of red nucleus microstimulation on the locomotor pattern and timing in the intact cat: a comparison with the motor cortex. *J. Neurophysiol.* 81, 2297, 1999.

Rossignol, S. Neural control of stereotypic limb movements. In: *Handbook of Physiology. Section 12. Exercise: Regulation and Integration of Multiple Systems.*, edited by: L. Rowell and J.T. Sheperd. New York, American Physiological Society, 1996.

Schaafsma, A., Otten, E., and Van Willigen, J.D. A muscle spindle model for primary afferent firing based on a simulation of intrafusal mechanical events. *J. Neurophysiol.* 65, 1297, 1991.

Schneider, C., Lavoie, B.A., and Capaday, C. On the origin of the soleus H-reflex modulation pattern during human walking and its task-dependent differences. *J. Neurophysiol.* 83, 2881, 2000.

Schubert, M., Curt, A., Colombo, G., Berger, W., and Dietz, V. Voluntary control of human gait: conditioning of magnetically evoked motor responses in a precision stepping task. *Exp. Brain Res.* 126, 583, 1999.

Sechenov, I.M., Reflexes of the brain, (Refleksy golovnogo mozga). In: *I.M. Sechenov, Selected Works*, edited by: A. A. Subkov. Moscow, State Publishing House, 264, 1863.

Serrien, D. and Wiesendanger, M. Role of the cerebellum in tuning anticipatory and reactive grip force responses. *J. Cogn. Neurosci.* 11, 672, 1999.

Sherrington, C.S. On the proprio-ceptive system, especially in its reflex aspects. *Brain* 29, 467, 1906.

Sherrington, C.S. Flexion-reflex of the limb, crossed extension-reflex, and reflex stepping and standing. *J. Physiol. (Lond.)* 40, 28, 1910.

Sherrington, C.S. *Integrative Action of the Nervous System.* Cambridge University Press (Cambridge) originally published in 1906 by Yale University Press, New Haven, CT, 1947.

Shik, M.L., Severin, F.V., and Orlovsky, G.N. Control of walking and running by means of electrical stimulation of the mid-brain. *Biophysics* 11, 756, 1966.

Shik, M.L., Severin, F.V., and Orlovsky, G.N. Control of walking and running by means of electrical stimulation of the mesencephalon. *Electroencephalography & Clin. Neurophysiol.* 26, 549, 1969.

Sinkjaer, T. Muscle, reflex, and central components in the control of the ankle joint in healthy and spastic man. *Acta Neurol. Scand. Supplementum* 170, 1, 1997.

Sinkjaer, T., Andersen, J.B., Ladouceur, M., Christensen, L.O., and Nielsen, J.B. Major role for sensory feedback in soleus EMG activity in the stance phase of walking in man. *J. Physiol.* 523, 817, 2000.

Sinkjaer, T., Andersen, J.B., and Larsen, B. Soleus stretch reflex modulation during gait in humans. *J. Neurophysiol.* 76, 1112, 1996.

Sinkjaer, T., Andersen, J.B., Nielsen, J.F., and Hansen, H.J. Soleus long-latency stretch reflexes during walking in healthy and spastic humans. *Clin. Neurophysiol.* 110, 951, 1999.

Smith, O.J.M. A controller to overcome dead time. *Instrum. Soc. of Americas J.* 6, 28, 1959.

Spencer, H. *Principles of Psychology (2nd Edition 1873).* New York, Appleton, 1855.

Stephens, J.A., Reinking, R.M., and Stuart, D.G. Tendon organs of cat medial gastrocnemius: responses to active and passive forces as a function of muscle length. *J. Neurophysiol.* 38, 1217, 1975.

Stephens, M.J. and Yang, J.F. Loading during the stance phase of walking in humans increases the extensor EMG amplitude but does not change the duration of the step cycle. *Exp. Brain Res.* 124, 363, 1999.

Stuart, D.G., Goslow, G.E., Mosher, C.G. and Reinking, R.M. Stretch responsiveness of Golgi tendon organs. *Exp. Brain Res.* 10, 463, 1970.

Stuart, D.G., Mosher, C.G., Gerlach, R.L., and Reinking, R.M. Mechanical arrangement and transducing properties of Golgi tendon organs. *Exp. Brain Res.* 14, 274, 1972.

Stuart, D.G., Pierce, P.A., Callister, R.J., Brichta, A.M., and McDonagh, J.C. Sir Charles S. Sherrington: Humanist, Mentor, and Movement Neuroscientist. In: *Classics in Movement Science*, edited by: M.L. Latash and V.M. Zatsiorsky. Champaign, IL, Human Kinetics, 2001.

Taga, G. A model of the neuro-musculo-skeletal system for human locomotion. I. Emergence of basic gait. *Biol. Cybern.* 73, 97, 1995a.

Taga, G. A model of the neuro-musculo-skeletal system for human locomotion. II. Real-time adaptability under various constraints. *Biol. Cybern.* 73, 113, 1995b.

Taga, G. A model of the neuro-musculo-skeletal system for anticipatory adjustment of human locomotion during obstacle avoidance. *Biol. Cybern.* 78, 9, 1998.

Taga, G., Yamaguchi, Y., and Shimizu, H. Self-organized control of bipedal locomotion by neural oscillators in unpredictable environment. *Biol. Cybern.* 65, 147, 1991.

Taylor, A., Ellaway, P.H., and Durbaba, R. Physiological signs of the activation of bag2 and chain intrafusal muscle fibers of gastrocnemius muscle spindles in the cat. *J. Neurophysiol.* 80, 130, 1998.

Taylor, A., Ellaway, P.H., and Durbaba, R. Why are there three types of intrafusal muscle fibers? *Prog. Brain Res.* 123, 121, 1999.

Taylor, A., Rodgers, J.F., Fowle, A.J., and Durbaba, R. The effect of succinylcholine on cat gastrocnemius muscle spindle afferents of different types. *J. Physiol.* 456, 629, 1992.

Terzuolo, C., Fohlmeister, J.F., Maffei, L., Poppele, R.E., Soechting, J.F., and Young, L. On the application of systems analysis to neurophysiological problems. *Arch. Ital. Bio.* 120, 18, 1982.

Tomovic, R., Anastasijevic, R., Vuco, J., and Tepavac, D. The study of locomotion by finite state models. *Biol. Cybern.* 63, 271, 1990.

Tomovic, R. and McGhee, R. A finite state approach to the synthesis of control systems. *IEEE Trans. Hum. Fac. Electron.* 7, 122, 1966.

Tracey, D.J. Characteristics of wrist joint receptors in the cat. *Exp. Brain Res.* 34, 165, 1979.

Valls-Sole, J., Rothwell, J.C., Goulart, F., Cossu, G., and Munoz, E. Patterned ballistic movements triggered by a startle in healthy humans. *J. Physiol.* 516, 931, 1999.

Vetter, P. and Wolpert, D.M. Context estimation for sensorimotor control. *J. Neurophysiol.* 84, 1026, 2000.

von Holst, E. Relations between the central nervous system and the peripheral organs. *Br. J. Anim. Behav.* 2, 89-94, 1954.

von Holst, E. and Mittelstaedt, H. Das Reafferenzprincip. *Naturwissenschaften* 37, 464, 1950.

Voss, H. Tabelle der absoluten und relativen Muskelspindelzahlen der menschlichen Skelett-muskulatur. *Anatomische Anzeiger* 129, 5562, 1971.

Wand, P., Prochazka, A., and Sontag, K.H. Neuromuscular responses to gait perturbations in freely moving cats. *Exp. Brain Res.* 38, 109, 1980.

Watt, H.J. Experimentelle Beiträge zu einer Theorie des Denkens. *Archiv gesamter Psychologie* 4, 289, 1905.

Wetzel, M.C., Atwater, A.E., Wait, J.V., and Stuart, D.G. Kinematics of locomotion by cats with a single hindlimb deafferented. *J. Neurophysiol.* 39, 667, 1976.

Wolf, H. and Pearson, K.G. Proprioceptive input patterns elevator activity in the locust flight system. *J. Neurophysiol.* 59, 1831, 1988.

Wolpert, D.M. and Ghahramani, Z. Computational principles of movement neuroscience. *Nat. Neurosci.* 3, 1212, 2000.

Yamazaki, N., Hase, K., Ogihara, N., and Hayamizu, N. Biomechanical analysis of the development of human bipedal walking by a neuro-musculo-skeletal model. *Folia Primatologica* 66, 253, 1996.

Zalkind, V.I. Method for an adequate stimulation of receptors of the cat carpo-radialis joint. *Sechenov Physiological Journal USSR* 57, 1123, 1971.

Zelenin, P.V., Deliagina, T.G., Grillner, S., and Orlovsky, G.N. Postural control in the lamprey: A study with a neuro-mechanical model. *J. Neurophysiol.* 84, 2880, 2000.

Zill, S. Proprioceptive feedback and the control of cockroach walking. In: *Feedback and Motor Control in Invertebrates and Vertebrates,* edited by: W.J.P. Barnes and M.H. Gladden. London, Croon Helm, 1985.

7 Sensory Learning and the Brain's Body Map

Mathew E. Diamond, Justin A. Harris, and Rasmus S. Petersen

CONTENTS

7.1 INTRODUCTION

One of the many ambitious goals set in our era of neuroscience is to understand the brain modifications that underlie learning — first, to identify the relevant neuronal circuits and second, to describe the cellular changes occurring within those circuits. For nearly three decades[1] long-term potentiation and long-term depression have been investigated, and synaptic plasticity of this sort is now believed to constitute the cellular change underlying most types of learning.[2] But progress in the first of the two issues has been slower: we still know relatively little about the locus of brain modifications during learning of even the simplest tasks. Exceptions do exist: the cerebellar cortex and interpositus nucleus clearly form a locus of plasticity during certain precisely-timed conditioned reflexes[3] and the amygdala, a critical locus during certain types of emotional conditioning.[4] Yet, consider how little is known about what parts of the brain are involved when a monkey learns to remember visual stimuli across a delay. Lesion experiments point to an important role of the hippocampus,[5] but even the claim of the hippocampal locus can be called into question.[6] Thus, while the hippocampus is the most systematically studied mammalian brain area when it comes to synaptic modification mechanisms, it remains difficult to pin down its involvement in any well-defined learning task.

Here, we will not address the problem of what learning consists of at the cellular level. Our purpose is another one, namely to consider *where* the neuronal modifications that underlie learning are stored and accessed in cerebral cortex. Our focus is on *sensory learning*, the improvement in sensory-perceptual ability that occurs with training in a specific task. Since the question "How are learning-related modifications distributed?" is poorly posed, we need to make the problem much more precise to have any hope of arriving at a concrete conclusion. To this end, we have focused on tactile learning, and have expressed the question in a more specific way: "Does the somatotopic organization that characterizes sensory cortex constitute a structural framework for perceptual learning and memory?"

7.2 CANDIDATE MODELS FOR THE DISTRIBUTION OF LEARNING-RELATED MODIFICATIONS

For almost a century, the presence of "maps" in mammalian sensory cortices has been recognized. The first clear demonstration was Inouye's mapping of human visual cortex from lesion-induced scotoma in 1909 (cited in Reference 7). Regarding the tactile modality, it was known long ago that both the primary and secondary somatosensory cortical regions (SI and SII) possess a map-like organization.[8] With the exception of a few notable discontinuities (e.g., the adjacency of the hand and face representations), neighboring peripheral receptors project to neighboring cortical sites, so that the spatial position of neuronal activity holds essential information about the nature of the event that is being represented.[9] Single-electrode physiological work through the 1980s revealed the exquisite organization of primate somatosensory cortex — distinct representations of the single pads of individual fingers were found in SI.[10] Recent fMRI work has shown distinct digit representations in human SI.[11] In SII, a less somatotopically precise map has been delineated.[12]

Still, the fact that these areas contain somatosensory maps does not prove that their map-like properties have a role in sensory-perceptual learning. Two alternative positions can be argued. First, SI and SII may truly be sites of somatotopic activity patterns, but these areas might serve only as "pre-cognitive" processing and distribution centers. In primates, the major intracortical sensory processing streams originating in SI and SII include a ventrally directed pathway passing through insular and entorhinal cortex before reaching the hippocampus[13] and a caudally directed pathway projecting to posterior parietal cortex before reaching prefrontal cortex.[14] It is conceivable that the conscious recognition of the features or identity of objects might be carried out exclusively by "higher-order" cortical regions based on sensory information received from SI and SII. In that case, learning might also be restricted to the same higher-order regions.[15]

Second, even if sensory cortical areas participate in sensory perception, the somatotopic map might be irrelevant. Topographic boundaries, which appear to be precise when the immobile human subject or anesthetized animal receives passive stimulation, might not be respected by activity patterns during natural usage of the sensory system. According to this argument, even if sensory cortex participates in learning and remembering, it does so through a widely distributed representation, overriding topographic boundaries. This view finds proponents among neurophysiologists.[16]

These two models will be critically examined in the next two sections. Our view is that critical evidence contradicts both models. Contrary to the first model, primary sensory cortex *does* appear to contribute to the cognitive experience of perception and even to sensory memory (the recall of previously perceived events). And, contrary to the second model, when sensory cortex participates in cognitive experiences, it appears that the map within which activity patterns reside is critical in the storage of the experience.

7.3 INVOLVEMENT OF SENSORY CORTEX IN PERCEPTION AND MEMORY

Recent studies indicate that meaningful sensations can be elicited by direct stimulation of primary sensory cortex. Romo and colleagues[17,18] trained monkeys to discriminate between two vibrations at different frequencies applied to the fingertips — the monkeys' task was to judge whether the frequency of the second vibration was higher or lower than that of the first vibration. Typically, the monkeys continued to make this discrimination with no loss in accuracy even if the first or the second tactile stimulus was substituted with direct electrical stimulation of neuronal populations in area 3b of SI (Figure 7.1). In other words, directly stimulating SI with electrical pulses at a given frequency produced a sensation that the monkeys treated as being identical to the real vibratory stimulus of the corresponding frequency. The meaning of this fascinating experiment is not yet certain. Does electrical stimulation simply succeed in activating the cortical targets of SI, which generate a perceptual correlate from the information relayed to them by SI neurons? Or is neuronal discharge in area 3b, by itself, the necessary and sufficient brain activation underlying perception, directly corresponding to sensation of the external event? More work is needed to distinguish between these possibilities; nonetheless, the data point to the idea that primary sensory cortical activity participates directly in the circuit whose activity underlies the percept.

With regard to sensory memory, there are reports suggesting that the circuitry activated during sensory recall overlaps that activated during sensory perception, and one of the specific sites of overlap is primary sensory cortex. The most important evidence comes from functional magnetic resonance imaging (fMRI) and positron emission tomography (PET) studies in subjects performing mental imagery *in the absence of* external sensory input. Although the number of studies is much greater for the visual modality (e.g., References 19 and 20), examples in the tactile realm do exist. For example, in an fMRI experiment, imagining the act of finger tapping led to activation of the same zone of SI cortex that had been originally engaged during actual finger tapping: memory of the tactile experience appeared to re-engage the same neurons involved in the initial experience.[21] It is intriguing that not only *memory*, but also *expectation*, of a tactile event activates SI cortex. A study using

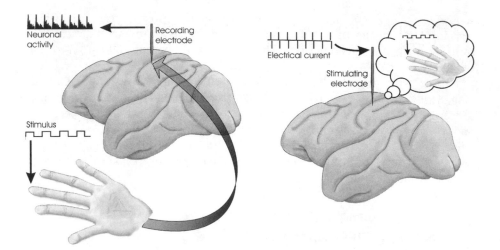

FIGURE 7.1 Interpretation of the recent work by Romo and colleagues.[17,18] In monkeys trained to estimate vibrations at different frequencies applied to the fingertips (left drawing), the vibratory stimulus evokes periodic firing among neurons in the topographically matched site of primary somatosensory cortex, as indicated by the regular peristimulus time histogram. Direct electrical stimulation of the same cortical site with the appropriate frequency parameters (right drawing) produces a percept that the monkey is able to evaluate as if it were a vibratory stimulus.

fMRI reported that subjects who knew they were about to receive a tickle-inducing stimulus showed increased activity in the same cortical regions, including SI and SII, that showed activation during actual tickling.[22]

7.4 TACTILE LEARNING IN A TOPOGRAPHIC FRAMEWORK

Having briefly argued that somatosensory cortex is a key component of the total neuronal ensemble that gives rise to the online experience, recall, and expectation of tactile events, we turn to the second issue: what is the role of somatosensory cortical *topography* in perceptual learning? One way of investigating the role of cortical topography is to require subjects to learn a task using a restricted set of sensory receptors. Later, they are tested on the same task, but are required to use a different set of receptors in the same sensory system. If the learning process is widely distributed, showing no respect for topographic boundaries, or if the learned information resides in "higher" areas, whose organization does *not* conserve the topographic arrangement of the sensory apparatus, then subjects will be able to transfer learning immediately to a second set of sensory receptors, regardless of the relative positions of the "trained" and "tested" sensory receptors. In contrast, if the learning process preferentially involves one part of the cortical topographic field, then learning transfer will be *incomplete*: subjects will require some retraining time before accomplishing the task using the second set of sensory receptors (Figure 7.2). This strategy has been applied to examine the distribution of learning in the visual

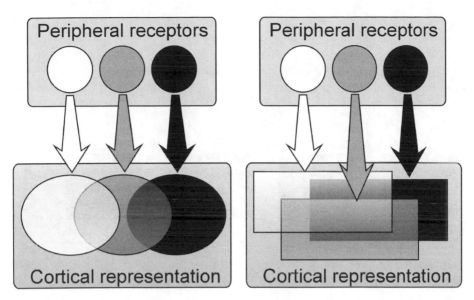

FIGURE 7.2 Schematic diagram illustrating two alternative models for the cortical distribution of sensory learning. On the left, an orderly array of sensory receptors projects to an orderly array of cortical processing zones, which correspond to storage sites for learned information. On the right, the cortical processing and storage zones are widely distributed, highly overlapping, and do not conserve topographic order. Experiments that probe the patterns of learning transfer to nearby sensory receptors after training with one restricted site can discriminate between these models.

modality[23-25] and we have used the same strategy to uncover the distribution of learning in the tactile modality.[26-29] Our experiments, reviewed below, show that when animals or human subjects learn a tactile discrimination task, the neural modifications associated with the learning are distributed according to the spatial arrangement of the cortical map.

7.5 EVIDENCE FROM LEARNING WITH THE WHISKERS

The rat whisker sensory system is well suited to the inquiry because individual whiskers on the snout, arranged in rows and arcs, project to contralateral cortical "barrel-columns" in a precise topographic manner.[30] Are all the barrel-columns of the cortical map bound together as a functional unit during learning or can different areas participate independently in sensory learning? Rats deprived of all but one whisker were trained in the "gap-crossing" task:[31] they learned to use their remaining intact whisker to locate a "goal platform" and to cross the intervening gap to reach a food reward. Training was carried out under red light, preventing the albino rats from using visual information. They learned to feel the opposite platform across gaps of up to 16 cm. Once a rat had learned to gap-cross reliably, the "trained" whisker was clipped and a "prosthetic" one attached to the stub of an "untrained" whisker, or to the stub of the previously trained whisker.

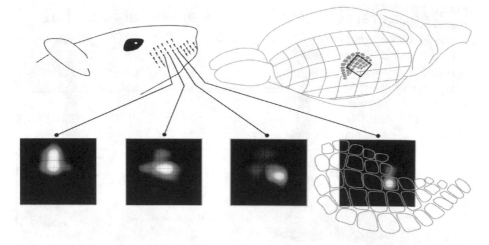

FIGURE 7.3 (See Color Figure 7.3 in color insert.) Topographically distributed "response maps" for four whiskers of the rat's snout (C_{1-4}) as revealed with large 10×10 microelectrode arrays inserted in somatosensory cortex. The four stimulus sites are shown on the drawing of the rat. To the right, the recording site is shown relative to the barrel field of the left hemisphere. The boxed area indicates the boundaries of the electrode array. At the bottom, spatial distributions of the four whiskers' representations are illustrated in relation to the barrel arrangement (right side). The center of each activated zone, where spike counts were highest, is given by the warmest color.

With the prosthetic whisker in place, they were immediately tested. The main result was that rats gap-crossed without delay *only* if the prosthetic whisker was re-attached to the stub of the previously trained whisker. For any other whisker transposition, a period of relearning was required: indeed, the number of trials necessary to reacquire the task increased systematically as a function of the distance along rows and arcs between the trained and the prosthetic whisker. Rats showed no benefit from their previous training if the prosthetic whisker was attached two or more positions away from the site of the original whisker.

Learning with a *single* whisker might seem to be a special case, but additional experiments demonstrated that the topographic learning rule was the same for multi-whisker learning. When rats were trained with a set of four adjacent whiskers, the learned ability quickly transferred to any of the whiskers immediately surrounding the trained ones, but transferred less readily to non-adjacent whiskers. Moreover, the topographic rule held up across the midline: rats could rapidly transfer learning to whiskers *symmetrically opposite* the trained whiskers (e.g., left C_3 to right C_3), but required additional training before successfully transferring to *non*-symmetric opposite whiskers. The corpus callosum, which connects homotopic sites in the left and right barrel field cortex,[32] could be the neural substrate for this transfer.

From large-scale electrophysiological recordings using 100-microelectrode arrays implanted in cortex (Figure 7.3), we found that the extent to which learning transferred across whisker positions was perfectly correlated with the degree of overlap between the representations of those whiskers in primary somatosensory "barrel" cortex.

Based on the behavioral evidence, it is difficult to hold that the cortical network that participated in learning and remembering the tactile task was uniformly or globally distributed; in that case, the rats would have utilized the prosthetic whisker to gap-cross without delay, even if it were attached far from its original site. Instead, the observed pattern points to a memory trace governed by the precise topography of sensory cortex. The neurophysiological evidence confirms the feasibility and simplicity of this explanation.

7.6 EVIDENCE FROM LEARNING WITH THE FINGERS

It is natural to wonder whether the topography rule discovered in rats would hold up for tactile learning in primates, particularly in humans. Interestingly, the literature does not fully support the view that tactile learning is topographically distributed; some studies have argued that subjects can transfer a learned tactile task immediately to any skin surface, regardless of its location relative to the trained site.[33-36] Because previous studies utilized extended training periods (weeks or months) during which topographic specificity could have been lost, we decided to examine the transfer of tactile discriminative abilities acquired during single training sessions. In each experiment, subjects were trained to use one fingertip ("I") to discriminate between two stimuli. Experiment 1 required identification of vibration frequency (on average, about 260 training trials to threshold performance), Experiment 2, punctate pressure (about 220 training trials), and Experiment 3, surface roughness (about 120 training trials). Immediately after reaching criterion with the trained fingertip, subjects were tested with the same fingertip, its first and second neighbors on the same hand (I_1 and I_2, respectively), as well as the three corresponding fingers on the contralateral hand (C, C_1, and C_2).

As expected, for all stimulus types, participants showed retention of learning with the trained fingertip. However, the transfer beyond the trained fingertip varied according the stimulus type (Figure 7.4). For vibration, learning did not transfer to other fingertips. For both the pressure and roughness stimuli, there was limited transfer, dictated by topographic distance: subjects performed well with the first neighbor of the trained finger and with the finger symmetrically opposite the trained one.

The objection might be raised that subjects did not learn *tactile information* — they simply learned to focus their attention on the trained finger. Perhaps, to the extent that there was some spatial spread in the "attention spotlight," subjects retained enough of the attention effect to show a benefit for untrained fingers during testing. The transfer of learning observed in Experiments 2 and 3 to a location that would hardly be expected to share the attention spotlight — that is, to the fingertip *contralateral* to the trained one — is inconsistent with the attention hypothesis. We carried out an additional test to show that what was learned was tactile information. In this final experiment, subjects were tested using the same pair of surfaces as in Experiment 3. The subjects had been trained, however, with a pair of surfaces different from those used during testing. If the training effect were due to spatial attention, subjects would have performed best with the trained fingertip even if the

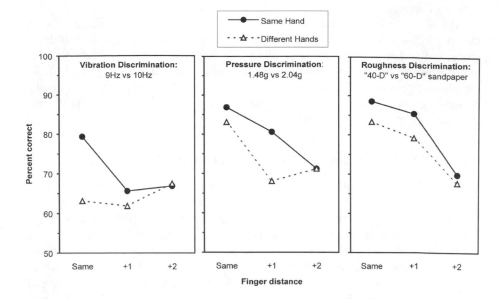

FIGURE 7.4 Topography of tactile learning in humans. Subjects were trained to use one fingertip to carry out a discrimination of vibration frequency (left panel), pressure (center panel), or roughness (right panel). After completion of training, they were tested with the trained fingertip ("same") and its first ("+1") and second ("+2") neighbors, as well as the corresponding fingertips on the opposite hand. The graphs show the performance during testing, and reveal topographic orderliness in the transfer of learning for each type of stimulus. Original data is reported in Reference 29.

test surfaces were unfamiliar. They did not: there was no advantage for the trained fingertip. This indicates that the tasks involved topographically distributed memory for *specific information.*

If we accept the topographic learning rule, then the finding that the precise pattern of the learning distribution depended upon the features of the acquired information (Figure 7.4) should not be surprising. Since sensory neocortex is parcelled into multiple areas, each specialized to process specific types of stimuli,[37] the spatial distribution of stored information should reflect the somatotopy present in the cortical area that processes and stores that type of stimulus. Although a detailed review is beyond the limits of this chapter, a survey of the literature gives a straightforward indication as to which cortical areas might be the critical substrates for the experiments described above. Area 3b, containing neurons with rapidly adapting response properties, single-digit receptive fields, and no callosal connectivity, is the best candidate for vibration frequency learning.[17,18,38-45] For roughness and punctate pressure learning, the functional organization of area 43 (also known as SII) appears to be well suited.[46-53] For example, the dense callosal connectivity of SII is likely to be the substrate for the transfer of roughness and punctate pressure learning to the opposite, symmetric fingertip.

7.7 TOPOGRAPHIC LEARNING: SHARED PRINCIPLES ACROSS SPECIES AND ACROSS MODALITIES

We are not aware of tests for frequency-specific learning in the auditory modality. However, transfer of visual learning has been studied extensively. Several studies have reported that human participants trained to recognize a visual stimulus (e.g., a bar with a specific orientation) at a specific retinotopic position were unable to identify that stimulus when presented at a different location.[24,25,54-57] Based upon the distribution of the learning effects with a sharp spatial gradient related to the receptor surface, primary visual cortex was taken to be the neural substrate for the learning of difficult visual discriminations. However, as task difficulty decreased, transfer of learning increased, suggesting that "higher-order" cortical areas (possessing less strictly topographical maps of the periphery) participate in the learning of easier tasks.[23]

In contrast, several studies have *failed* to detect a topography for tactile learning in humans: when subjects learned to use one finger to discriminate between two stimuli, their ability readily transferred to other fingers.[34-36] These findings were interpreted as showing "complete transfer" of perceptual learning, suggesting that a fundamental difference exists between visual and tactile learning. However, most of these studies only examined transfer over a limited spatial extent (e.g., transfer between neighboring fingers or the corresponding fingers on each hand). Our study shows that, when a greater range of sites is tested, a topographic distribution of tactile learning is revealed. This conclusion also resolves the apparent discrepancy between visual and tactile learning in humans. Moreover, the existence of the principle of topographic learning in species as distinct as humans and rats implies that the storage of sensory information within the framework of a body map is a fundamental and successful strategy adopted by the mammalian cortex.

7.8 CONCLUSIONS

Behavioral and physiological studies suggest that the internal, map-like organization of sensory cortical areas shapes the way sensory information is learned. When we assert that information is stored within a topographic framework, we must clarify the sort of information to which we are referring. In most situations, a large part of what is learned is the "behavioral strategy" (consider, for example, the controversy as to whether the Morris water maze best measures behavioral strategy learning as opposed to spatial map learning[58]). An animal placed in an apparatus to make specific sensory discriminations must first solve problems concerning the overall context of the situation. One would hardly expect the learning of behavioral strategy to be localized to a few cortical columns or even to a few cortical processing regions. Indeed, Lashley is famous for showing (inadvertently) that learning of behavioral strategy is poorly localized in the brain.[59] An additional complicating factor is that, during regimes of sensory training extending across weeks or months, the distribution of learning may change. The idea of a "subcorticalization" of motor procedures during long periods of practice is common, and similar redistributions may occur in sensory learning.

Therefore, an attempt to uncover possible topographic components of learning can only be successful if the investigator separates the learning of behavioral strategy from the learning of a specific sensory task. We did so by allowing the subject (rat or human), upon testing, to persevere with the same behavioral strategy used during previous training, but now with the requirement of gathering sensory information with an "untrained" part of the sensory apparatus. This approach uncovered the fact that sensory information processed by a "naive" part of the sensory system could not be used immediately to solve the task.

If vague concepts such as "widely distributed" or "parallel" sensory processing are ever to be properly defined, the new definition will have to take into account the empirical observation that, under certain experimental conditions, what was learned with one fingertip can be of essentially no benefit to the performance with the adjacent fingertip. As an alternative to the idea of globally distributed processing, we propose a model for sensory learning that emphasizes modularity. (1) The features of the sensory information determine which cortical area is involved in information processing and storage (thus, vibration frequency and surface texture would be explicitly represented in different cortical regions). (2) Task difficulty also influences the site of cortical modification. Easier discriminations do not entail modifications in "lower-order" cortical areas, but may entail the learning of a behavioral strategy (also see Reference 23). (3) For sensory learning tasks that involve topograpically organized areas, a critical component of the relevant sensory information is stored within a restricted region of the sensory map by the same neuronal circuits that process the sensory signal during training. (4) If subjects are subsequently tested with untrained parts of the sensory system, then performance on the task will be proportional to the extent to which the topographic region activated by the new sensory signal overlaps the region where the sensory information has been stored. In this model, the brain's body maps indeed play a fundamental role in sensory learning.

ACKNOWLEDGMENTS

Supported by the Telethon Foundation, the J.S. McDonnell Foundation, the European Community, the Ministero dell'Università e della Ricerca Scientifica e Tecnologica (MURST), Consiglio Nazionale delle Ricerche, and the Regione Friuli Venezia Giulia.

REFERENCES

1. Bliss, T.V. and Lomo, T., Long-lasting potentiation of synaptic transmission in the dentate area of the anaesthetized rabbit following stimulation of the perforant path, *J. Physiol.*, 232, 331, 1973.
2. Bailey, C.H., Bartsch, D., and Kandel, E.R., Toward a molecular definition of long-term memory storage, *Proc. Natl. Acad. Sci. U.S.A.*, 93, 13445, 1996.
3. Thompson, R.F., Swain, R., Clark, R., and Shinkman, P., Intracerebellar conditioning — Brogden and Gantt revisited, *Behav. Brain Res.*, 110, 3, 2000.

4. Fanselow, M.S. and LeDoux, J.E., Why we think plasticity underlying Pavlovian fear conditioning occurs in the basolateral amygdala, *Neuron.* 23, 229, 1999.
5. Zola-Morgan, S., Squire, L.R., and Amaral, D.G., Lesions of the hippocampal formation but not lesions of the fornix or the mammillary nuclei produce long-lasting memory impairment in monkeys, *J. Neurosci.*, 9, 898, 1989.
6. Horel, J.A., Some comments on the special cognitive functions claimed for the hippocampus, *Cortex*, 30, 269, 1994.
7. Glickstein, M., The discovery of the visual cortex, *Sci. Am.*, 259 118, 1988.
8. Woolscy, C.N., In: Harlow, H.F. and Woolsey, C.N. Eds., *Biological and Biochemical Bases of Behavior*, University of Wisconsin Press, Madison, pp. 63, 1958.
9. Petersen, R.S. and Diamond, M.E., Topographic maps in the brain. In: *Encyclopedia of Life Sciences*, Macmillan, 2001.
10. Kaas, J.H., Topographic maps are fundamental to sensory processing, *Brain Res. Bull.*, 44, 107, 1997.
11. Maldjian, J.A., Gottschalk, A., Patel, R.S., Detre, J.A., and Alsop, D.C., The sensory somatotopic map of the human hand demonstrated at 4 Tesla, *Neuroimage.*, 10,55, 1999.
12. Gelnar, P.A., Krauss, B.R., Szeverenyi, N.M., and Apkarian, A.V., Fingertip representation in the human somatosensory cortex: An fMRI study, *Neuroimage*, 7, 261, 1998.
13. Friedman, D.P., Murray, E.A., O'Neill, J.B., and Mishkin, M., Cortical connections of the somatosensory fields of the lateral sulcus of macaques: evidence for a corticolimbic pathway for touch, *J. Comp. Neurol.*, 252, 323, 1986.
14. Preuss, T.M. and Goldman-Rakic, P.S., Connections of the ventral granular frontal cortex of macaques with perisylvian premotor and somatosensory areas: anatomical evidence for somatic representation in primate frontal association cortex, *J. Comp. Neurol.*, 282, 293, 1989.
15. Mesulam, M.M., From sensation to cognition, *Brain*, 121, 1013, 1998.
16. Nicolelis, M.A., Fanselow, E.E., and Ghazanfar, A.A., Hebb's dream: the resurgence of cell assemblies, *Neuron*, 19, 219, 1997.
17. Romo, R., Hernandez, A., Zainos, A., and Salinas, E., Somatosensory discrimination based on cortical microstimulation, *Nature*, 392, 387, 1998.
18. Romo, R., Hernandez, A., Zainos, A., Brody, C.D., and Lemus, L., Sensing without touching: psychophysical performance based on cortical microstimulation, *Neuron*, 26, 273, 2000.
19. Kosslyn, S.M., Thompson, W.L., Kim, I.J., and Alpert, N.M., Topographical representations of mental images in primary visual cortex, *Nature*, 378, 496, 1995.
20. Kosslyn, S.M., Pascual-Leone, A., Felician, O., Camposano, S., Keenan, J.P., Thompson, W.L., Ganis, G., Sukel, K.E., and Alpert, N.M., The role of area 17 in visual imagery: convergent evidence from PET and rTMS, *Science*, 284, 167, 1999.
21. 21.Porro, C.A., Francescato, M.P., Cettolo, V., Diamond, M.E., Baraldi, P., Zuiani, C., Bazzocchi, M., and di Prampero, P.E., Primary motor and sensory cortex activation during motor performance and motor imagery: a functional magnetic resonance imaging study, *J. Neurosci.*, 16, 7688, 1996.
22. Carlsson, K., Petrovic, P., Skare, S., Petersson, K.M., and Ingvar, M., Tickling expectations: neural processing in anticipation of a sensory stimulus, *J. Cogn. Neurosci.*, 12, 691, 2000.
23. Ahissar, M. and Hochstein, S., Task difficulty and the specificity of perceptual learning, *Nature*, 387, 401, 1997.

24. Dill, M. and Fahle, M., The role of visual field position in pattern-discrimination learning, *Proc. R. Soc. Lond. B Biol. Sci.*, 264, 1031, 1997.
25. Karni, A. and Sagi, D., Where practice makes perfect in texture discrimination: evidence for primary visual cortex plasticity, *Proc. Natl. Acad. Sci. U.S.A.*, 88, 4966, 1991.
26. Diamond, M.E., Petersen, R.S., and Harris, J.A., Learning through maps: functional significance of topographic organization in primary sensory cortex, *J. Neurobiol.*, 41, 64, 1999.
27. Harris, J.A., Petersen, R.S., and Diamond, M.E., Distribution of tactile learning and its neural basis, *Proc. Natl. Acad. Sci. U.S.A.*, 96, 7587, 1999.
28. Harris, J.A. and Diamond, M.E., Ipsilateral and contralateral transfer of tactile learning, *Neuroreport*, 11, 263, 2000.
29. Harris, J.A., Harris, I.M., and Diamond, M.E., The topography of tactile learning in humans, *J. Neurosci.*, 21, 1056, 2001.
30. Woolsey, T.A. and Van der Loos, H., The structural organization of layer IV in the somatosensory region SI of mouse cerebral cortex. The description of a cortical field composed of discrete cytoarchitectonic units, *Brain Res.*, 17, 205, 1970.
31. Hutson, K.A. and Masterton, R.B., The sensory contribution of a single vibrissa's cortical barrel, *J. Neurophysiol.*, 56, 1196, 1986.
32. Olavarria, J., Van Sluyters, R.C., and Killackey, H.P., Evidence for the complementary organization of callosal and thalamic connections within rat somatosensory cortex, *Brain Res.*, 291, 364, 1984.
33. Nagarajan, S.S., Blake, D.T., Wright, B.A., Byl, N., and Merzenich, M.M., Practice-related improvements in somatosensory interval discrimination are temporally specific but generalize across skin location, hemisphere, and modality, *J. Neurosci.*, 18, 1559, 1998.
34. Sathian, K. and Zangaladze, A., Tactile learning is task specific but transfers between fingers, *Percept. Psychophys.*, 59, 119, 1997.
35. Sathian, K. and Zangaladze, A., Perceptual learning in tactile hyperacuity: complete intermanual transfer but limited retention, *Exp. Brain Res.*, 118, 131, 1998.
36. Spengler, F., Roberts, T.P., Poeppel, D., Byl, N., Wang, X., Rowley, H.A., and Merzenich, M.M., Learning transfer and neuronal plasticity in humans trained in tactile discrimination, *Neurosci. Lett.*, 151, 232, 1997.
37. Kaas, J.H., Evolution of multiple areas and modules within neocortex, *Perspect. Dev. Neurobiol.*, 1, 101, 1993.
38. Hernandez, A., Zainos, A., and Romo, R., Neuronal correlates of sensory discrimination in the somatosensory cortex, *Proc. Natl. Acad. Sci. U.S.A.*, 97, 6191, 2000.
39. Iwamura, Y., Tanaka, M., Sakamoto, M., and Hikosaka, O., Rostrocaudal gradients in the neuronal receptive field complexity in the finger region of the alert monkey's postcentral gyrus, *Exp. Brain Res.*, 92, 360, 1993.
40. Killackey, H.P., Gould, III, H.J.,Cusick, C.G., Pons, T.P., and Kaas, J.H., The relation of corpus callosum connections to architectonic fields and body surface maps in sensorimotor cortex of New and Old World monkeys, *J. Comp. Neurol.*, 219, 384, 1983.
41. Merzenich, M.M., Kaas, J.H., Sur, M., and Lin, C.S., Double representation of the body surface within cytoarchitectonic areas 3b and 1 in "SI" in the owl monkey *Aotus trivirgatus*, *J. Comp. Neurol.*, 181, 41, 1978.
42. Mountcastle, V.B., Talbot, W.H., Sakata, H., and Hyvarinen, J., Cortical neuronal mechanisms in flutter-vibration studied in unanesthetized monkeys. Neuronal periodicity and frequency discrimination, *J. Neurophysiol.*, 32, 452, 1969.

43. Mountcastle, V.B., Steinmetz, M.A., and Romo, R., Frequency discrimination in the sense of flutter: psychophysical measurements correlated with postcentral events in behaving monkeys, *J. Neurosci.*, 10, 3032, 1990.

44. Recanzone, G.H., Merzenich, M.M., Jenkins, W.M., Grajski, K.A., and Dinse, H.R., Topographic reorganization of the hand representation in cortical area 3b owl monkeys trained in a frequency-discrimination task, *J. Neurophysiol.*, 67, 1031, 1992.

45. Wang, X., Merzenich, M.M., Sameshima, K., and Jenkins, W.M., Remodelling of hand representation in adult cortex determined by timing of tactile stimulation, *Nature*, 378, 71, 1995.

46. Caselli, R.J., Rediscovering tactile agnosia, *Mayo Clin. Proc.*, 66, 129, 1991.

47. Jiang, W., Tremblay, F., and Chapman, C.E., Neuronal encoding of texture changes in the primary and the secondary somatosensory cortical areas of monkeys during passive texture discrimination, *J. Neurophysiol.*, 77, 1656, 1997.

48. Ledberg, A., O'Sullivan, B.T., Kinomura, S., and Roland, P.E., Somatosensory activations of the parietal operculum of man. A PET study, *Eur. J. Neurosci.*, 7, 1934, 1995.

49. Murray, E.A. and Mishkin, M., Relative contributions of SII and area 5 to tactile discrimination in monkeys, *Behav. Brain Res.*, 11, 67, 1984.

50. Pons, T.P., Garraghty, P.E., and Mishkin, M., Serial and parallel processing of tactual information in somatosensory cortex of rhesus monkeys, *J. Neurophysiol.*, 68, 518, 1992.

51. Roland, P.E., O'Sullivan, B., and Kawashima, R., Shape and roughness activate different somatosensory areas in the human brain, *Proc. Natl. Acad. Sci. U.S.A.*, 95, 3295, 1998.

52. Sinclair, R.J. and Burton, H., Neuronal activity in the second somatosensory cortex of monkeys *Macaca mulatta* during active touch of gratings, *J. Neurophysiol.*, 70, 331, 1993.

53. Tremblay, F., Ageranioti-Belanger, S.A., and Chapman, C.E., Cortical mechanisms underlying tactile discrimination in the monkey. I. Role of primary somatosensory cortex in passive texture discrimination, *J. Neurophysiol.*, 76, 3382, 1996.

54. Ahissar, M. and Hochstein, S., Learning pop-out detection: specificities to stimulus characteristics, *Vision Res.*, 36, 3487, 1996.

55. Dill, M. and Fahle, M., Limited translation invariance of human visual pattern recognition, *Percept. Psychophys.*, 60, 65, 1998.

56. Fahle, M., Human pattern recognition: parallel processing and perceptual learning, *Perception*, 23, 411-427 1994.

57. Schoups, A.A. and Orban, G.A., Interocular transfer in perceptual learning of a pop-out discrimination task, *Proc. Natl. Acad. Sci. U.S.A.*, 93, 7358-7362, 1996.

58. Hoh,T., Beiko, J., Boon, F., Weiss, S., and Cain, D.P., Complex behavioral strategy and reversal learning in the water maze without NMDA receptor-dependent long-term potentiation, *J. Neurosci.*, 19, RC2, 1999.

59. Lashley, K.S. Psychological Mechanisms in Animal Behavior: Society of Experimental Biology Symposium, No. 4, Cambridge, U.K., Cambridge University Press, 1950.

8 Attention in the Somatosensory System

Steven S. Hsiao and Francisco Vega-Bermudez

CONTENTS

8.1 INTRODUCTION

In this chapter, we review the psychophysical and neurophysiological basis of tactile selective attention. As in other sensory systems, attention plays a major role in the way that sensory inputs are processed and perceived. The importance of attention in touch can easily be demonstrated by switching one's focus of attention to different locations on the body. For example, if you switch your focus of attention to your foot, you immediately become conscious of sensations arising from receptors in your foot that were non-existent a moment earlier. This simple observation demonstrates the power of selective attention and emphasizes two important aspects of sensory processing. First, it shows that we have a limited information processing capacity and that, under normal circumstances, we ignore most of the sensory inputs that impinge on our bodies. Second, it shows that attention is under cognitive control and, like a lens, focuses our mental efforts onto specific sensory inputs at selected body locations.

Defining attention has been difficult. While each of us has our own subjective sense of what attention is, scarcely anyone is able to express exactly what is meant by the term attention. The difficulty lies in the fact that attention is a property of our internal mental states and, as such, it is a property that is unique to the mind; and, nobody has a definition of what constitutes the mind. However, if one assumes that the mind and brain are the same then one definition of attention is that it is a neural mechanism that allows observers to direct their mental efforts onto specific objects or events. In this definition, the specific objects and events can be practically anything, including external stimuli such as visual scenes, sounds, or locations on the body surface, or internal mental states such as stored memories.

There have been numerous studies that have attempted to characterize attention. In these studies, differences in human sensory performance and their underlying neural processes are monitored while selective attention is switched between different sensory modalities, different locations on the body, and different aspects of the sensory stimulus. These studies show that attention is rapidly engaged and withdrawn from particular stimuli, and when engaged, stimuli that are under the attentional focus are perceived more rapidly and accurately.

There are two main approaches that have been used to characterize attention in the somatosensory system. Psychophysical studies on humans demonstrate that the amount of attention needed to process different kinds of tactile stimuli differ and that some tactile stimuli are processed pre-attentively. These stimuli are processed by "bottom–up," or ascending, mechanisms that are thought to require minimal attentional resources. Bottom–up processing mechanisms are important because they capture and draw our attention to stimuli such as a pinprick on the skin that require immediate attention. There is also a selective component of attention that allows us to focus or concentrate our efforts onto a specific stimulus at a specific location on the body. These "top–down," or descending, mechanisms provide us with the ability to selectively filter out and suppress irrelevant information and enhance the central representations of sensory stimuli that are immediately relevant. Some of the issues that are important here are to determine the capacity of selective attention, the degree that attention can be focused to particular body sites and modalities, and the effect that selective attention has on information processing.

The second approach to understanding attention is to study the effects directly on the responses of neurons in the nervous system. There are two ways that this is done. In the first, animals are trained to perform specific behavioral tasks that require them to switch their focus of attention (or cognitive efforts) between different stimuli presented within or between sensory systems. The assumption is then made that changes in neural activity that occur with changes in the animal behavior is a reflection of the attentional effort required to perform the task. In most of these animal studies, animals are trained in tactile tasks that require them to switch their focus of attention back and forth between a tactile task at a specific location on the body and a control task that diverts the animal's focus of attention to either a different body site, or to an auditory or visual stimulus. This paradigm controls for differences in arousal since the animals are continuously performing a behavioral task. The other way that is now commonly used to study attention is to perform imaging studies directly on human subjects. In this approach, humans are asked to perform a variety of tactile tasks and the effects of attention are assessed by examining the change in activation of different cortical regions. This method has the advantage of allowing one to simultaneously study many areas of the nervous system involved in attention. However, the approach is limited because it only provides information about which anatomical locations are affected by attention.

Here, we describe psychophysical studies on humans, neurophysiological studies of monkeys, and imaging studies of humans performing various selective attention tasks.

8.2 PSYCHOPHYSICAL STUDIES OF TACTILE ATTENTION

8.2.1 Ascending or Bottom–Up Mechanisms of Attention

There have been several psychophysical studies in which humans performed tactile search tasks. These studies are based on a paradigm originally developed by Treisman,[41] who demonstrated in subjects performing visual search tasks that the ability to perceive certain types of target stimuli is unaffected by the number of background distractors. She hypothesized that these stimuli stand out because they contain features that are processed pre-attentively and, as such, require limited attentional resources. In tactile search studies, performance is measured by the response time required to locate a target stimulus presented to one of the finger pads as a function of the number of distractors placed on the finger pads of both hands. When these reaction time functions are relatively flat, the stimulus is considered to "pop out." Using this paradigm, Lederman and Klatzky[30] found that material properties, such as rough-smooth, hard-soft, cool-warm, and surface discontinuities, such as vertical or horizontal bars vs. a flat surface or a hole vs. no-hole, tended to pop out (reaction time slopes RT < 30 ms per digit). In contrast, surface features related to the relative two and three dimensional spatial position of objects on the finger pads or to the orientation of stimuli, such as a horizontal vs. vertical bars, did not. The range of RT slopes varied greatly, with roughness having a slope of about 3.7[28] to 9 msec/digit[28,30] and stimuli that tested the relative position of different shapes having reaction time slopes close to 440 msec/digit (see Figure 8.1). Surfaces composed of three-dimensional (3D) contours such as curved vs. flat had intermediate reaction times. Similarly, recent studies showed that selective attention is only minimally required to detect changes in vibration amplitude[43,49] or texture.[43] These results suggest that some surface features, such as texture or vibration, are processed pre-attentively, whereas more complex features that require information about the spatial positions between stimulus features on different fingers are not.

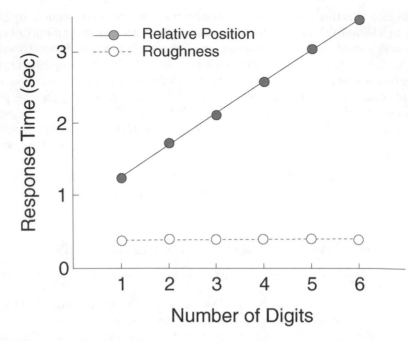

FIGURE 8.1 Reaction times in humans performing tactile search tasks. In one study, subjects were presented with a rough target stimulus on one digit and smooth distracting stimuli on the other digits. In the other study, the target stimulus was a raised dot positioned on the right side of the finger pad. The distractor stimuli were dots on the left side of the pads. The abscissa is the number of finger pads being stimulated. The ordinate is the reaction time to detect the target stimulus. The slope of the roughness task was about 3.7 msec/digit, indicating that the sensation of roughness "pops out" from the distractor stimuli. The slope of the relative position task was about 450 msec/digit, indicating that the subjects needed to perform a serial search to locate the target stimulus. Adapted from Reference 28.

8.2.2 DESCENDING MECHANISMS OF ATTENTION

Our sensory systems are being constantly stimulated by sights, sounds, and a wide range of tactile stimuli. It is only through selective attention mechanisms that we filter out irrelevant information and prevent sensory overload. Studies that directed attention between sensory modalities suggest that attention has a limited capacity for processing information presented simultaneously to different modalities. In one series of studies, Chapman and her colleagues[41,52] had subjects perform cross-modal tactile visual tasks. In these studies, reaction times were measured while subjects directed their focus of attention toward or away from a stimulus or directed their attention to both the tactile and visual stimuli (divided attention task). In the first study,[41] subjects were required to detect the onset of a 40 Hz vibration on their hands or the onset of a light. In the second study,[52] subjects were required to detect a change in surface texture on the tip of one digit or the change in intensity of a light. In both studies, they found that subjects had the shortest reaction times when cued correctly (440 msec), intermediate reaction times (490 msec) when given the neutral cue (divided attention task) and longest reaction times when cued incorrectly (575 msec). In addition, the reaction times increased as the tasks were made more difficult. Driver and Spence[14] also found that there are extensive interactions between modalities and that a salient event (either relevant or irrelevant) in one modality at the correct location improves performance in detecting events in another modality. For example, tactile stimulation of the left hand with a set of vibrators led to faster discrimination of the location of visual stimuli in the left but not the right visual field. The improvement was based on external body coordinate systems; when the hands were crossed so that the right hand was then in the left visual field, improvement occurred with the right rather than the left hand. These studies, together with Chapman's studies,[41,52] suggest that the sensory modalities do not function independent of each other and that there are limited attentional resources for processing information across sensory systems.

There are now at least three different studies suggesting that the ability to attend to multiple stimuli attention is greater when subjects are asked to attend to multiple somatosensory stimuli at different body locations than when attending to multiple stimuli presented to different sensory modalities. In all three studies, tactile stimuli are presented simultaneously to different locations on the body surface and subjects are asked to perform directed or divided attention tasks. In two of the studies, subjects were presented with salient stimuli that consisted of vibrations to the thenar eminence of the right hand, left fingertip, and the forearms or taps to the right and left index fingers.[40,44] In the other study, Craig[9] directed subjects to attend to stimuli on either hand or to divide their attention to patterned stimuli presented on both hands. All three studies found that there is essentially no difference in performance when attention is directed or divided between stimuli presented to different locations on opposite sides of the body, suggesting that there are large attentional resources when attention is confined to the somatosensory system. The results from the Craig study are of particular interest since they did not investigate simply whether subjects could detect stimuli but also tested the ability of subjects to identify vibrotactile patterns presented to different fingers. In his study, spatially patterned vibrotactile stimuli were presented to pairs of fingers using two Optacons, which consists of 6×24 array of vibrating pins used as a reading aid for the blind. Subjects were then asked to identify vibrotactile patterns that were presented with different stimulus onset asynchonies to the different digits. The study reported that subjects performance was about 90 to 98% in identifying patterns when attention is directed to a single digit and that there was a small reduction in performance when the patterns were presented to two digits bilaterally (see Figure 8.2). This is essentially the same result reported by Shiffrin and Posner. Contrary to expectations, Craig also found that there was a large decline in performance when the patterns were presented ipsilaterally to the subjects' index and middle fingers and that the decline in performance disappears when the onset asynchrony is greater than about 200–400 msec (see Figure 8.2). One explanation for these results is that the capacity of attention may be unlimited[44] when stimuli are presented to locations on the body surface that are widely separated from each other (i.e., the two hands) but stimulus interference occurs when stimuli are placed close to each other in space (same hand) and time (within 400 msec).

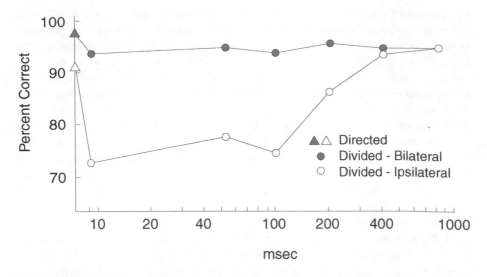

FIGURE 8.2 Directed vs. divided attention. In this experiment, subjects were presented with Optacon-generated patterns of lines on adjacent digits on the same hand (index and middle) or to the index finger on one hand and middle finger on the opposite hand. In the directed attention studies, subjects were cued to attend to a single digit and the two letters were presented simultaneously to the fingers. In the divided attention task, the subjects were presented with the two letters and 500 msec later they were asked to report the letter that was presented on one of the digits chosen at random. There is a small decrement in performance in the directed attention task when the two are presented to the ipsilateral hand, suggesting that the patterns presented to the same hand interfere with each other. This decrement in performance increases for stimulus onset asynchronies up to 400 msec (abscissa), suggesting that the interference disappears when two patterns are presented separately at sufficiently long intervals. Adapted from Craig, J.C., Psychophys, 38, 496–511, 1985. With permission.

8.2.3 FOCUS OF ATTENTION

There are now several studies that confirm the subjective impression that attention functions like a cognitive "spotlight" that can be focused and moved to selected body locations. In one recent study, Lakatos and Shepard[29] showed that the time it takes to shift one's focus of attention is dependent on the distance between the cued site and the test location. In their study, subjects were asked to attend to a specific location on the body. Then, two seconds later, a second location was announced and, simultaneously, the subjects received air puffs at four of eight different body locations. Subjects were required to respond as rapidly as possible if an air puff occurred at the second-announced body location. The study found that reaction times increased as the distance between the two sites increased and that the critical distance is not related to the somatotopic distance between body locations but to the straight line distance in external space between the two test sites. Driver and Grossenbacher[13] obtained similar results when they presented vibrotactile stimuli on the hands, which were placed either together or spread apart at different distances. Interestingly, they also showed that the reaction times decreased when subjects oriented their heads toward the relevant hand, independent of whether the subjects were looking at the hand or were blindfolded. These studies indicate that attention is based on an extra-personal coordinate system that is affected by proprioceptive inputs (e.g., changes in head orientation), and is not organized somatotopically. In this coordinate system, attention shifts between body parts like a spot of light moving directly between body sites.

Studies that attempt to characterize the spotlight of attention indicate that the focus of attention has an aperture size that is modality dependent. As discussed earlier, Craig[9] had subjects attend to patterns presented to fingers on the same and opposite hands and found that the patterns were more easily perceived when pre-sented to fingers of the opposite hand than when presented to adjacent fingers on the same hand. He also found that the ability to integrate patterns on the two hands was unaffected by the spatial separation between hands. The results from this study suggest that the hand is under a single focus of attention, and that stimuli presented to digits of the same hand interfere with one another. In a similar study, Evans et al.[16] had subjects report the direction of movement of a pattern presented on one fingertip while another moving (non-target) pattern was presented to an adjacent finger pad on the same or opposite hand. When the target and non-target patterns were presented on the same hand, they found that the accuracy was higher and reaction times were faster when the two patterns moved in the same direction and that performance declined when the non-target stimulus moved in the opposite direction. This was true even though subjects were told to ignore stimuli on the non-target finger. In contrast, the non-target stimulus had no effects on performance when presented to a finger on the opposite hand.

Although these studies strongly suggest that the entire hand is under a single attentional focus, a recent study in our laboratory suggests otherwise. In those studies,[11] we presented sandpapers of varying grit numbers to individual fingers and to pairs of fingers on the same and opposite hands. We found that when a single finger is used, there is a monotonic decrease in roughness magnitude judgments as grit number increased, confirming previous roughness studies that used sandpapers, and that roughness magnitude estimates were identical across fingers on the two hands. When pairs of fingers were used, subjects scanned different grit sandpapers with the two fingers and were asked to report their roughness judgments on the target finger from only the target finger. We found that roughness judgments on the target finger were unaffected by sandpapers presented to non-target fingers on either hand and that the roughness estimates obtained were identical to the estimates obtained during the single finger task. These results suggest that, at least for rough-ness stimuli, the minimum aperture size is the finger pad and not the entire hand as reported by Craig.[9] One explanation for this inconsistency is that the degree that attention can be focused may differ for different tactile modalities. Thus, roughness perception, which is encoded by the SA1 afferent system, may have a smaller focus of attention than vibration, which is encoded by the RA and PC afferent systems.[27] This hypothesis is supported by psychophysical studies showing that there is little integration of spatial information across finger pads in subjects performing two-dimensional pattern identification tasks of large patterns that extend across multiple digits.[31]

8.3 NEUROPHYSIOLOGICAL STUDIES OF TACTILE ATTENTION

8.3.1 ANATOMY

In this section, we briefly review the ascending and central pathways of the somatosensory system. We will later show that neurons at practically all stages within the system, starting at the level of the brain stem nuclei, have responses that are affected by the animal's focus of attention; with neurons at higher levels of the central nervous system showing greater effects than neurons closer to the peripheral inputs.

Afferent fibers entering the spinal cord separate into two parallel paths (for a review of the afferent fibers, see Reference 27 for a review. One pathway, called the dorsal-column-medial leminiscal pathway, contains axons from large and medium diameter fibers that synapse on neurons in the ipsilateral dorsal column nuclei (DCN). Afferent fibers in this pathway carry information to the central nervous system about mechanoreceptive and proprioceptive events at the periphery. The other main pathway is called the spinothalamic tract (STT). This pathway receives fibers originating from neurons in the contralateral dorsal horn and conveys information mainly from the small myelinated and unmyelinated afferents that enter the spinal cord. It is responsible primarily for conveying information about pain and temperature to the central nervous system. Fibers from the second order neurons in the DCN and Dorsal horn project to neurons in the ventroposterior lateral nucleus of the thalamus (VPL), where information is further segregated into nuclei that respond to deep and cutaneous tissues. Neurons in VPL send their projections in two directions, a large projection to primary somatosensory cortex (SI) and a smaller projection to secondary somatosensory cortex (SII).[23]

SI cortex is composed of four separate areas: 3a, 3b, 1, and 2. These areas receive specific projections from neurons in VPL and from neurons in other areas of SI cortex. Neurons in areas 3a and 2 respond to inputs from deep tissues and neurons in areas 3b and 1 respond to cutaneous mechanoreceptive inputs. It can be inferred from the anatomical connectivity that areas 3b and 1 are important for mechanoreceptive functions, such as form and texture perception, and neurons in areas 3a and 2 are important for functions that require information from deep tissues, such as joint angle and stereognosis (area 2).

Neurons in SI cortex project in two main directions. One projection is toward neurons in areas 5, 7b, and Ri. This pathway is concerned primarily with processing information about where our limbs are in space and how our hands are positioned when grasping objects. The other projection is toward SII. SII cortex is a large cortical region located in the upper bank of the lateral sulcus and is composed of two or more somatosensory areas. The pathway toward SII is thought to be responsible for processing information about tactile form and texture perception. While neurons in SI tend to have small receptive fields that are within one (3b) or a few digits (1, 2), neurons in SII cortex have a wide range of receptive field sizes and shapes, ranging from single digit receptive fields to bilateral receptive fields that cover all or part of both hands. SII cortex is heavily interconnected with many cortical areas.

Among those connections that are of consequence in relation to attention are projections to areas in the frontal cortex, insula, and a weak projection to the anterior cingulate, which also receives inputs from frontal cortical areas.

8.3.2 Effects of Attention on Neuronal Responses

8.3.2.1 Sub-Cortical Areas

Neurons at all levels of processing appear to be affected by the animal's attentional state. In one study, Hayes et al.[21] trained monkeys to perform thermal and light discrimination tasks. They found that wide dynamic range and nociceptive neurons in the dorsal horn showed a greater response to behaviorally relevant thermal stimuli than equivalent irrelevant thermal stimuli. Similar results were reported by Bushnell et al.[6] and Duncan et al.[15] who trained animals to detect either a change in noxious or innocuous thermal stimuli, or to detect the onset of a light. In the psychophysical study, they reported that the reaction times (in both man and monkey) needed to detect the stimuli were shortest when the subjects were correctly cued and that the effects were greater for noxious than for warm stimuli.[7] In the neurophysiological study, they reported that about 45% of the neurons in the medullary dorsal horn had "task-related" responses.[6]

Although initial reports by Poranen and Hyvarinen[39] suggested that neurons in VPL are not affected by the animal's focus of attention, two other studies report that attention has small effects. Tremblay et al.[48] trained animals to perform an air puff detection task and a light discrimination task. They found that 2 of 23 neurons in VPM showed clear attention modulated responses. Recently, Morrow and Casey[34] reported that 6 of 18 neurons that they studied in VPL showed attention-modulated responses. In their study, animals were trained to detect the cessation of an air puff or mechanical taps with a Von Frey hair on the skin. These studies suggest that attention does not play a large role at the sub-cortical level and that the "spotlight" of tactile attention most likely does not originate from neurons in VP. Whether or not other areas of thalamus, such as the reticular formation,[10] are involved in tactile attention is not known.

8.3.2.2 SI Cortex

Initial studies suggested that attention had minimal effects on the response properties of neurons in SI cortex.[18] A later study by Hyvarinen et al.[25] suggested that attention had a small effect. They trained animals to perform both a vibration detection task and to sit passively while the same stimuli were presented to the hand. They found that about 8% of the neurons in area 3b and about 22% of the neurons in area 1 were affected by the animal's focus of attention. More recently, two studies have shown that attention has much greater effects on neurons in SI cortex.

Hsiao et al.[24] trained animals to perform a tactile letter discrimination task and a visual light detection task. During the recording sessions, the animal's focus of attention was switched back and forth every 5 to 10 minutes between the two tasks while the same tactile stimulus was scanned across the distal pads of the animal's hand. The effects of attention were assessed by comparing a neuron's responses to

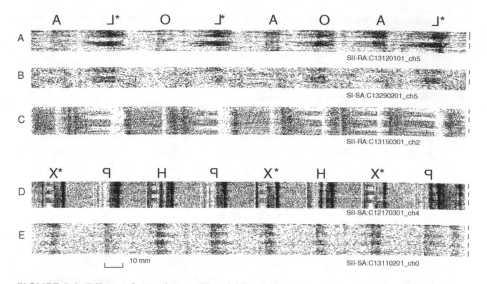

FIGURE 8.3 Effects of attention on SI and SII neurons. Raster plots showing the responses of five neurons from SI (B) and SII cortex (A, C–E). Each raster represents the response of the neuron while the animal's focus of attention was switched back and forth between performing a tactile letter discrimination task and a visual task. Periods where the animal was performing the tactile task are marked by a vertical line on the right side of each raster. The top three rasters represent the responses to a stimulus that contained the letters A, L, and O and the bottom two rasters represent the responses to the letters X, P, and H. The stimuli were mounted on a rotating drum that scanned the letters repeatedly across the receptive field (letters moved from right to left). In the top three rasters, the target letter (*) was the letter L and in the bottom two rasters the target letter was the letter X. Each black tick represents the occurrence of an action potential. Adapted from Reference 24.

the embossed letters for the two tasks. They reported that about 50% of the neurons in areas 3b and 1 of SI cortex had increased firing rates when the animal performed the tactile task; this study differed from the first two studies in that it required the animal to perform a visual distracting task as the control. Having the animal perform this task served two functions. One was to control for arousal effects since the animal received rewards at about the same rate during both tasks. The other was to draw the animal's focus of attention away from the tactile task, which may explain why these authors found greater effects in SI cortex.

An example of the response of a typical SA neuron in area 3b is shown in Figure 8.3B. In this figure, the letters above the raster plots represent the letters of the alphabet that were scanned across the animal's finger and the target letter. Although attention affected fewer neurons and had smaller effects on neurons in SI than SII cortex, there was a clear enhancement of the neural responses when the animal performed the tactile task relative to when it performed the visual task. The enhanced responses were not only observed on the target letters but also on non-reward letters that did not directly follow a target letter (e.g., see the response to the letter O and the second letter A following the second target letter L in Figure 8.3A). Attentional effects were not observed during two types of periods. One was time-

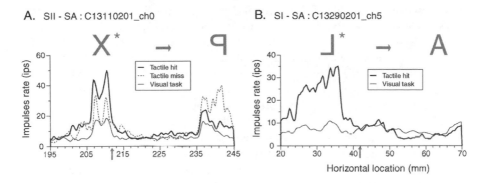

FIGURE 8.4 Effects of attention on target and non-target letters. Response of two neurons from SI (B) and SII (A) cortex to target letters and letters following target letters. The ordinate shows the mean rates evoked by the letters. The abscissa represents the circumferential distance in mm from the start of the pattern (see Figure 8.3). Letters scanned from right to left at 15 mm/sec across the receptive field. Thick lines show responses during hits (animal correctly identified the letters), thin lines represent the response during the visual task, and dashed lines represent the response during missed trials. The arrows along the abscissa represent the mean location of the animals response. Adapted from Reference 24.

out periods that were triggered by false-positive responses. The other was reward periods, which corresponded to non-target letters following target letters in which the animal correctly identified the letter. The most likely explanation for the lack of attention effects on these letters is that the animal learned that certain letters were behaviorally irrelevant (e.g., two target letters never appeared in succession on the drum).

The effects of attention on the mean impulse rates evoked by a target letter and the following letter, while the animal was performing the tactile and visual tasks is shown in Figure 8.4. Figure 8.4A shows a response that was typical of neurons in SI and SII. Neurons in these areas showed a vigorous response to non-target letters (e.g., letter P) following misses and minimal responses to these same letters following hits. Figure 8.4B does not show the responses during tactile misses since the animal had too few misses during these trials.

This study demonstrates several aspects concerning the effects of attention on neurons in SI cortex. Besides showing that the neural response of a large fraction of neurons in SI are affected by attention, it demonstrates that attention has large effects on the responses of neurons and that these effects cannot be explained by simple changes in neuronal gain. For example, the response to the target letter L (Figure 8.4B) became weaker when the same animal performed the tactile task and almost non-existent when the same letter passed over the receptive field and the animal was performing the visual task. This suggests that attention modifies not only the sensitivity of the neuron but also the spatiotemporal form of the response. Second, this study demonstrates that attention is turned on and off quickly. In Figure 8.4, the small arrows represent the mean time that the animal turned the response switch. These responses suggest that the effects of attention on the neural responses are engaged and disengaged within 100 msec, which indicates that the effects of attention may be dynamically changing throughout the recording period. A third finding from this study was that neurons in SI only showed enhanced responses during the tactile task; however, more recent studies described below have shown that neurons in SI have both enhanced and suppressed responses.

Recently, Burton and Sinclair[4] reported results from neurons in SI cortex in animals trained to detect a change in vibratory amplitude. In agreement with the findings by Hsiao et al.,[24] they reported that about 50% of the neurons in SI are affected by the animal's focus of attention when switching between two modalities. In their study, the animal's focus of attention was directed to either one location from mirror image sites on both hands or to an auditory tone. They found that in trials where the animal was cued to respond to the contralateral hand, some neurons in SI showed enhanced responses while others showed suppressed responses.

8.3.2.3 Memory in SI

There is evidence that neurons in SI may play a role in the memory of tactile stimuli. Zhou et al.[50,51] trained animals to perform either a tactile-tactile matching task or a cross-modal visual-tactile matching task. In these studies, animals were given either a tactile cue (e.g., reach out and touch a bar containing horizontal bars) or a visual cue (e.g., display with horizontal bars). Following a short delay, the animal was trained to touch a bar that either contained a matching (horizontal bars) or non-matching (vertical bars) stimulus. They found that, during the delay period between the stimuli, many neurons in all four areas of SI cortex showed elevated rate changes. While these rate changes could be attributed to attentional effects, they are also indicative of neurons that may play a role in the short-term memory of tactile stimuli.

8.3.2.4 SII Cortex

There have been several attention studies in SII cortex and all of them have shown that selective attention has profound effects on the responses of neurons. In the initial studies, Poranen and Hyvarinen[39] trained animals to detect the cessation of a vibration and reported that, while attention had minimal effects on the responses of neurons in SI, all of the neurons in SII were greatly affected. Similarly, Hsiao et al.,[24] recording from animals performing the same task described earlier, reported that more than 80% of the neurons in SII were affected by attention and that the attention to the tactile stimulus caused both increases (58%) and decreases (22%) in the firing rates of SII neurons. In general, the effect of attention on the responses of neurons was similar in form but greater in magnitude to what was seen in SI cortex. Examples of four neurons in SII are shown in Figure 8.3. The neurons shown in Figures 8.3A 8.3D, and 8.3E showed enhanced responses during the tactile tasks while the neuron shown in Figure 8.3C showed suppressed responses. Figure 8.4A shows that the spatial structure of the responses to the target letter X during "hits" and "misses" is not a simple multiplicative gain from the response recorded while the animal performed the visual task. These results suggest that attention alters the responses in more complex ways than by simply changing the "neural gain" of the signal.

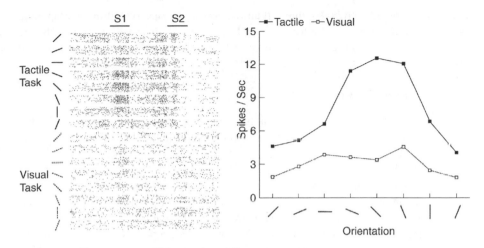

FIGURE 8.5 Effects of attention on orientation tuning curves. Response of a neuron in SII to oriented bars while performing a tactile and visual task. During the tactile task, the bar indented the skin at one of eight randomly chosen orientations (S1) for 500 msec. After a 1 second delay, the bar indented the skin for an additional 500 msec at either the same orientation or at the orthogonal orientation (S2). The animal's task was to pull a switch with its foot if the orientations of the two bars were the same or to push the switch if the orientations were different. Raster plots on the left show the responses sorted according to bar orientation and task. The curves to the right show the mean impulse rates as a function of orientation. (Hsiao, unpublished data.)

Recently, Hsiao and his colleagues found that about 30% of the neurons in SII cortex show orientation-tuned responses and that the tuning is greatly affected by attention.[17] In their studies, they trained animals to perform a visual task and a match-to-sample tactile orientation discrimination task. In the tactile task, the animal was presented with a small bar at one of eight orientations on the distal finger pad for 500 msec. Then, one second later, the animal was presented with the bar at either the same orientation or at the orthogonal orientation. The animal's task was to push a switch with its foot if the orientations of the two bars were the same and to pull the switch if the orientation of the two bars differed. In the visual task, the animal pulled the switch with its foot when a white square dimmed on a video screen. Figure 8.5 shows an example of how attention affects the tuning properties of neurons in SII cortex. In this example, the neuron shows minimal orientation tuning during the visual task but shows strong tuning during the tactile task. This figure also shows that neurons in SII exhibit attention modulated responses during the delay period — suggesting that SII may also play a role in tactile memory. For this neuron, the delay period response shows the same tuning characteristics as the stimulus-driven responses.

In another study of attention in SII cortex, Burton et al.[5] studied animals performing a vibration detection task similar to the one described earlier. They found that 45% of the neurons in SII cortex were affected by the animal's focus of attention.

Single neurons showed both enhanced and suppressed responses with the responses being generally suppressed during the early phases of a trial and enhanced during the late phases of a trial. These findings suggest that attention may improve the signal-to-noise ratio of neurons in SII cortex.

8.3.2.5 Prefrontal Cortex

Recently, Romo et al.[42] have shown that neurons in the prefrontal cortex are affected by the animal's focus of attention. In their study, they trained the animals to perform a vibrotactile match-to-sample task in which the animals were presented with two sequential vibrations to the finger pads. The animal was required to press a key indicating whether the second vibration was at a higher or lower frequency than the first. They found that neurons showed a continuous discharge during the delay period (up to 6 seconds) and that 65% of the neurons had discharge rates during the delay period that varied as a monotonic function of the frequency of the base stimulus. These results suggest that neurons in prefrontal cortex play a role in tactile working memory.

8.3.3 MECHANISMS OF ATTENTION

While all studies of attention that we reviewed demonstrated that neurons either increase or decrease their firing rates with attention, a recent study has shown that attention also modulates the temporal firing characteristics of neurons. Steinmetz et al. (2000) recorded, using a multiple electrode array, from three animals trained to perform three different tactile tasks and a visual task. During the experiments, the animal's focus of attention was switched back and forth every few minutes between the tactile and visual tasks. They found that a large fraction of the neuron pairs in SII cortex fired synchronously and, more importantly, found that the degree of synchronous firing changed with the animal's focus of attention (Figure 8.6). They reported that, on average, about 17% of the neuron pairs showed changes in synchrony, when the animal's focus of attention was switched from the visual task to the tactile task. About 80% of these neuron pairs showed increases in synchrony while the remaining 20% showed decreases in synchrony. In addition, they found that the percentage of neurons that showed changes in synchrony differed between the three animals; about 35% of the neuron pairs from the animal that performed the most difficult task showed changes, whereas only 9% of the neuron pairs were affected in the animal that performed the easiest task.

These results indicate that increasing the degree of synchrony between neurons may be a mechanism that underlies selective attention. Increasing the degree of synchronous firing between pairs or populations of neurons produces larger excitatory post-synaptic potentials in target neurons and is, therefore, more effective at driving target neurons. This results in the message contained in the population of neurons that are firing more synchronously to be selected for further processing at higher stages and the message contained in neurons firing asynchronously to be diminished.

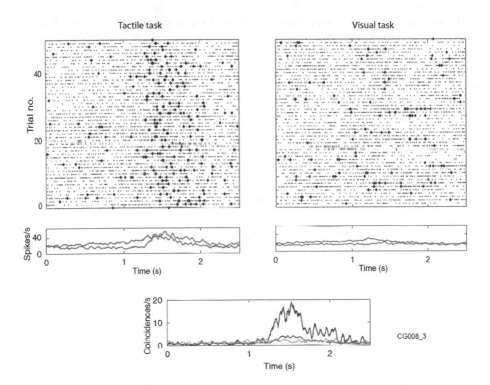

FIGURE 8.6 (See Color Figure 8.6 in color insert.) Effects of attention on synchrony. Response of a typical neuron pair (red and green) in monkey cortical area SII. The response rasters are triggered at the onsets of 50 tactile stimulus periods while the monkey performs the tactile letter discrimination task and the visual dim detection task. Each row in the top two rasters represents one stimulus period, 2.5 s long, corresponding to the presentation of one letter. Red and green dots represent the action potentials of the two neurons. Peristimulus time histograms are shown below each raster plot with corresponding colors. Synchronous events, defined as spikes from each neuron within 2.5 ms of each other, are represented as blue diamonds. This figure shows that the number of synchronous events is much higher when attention is directed towards the tactile stimuli. The number of coincident events is shown in the bottom figure — blue curve — tactile task, red curve — visual task, violet curve — coincidences expected by chance. Adapted from Steinmetz, P.N., Roy, A., Fitzgerald, P.J., Hsiao, S.S., Johnson, K.O., and Niebur, E., Nature, 404: 187–190, 2000. With permission.

8.4 ATTENTION AND IMAGING IN HUMANS

Recently, there have been a number of studies that have used imaging techniques to determine what areas of the human brain are affected by tactile selective attention. Below, we briefly review the results from several of those studies.

In agreement with the neurophysiology studies discussed earlier, imaging studies on SI cortex suggest that attention plays a small but significant role in tactile perception. The effects have been shown using four different imaging techniques (PET, fMRI, magnetic source imaging (MEG), and somatosensory evoked potentials [SSEP]) and a variety of tactile stimuli that include simple touch, vibration, electrical stimuli, and roughness patterns.[2,3,12,20,26,32,37,38] The basic finding from all of these studies is that there are modest changes (increases or decreases) in the neural images when subjects are required to perform a task that requires them to attend to the specific stimuli. The changes in the strength of the neural image are attributed to changes in blood flow and thereby to the corresponding neural activity related to selective attention for PET and fMRI, and to changes in the magnetic and electrical fields related to attention in the MEG and SSEP studies. These results, however, are not universal since three other imaging studies found little or no modulation of activity in SI cortex with attention.[19,22,23] There are several reasons why these studies may not have found attention modulated responses. One is that two of these studies did not have the subjects perform an active control task (e.g., attend to a distracting visual task), but instead compared the responses to periods when the subjects sat quietly.[22,23] Another possibility is that the degree of modulation in SI is small and the results may not have been large enough to be significant.

Although most studies show an increase in blood flow in SI with attention (see Reference 3 as an example), others show that there may be a decrease. A PET study by Drevets et al.[12] reported that the changes in activation of SI were due to decreases in blood flow outside the attended location. For example, attending to the toe resulted in a decrease in blood flow in the area represented by the hand and face. Based on their results, they hypothesized that one of the mechanisms of attention is to suppress background activity.

The effects of attention on SII cortex are greater and much more pronounced than on SI cortex[3,19] agreeing with the large changes in responses that are observed in neurophysiological studies in attending animals. Specifically, the imaging studies show that the effects of attention on SII are more consistent irrespective of the tactile tasks or the method that was used to measure brain activity.[1,3,12,18–20,22,26,32,33] These studies also show that the attentional effects of SII are bilateral with the strength of the signals being larger in the hemisphere contralateral to the stimulus.[3,22,33] In addition, the magnitude of the effects in SII were independent of whether attention was divided between hands or was directed toward a single hand.[26]

Several other cortical areas have also been shown to be affected by tactile attention. These areas include the superior parietal area, anterior cingulate, and frontal and insular cortex. In particular, the frontal cortex and superior parietal area, corresponding to areas 5 and/or 7, have been shown in several studies to be affected by attention in humans performing touch detection, roughness discrimination, vibratory detection, and 3-D shape discrimination tasks.[1,3,32,38] In addition, Burton et al.[3] reported that the cingulate is activated when subjects are asked to perform difficult roughness discrimination tasks. Finally, bilateral activation of the insula has been shown in two different studies.[3,36] Surprisingly, these studies also show that the strength of activation in the insula is lower than for both SI and SII cortex, which suggests that neurons in the insula may not be driven well during selective attention tasks.

8.5 NEGLECT SYNDROME IN HUMANS

Studies in patients with unilateral lesions of left and, particularly, right parietal cortex suffer from an attentional disorder in which they neglect or deny the existence of the body part that is contralateral to the injured hemisphere.[36] The deficits can be severe, with patients refusing to move their limbs in the contralateral space. Patients with a mild form of hemi-neglect tend to have normal sensory functions that allow them to localize and perceive tactile stimuli on the affected side. However, if they are touched simultaneously on both sides of the body, they are unable to localize or perceive the stimuli on the side contralateral to the lesion. This effect, which is called tactile extinction, is attributed to subjects being unable to disengage attention from the ipsilateral side of the body and to re-engage attention on the competing or contralateral tactile stimulus.[35] The mechanisms of the parietal cortex in redirecting attention in the tactile system may be similar to the mechanisms used by parietal cortical neurons in redirecting visual attention.[45]

Tactile extinction appears to be based on an extrapersonal frame of reference rather than on somatotopic organization. Patients tend to neglect stimuli on the side of space contralateral to the injured hemisphere.[35] However, the area of neglect changes depending on where attention is being focused. For example, subjects with right hemisphere injury will neglect tactile stimuli on the left arm. However, if attention is focused on the right arm, they neglect the left part of the right arm, independent of whether the palm is facing up or facing down.[35] Patients with severe hemi-neglect tend to ignore all somatosensory information from the opposite side of the body.

8.6 SUMMARY AND CONCLUSIONS

There is now strong evidence that selective attention is an emergent property of the mind. Neurophysiological studies in monkeys and imaging studies in humans show that the effects of attention can be observed at all levels of processing in the central nervous system, and that the percentage of neurons activated by attention becomes higher and the effects become stronger as one moves more centrally. The studies show that attention has minimal effects on neurons in the brain stem and thalamus, modest effects on neurons in SI cortex, and very large effects on neurons in SII cortex. In SII cortex, more than 80% of the neurons are affected by selective attention.

Psychophysical studies have characterized attention effectively. These studies have demonstrated several important findings. First, they confirm our intuition that our sensory systems have a limited capacity for processing information. Second, they reveal that material properties of surfaces and vibration tend to be processed pre-attentively. Third, they show that selective attention functions like a spotlight that is under cognitive control. This attentional spotlight can be directed at will to a specific location on the body and can be focused down to a minimum size; stimuli that are within the focus are enhanced and stimuli outside the focus are ignored. Whether attention can be further focused down to specific stimulus attributes (i.e., attend to the roughness and not the vibration within the spotlight) remains to be seen.

While the mechanisms of attention are not understood, there are clues to suggest that directing attention is controlled by neurons in the posterior parietal cortex and that the focusing of attention is accomplished by changing the degree of synchronous firing between neurons. The evidence for the posterior parietal cortex being important for attentional control comes from studies on patients with damaged cortex, as these patients are unable to direct their attention to specific tactile (or visual) stimuli in the space contralateral to their lesion. The evidence for synchronous firing being the neural correlate of selective attention is that changing synchrony has two potential effects. One is to make those neurons that are firing in synchrony stand out from the background neural noise. The second is that synchrony increases the size of the EPSP in downstream neurons and therefore enhances the message conveyed by those neurons at higher structures in the brain.

ACKNOWLEDGMENTS

We would like to thank Dr. Dianne Pawluk for her insightful comments and John Lane and Sapna Prasad for technical assistance. Supported by NIH R01-NS34086 and Robert Wood Johnson Foundation, grant number: 037218.

REFERENCES

1. Binkofski F, Buccino G, Stephan KM, Rizzolatti G, Seitz RJ, and Freund HJ (1999) A parieto-premotor network for object manipulation: evidence from neuroimaging. Exp Brain Res 128: 210-213.
2. Buchner H, Reinartz U, Waberski TD, Gobbele R, Noppeney U, and Scherg M (1999) Sustained attention modulates the immediate effect of de-afferentation on the cortical representation of the digits: source localization of somatosensory evoked potentials in humans. Neurosci Lett 260: 57-60.
3. Burton H, Abend NS, MacLeod AM, Sinclair RJ, Snyder AZ, and Raichle ME (1999) Tactile attention tasks enhance activation in somatosensory regions of parietal cortex: a positron emission tomography study. Cereb Cortex 9: 662-674.
4. Burton H and Sinclair RJ (2000) Tactile-spatial and cross-modal attention effects in the primary somatosensory cortical areas 3b and 1–2 of rhesus monkeys. Somatosens Mot Res 17: 213-228.
5. Burton H, Sinclair RJ, Hong SY, Pruett JR, and Whang KC (1997b) Tactile-spatial and cross-modal attention effects in the second somatosensory and 7b cortical areas of rhesus monkeys. Somatosens Mot Res 14: 237-267.
6. Bushnell MC, Duncan GH, Dubner R, Fang L, and He LF (1984) Activity of trigeminothalamic neurons in medullary dorsal horn of awake monkeys trained in a thermal discrimination task. J Neurophysiol 52: 170-187.
7. Bushnell MC, Duncan GH, Dubner R, Jones RL, and Maixner W (1985) Attentional influences on noxious and innocuous cutaneous heat detection in humans and monkeys. J Neurosci 5: 1103-1110.
8. Carli G, LaMotte RH, and Mountcastle VB (1971) A simultaneous study of somatic sensory behavior and the activity of somatic sensory cortical neurons. Fed Proc 30: 664.
9. Craig JC (1985) Attending to two fingers: two hands are better than one. Percept Psychophys 38: 496-511.
10. Crick F (1984) Function of the thalamic reticular complex: The searchlight hypothesis. Proc Natl Acad Sci U S A 819: 4586-4590.
11. Dorsch, AK, Hsiao, SS, Johnson, KO, and Yoshioka, T (2001) Tactile attention: Subjective magnitude estimates of roughness using one or two fingers. Society for Neuroscience Abstracts.

12. Drevets WC, Burton H, Videen TO, Snyder AZ, Simpson JR, Jr., and Raichle ME (1995) Blood flow changes in human somatosensory cortex during anticipated stimulation. Nature 373: 249-252.

13. Driver J and Grossenbacher PG (1996) Multimodal Spatial constraints on Tactile Selective Attention. In: Attention and Performance XVI: Information Integration in Perception and Communication (Attention and Performance) (Inui T and McClelland J, eds), Cambridge, Mass: MIT Press.

14. Driver J and Spence C (1998) Cross-modal links in spatial attention. Philos Trans R Soc Lond B Biol Sci 353: 1319-1331.

15. Duncan GH, Bushnell MC, Bates R, and Dubner R (1987) Task-related responses of monkey medullary dorsal horn neurons. J Neurophysiol 57: 289-310.

16. Evans PM and Craig JC (1991) Tactile attention and the perception of moving tactile stimuli. Percept Psychophys 49: 355-364.

17. Fitzgerald, P.J, Lane, JW, and Hsiao, SS (1998) Attentional Effects in Somatosensory Cortex During an Orientation Discrimination Task. Society for Neuroscience Abstracts 25.

18. Francis ST, Kelly EF, Bowtell R, Dunseath WJ, Folger SE, and McGlone F (2000) fMRI of the responses to vibratory stimulation of digit tips. Neuroimage 11: 188-202.

19. Hamalainen H, Hiltunen J, and Titievskaja I (2000) fMRI activations of SI and SII corticces during tactile stimulation depend on attention. Neuroreport 11: 1673-1676.

20. Hansson T and Brismar T (1999) Tactile stimulation of the hand causes bilateral cortical activation: a functional magnetic resonance study in humans. Neurosci Lett 271: 29-32.

21. Hayes RL, Dubner R, and Hoffman DS (1981) Neuronal activity in medullary dorsal horn of awake monkeys trained in a thermal discrimination task. II. Behavioral modulation of responses to thermal and mechanical stimuli. J Neurophysiol 46: 428-443.

22. Hoechstetter K, Rupp A, Meinck HM, Weckesser D, Bornfleth H, Stippich C, Berg P, and Scherg M (2000) Magnetic source imaging of tactile input shows task-independent attention effects in SII. Neuroreport 11: 2461-2465.

23. Hsiao SS, Johnson KO, and Yoshioka T (2001) Neural mechanisms of tactile perception. In: Comprehensive Handbook of Psychology: Volume 3: Biological Psychology (Gallagher M. and Nelson RJ, eds.), John Wiley and Sons, Inc.

24. Hsiao SS, O'Shaughnessy DM, and Johnson KO (1993) Effects of selective attention of spatial form processing in monkey primary and secondary somatosensory cortex. J Neurophysiol 70: 444-447.

25. Hyvärinen J, Poranen A, and Jokinen Y (1980) Influence of attentive behavior on neuronal responses to vibration in primary somatosensory cortex of the monkey. J Neurophysiol 43: 870-882.

26. Johansen-Berg H, Christensen V, Woolrich M, and Matthews PM (2000) Attention to touch modulates activity in both primary and secondary somatosensory areas. Neuroreport 11: 1237-1241.

27. Johnson KO, Yoshioka T, and Vega-Bermudez F (2001) Sensory functions of the cutaneous mechanoreceptive afferents innervating the hand. J Clin Neurophysiology 17(6) 539-558.

28. Klatzky RL, Lederman SJ, and O'Neil CO (1996) Haptic object processing I: Early perceptual features. In: Somesthesis and the Neurobiology of the Somatosensory Cortex (Franzén O, Johansson RS, and Terenius L, eds), pp 147-152. Basel: Birkhäuser Verlag.

29. Lakatos S and Shepard RN (1997) Time-distance relations in shifting attention between locations on one's body. Percept Psychophys 59: 557-566.

30. Lederman SJ and Klatzky RL (1997) Relative availability of surface and object properties during early haptic processing. J Exp Psychol 23: 1-28.
31. Loomis JM, Klatzky RL, and Lederman SJ (1991) Similarity of tactual and visual picture recognition with limited field of view. Perception 20: 167-177.
32. Macaluso E, Frith CD, and Driver J (2000) Modulation of human visual cortex by cross-modal spatial attention. Sci 289: 1206-1208.
33. Mima T, Nagamine T, Nakamura K, and Shibasaki H (1998) Attention modulates both primary and second somatosensory cortical activities in humans: a magnetoencephalographic study. J Neurophysiol 80: 2215-2221.
34. Morrow TJ and Casey KL (2000) Attention-related, cross-modality modulation of somatosensory neurons in primate ventrobasal (VB) thalamus. Somatosens Mot Res 17: 133-144.
35. Moscovitch M and Behrmann M (1994) Coding of spatial information in the somatosensory system: Evidence from patients with neglect following parietal lobe damage. J Cogn Neurosci 6: 151-155.
36. Mountcastle VB (1978) Some neural mechanisms for directed attention. In: Cerebral Correlates of Conscious Experience (Buser PA and Rougeul-Buser A, eds), pp 37-51. Elsevier/North-Holland Biomedical Press.
37. Noppeney U, Waberski TD, Gobbele R, and Buchner H (1999) Spatial attention modulates the cortical somatosensory representation of the digits in humans. Neuroreport 10: 3137-3141.
38. Pardo JV, Fox PT, and Raichle ME (1991) Localization of a human system for sustained attention by positron emission tomography. Nature 349: 61-64.
39. Poranen A and Hyvärinen J (1982) Effects of attention on multiunit responses to vibration in the somatosensory regions of the monkey's brain. Electroencephalogr Clin Neurophysiol 53: 525-537.
40. Posner MI (1978) Chronometric explorations of mind. Hillsdale, NJ: Lawrence Erlbaum.
41. Post LJ and Chapman CE (1991) The effects of cross-modal manipulations of attention on the detection of vibrotactile stimuli in humans. Somatosens Mot Res 8(2): 149-157.
42. Romo R, Brody CD, Hernández A, and Lemus L (1999) Neuronal correlates of parametric working memory in the prefrontal cortex. Nature 399: 470-473.
43. Sathian K and Burton H (1991) The role of spatially selective attention in the tactile perception of texture. Percept Psychophys 50: 237-248.
44. Shiffrin RM, Craig JC, and Cohen E (1973) On the degree of attention and capacity limitation in tactile processing. Percept Psychophys 13: 328-336.
45. Steinmetz MA and Constantinidis C (1995) Neurophysiological evidence for a role of posterior parietal cortex in redirecting visual attention. Cereb Cortex 5: 448-456.
46. Steinmetz PN, Roy A, Fitzgerald PJ, Hsiao SS, Johnson KO, and Niebur E (2000) Attention modulates synchronized neuronal firing in primate somatosensory cortex. Nature 404: 187-190.
47. Treisman A and Gormican S (1988) Feature analysis in early vision: evidence from search asymmetries. Psychol Rev 95: 15-48.
48. Tremblay N, Bushnell MC, and Duncan GH (1993) Thalamic VPM nucleus in the behaving monkey. II. Response to air-puff stimulation during discrimination and attention tasks. J Neurophysiol 69: 753-763.
49. Whang KC, Burton H, and Shulman GL (1991) Selective attention in vibrotactile tasks: Detecting the presence and absence of amplitude change. Percept Psychophys 50: 157-165.

50. Zhou YD and Fuster JM (1996) Mnemonic neuronal activity in somatosensory cortex. Proc Natl Acad Sci U S A 93: 10533-10537.
51. Zhou YD and Fuster JM (1997) Neuronal activity of somatosensory cortex in a cross-modal (visuo-haptic) memory task. Exp Brain Res 116: 551-555.
52. Zompa IC and Chapman CE (1995) Effects of cross-modal manipulations of attention on the ability of human subjects to discriminate changes in texture. Somatosens Mot Res 12: 87-102.

9 New Applications of Digital Video Technology for Neurophysiological Studies of Hand Function

Daniel J. Debowy, K. Srinivasa Babu,
Edward H. Hu, Michelle Natiello, Shari Reitzen,
Maria Chu, Jill Sakai, and Esther P. Gardner

CONTENTS

9.1 INTRODUCTION

The brain's own body image is normally expressed in both a cognitive and pragmatic manner. By projecting the indentation patterns impressed upon the skin onto external objects that touch its surface, the tactile sense in the hand serves a cognitive function. Information derived from tactile receptors about an object's intrinsic properties — its size, shape, and surface texture — is used to identify and classify it. However, the hand is not merely a sensor; it is also a skillful motor device that can manipulate objects to achieve a desired goal. A functional body image is essential for planning and executing movements, especially the skilled actions of the hand that distinguish primates from lower species. Cutaneous receptors in both glabrous and hairy skin provide information about the kinematics and posture of the hand, as well as the grip and load forces used during prehension and object manipulation (Edin and Abbs, 1991; Edin and Johansson, 1995; Hulliger et al., 1979; Johansson, 1996; Westling and Johansson, 1987). In order to understand coordinated skilled movements of the hand, we need to examine the central representation and dynamics of the pragmatic functions monitored by the tactile system.

Tactile information processing, like that of vision, includes a prominent spatial component. Whether we acquire information by active touch, or by passive stimulation, it is important that we define where and how the skin is stimulated. The spatial dimension is of particular significance during skilled actions of the hand, as sensory feedback from the skin is used to guide hand behavior. Visualization of the hand dynamics would therefore provide essential information for interpreting the neurophysiological correlates of manipulative actions.

Studies in our laboratory have focused in recent years on hand behavior during prehension. In the course of these investigations, we developed various tools using digital video to correlate the hand kinematics with activity of neurons in parietal cortex. Digital multimedia technology has not only revolutionized telecommunications, but has proven to be a useful tool in neuroscience. Digital imaging of motor activity has been linked with neuronal spike trains to analyze neural correlates of both trained and spontaneous behaviors in experimental animals (Debowy et al., 2001; Fenton and Muller, 1996; Gardner et al., 1999; Ro et al., 1998, 2000). In this chapter, we describe how digital video (DV) can be used to visualize the actions of the hand during acquisition and manipulation of objects. We also explain how the video images are linked to the simultaneously acquired spike trains of cortical neurons and how the neurophysiological responses and kinematic actions are quantified.

9.2   WHAT STUDIES CAN DIGITAL VIDEO ENABLE?

In earlier reports, we showed that video camcorders could be used to capture hand kinematics: the actions of the hand when monkeys performed a trained prehension task (Debowy et al., 2001; Gardner et al., 1999; Ro et al., 1998, 2000). Consumer-grade video recorders that incorporated zoom lenses, autofocus, and automatic exposure controls were easily adapted for use in ordinary laboratory lighting conditions and yielded good quality color images (Figure 9.1). Remote control of camcorder record and zoom functions enabled us to place the cameras at a convenient distance from both the animal subject and electrophysiological recording equipment, so that they did not interfere with the normal progress of experiments. These cameras proved particularly useful for close-up examination of the hand by several investigators during recording sessions, allowing groups to observe live action in real-time on a large-screen color-monitor while recording single-unit activity from cortical

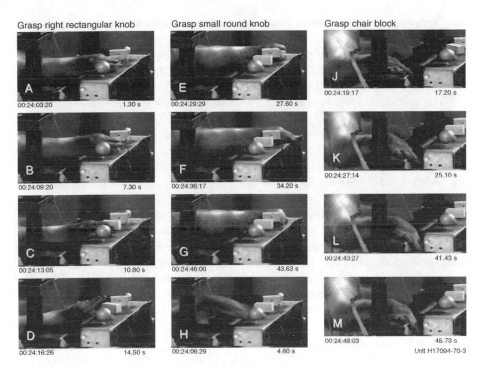

FIGURE 9.1 (See Color Figure 9.1 in color insert.) Digitized Hi-8 images of the hand kinematics during grasp of various objects. These pictures were cropped from full-frame views captured at the peak of bursts flagged in Figure 9.3; time code (left) and time in clip (right) are indicated below each image. Note that in D, the neuron failed to respond because the animal did not properly grasp the rectangular knob, and instead lifted it with the fingers extended.

neurons. In this manner, we could accurately map receptive fields and qualitatively assess whether the neuron under study responded to the behavioral task.

In addition, and most important for these studies, the camcorder also recorded the spike trains of cortical neurons on its audio channels, providing a simple method for linking neurophysiological and behavioral data. The video images and simultaneously recorded spike trains were stored together on videocassettes for subsequent analysis. Although we could play back these videotapes at a later time, simply viewing tapes of the animal and listening to the spike trains (or displaying them on an oscilloscope) yielded little beyond the same qualitative information that could be gleaned during the experiment itself. In order to make direct frame-by-frame quantitative correlations between video images of behavior and spike data, we digitized both sets of recordings.

Digital editing software was used to download these files to a computer and to examine details of the hand kinematics in digitized images displayed on a computer monitor. The hand behavior could be viewed in real time, at high speed, or by selected individual frames. Forward and backward bracketing of sequential frames was particularly useful for visualizing how the hand or the stimulus changed position over time. As each image was labeled during the initial data acquisition with time code specifying the hour, minute, second, and frame number at 30 frames per second (fps), the editing software displayed precise time markers accurate to the 33.3 ms frame interval. An event log of relevant time codes enabled us to identify the onset and duration of specific behaviors whose neural correlates could be analyzed using the software tools described below.

The spike train recorded on the audio channels was digitized together with the video images, providing a set of high-resolution electrophysiological records matched to the video images. These signals are synchronized because each 33.3 ms segment of the audio waveform is labeled with the same time code as the concurrent video image. The action potential waveform is preserved because the spike records are sampled at 32 or 48 kHz with 12- or 16-bit accuracy, respectively. Spikes fired by individual neurons can therefore be distinguished with standard electrophysiological techniques based on the action potential amplitude and duration (Ro et al., 1998); the digitized audio signal can substitute for an analog-to-digital (A/D) converter when capturing unit data. The editing software also displays the spike train audio signal as a strip chart in a separate window that can be advanced and viewed independently. By matching the time codes of the video and audio signals, qualitative observations can be made about the relevance of specific behaviors, or stimulation paradigms to neuronal firing patterns.

Quantification of firing rates enabled two-way analyses of neurophysiological data. We used the spike train as an index to the images, examining the animal's behavior during periods of high or low firing to see what features were shared (burst analysis). Conversely, images of specific behaviors served as alignment points to construct rasters and peristimulus time histograms (PSTHs) to assess strength and reproducibility of firing patterns evoked by these actions. Finally, features of the hand kinematics were abstracted, from the digitized images, and superimposed, providing a time series of the behaviors that accompanied individual spike trains.

9.3 HOW IS DIGITAL VIDEO IMPLEMENTED?

Two major advances in hardware have made the creation of DV movies literally a matter of "plug-and-play:" consumer-grade DV format camcorders and FireWire (IEEE 1394) digital interfaces to computers. Our original implementation of digital video required a dedicated compression–decompression board (codec) installed in the computer to both capture and view digital movies (Ro et al., 1998). Newer DV format camcorders digitize and compress the video images as they are acquired. This has multiple advantages. First, the DV standard compresses video with lower loss and degradation than older hardware–software schemes. DV also improves efficiency of data analysis because any computer equipped with QuickTime could be used for acquisition or analysis of DV files.

The DV format was developed by the international High Definition (HD) Digital VCR consortium to improve video image quality (Bovik, 2000). It achieves excellent image fidelity because the luminance (Y) is sampled at 13.5 MHz, and the color difference signals (R-Y, B-Y), at 3.375 MHz each; the 4:1:1 sample rate is comparable to the Betacam SP standard used for broadcast television (see e.g., Figure 9.2). The video image is further compressed 5:1 to 3.1 MB with a type of intraframe

MPEG-2 sampling that digitizes only the 720 × 480 pixels of "active video," i.e., those pixels that change between frames. Image compression reduces both file size and data storage rates.

DV camcorders also include two or four digital audio tracks that we use to record spike trains or other electrophysiological data such as EMG or EEG. The typical high-end DV camcorder offers two-channel sampling at 48 kHz with 16-bit resolution, and some allow four-channel data acquisition at 12-bit, 32 kHz/channel rates. These fast sampling rates allow continuous, high-fidelity recording of the digitized neural data trace, permitting the use of sophisticated clustering or template matching algorithms for spike separation offline. As the bandwidth of electrophysiological recordings is small compared to the video signal, there is no cost of storing the actual raw spike records. Some DV camcorders have built-in level controls for adjusting the input gain; others require an external adapter (see hardware description, Section 9.6.1).

A final advantage of the DV hardware is loss-less data transfer from the source camcorder to the computer. DV supports direct digital input and output of both video and audio signals using a single FireWire (IEEE 1394 standard) cable. DV images and audio files are recorded on Mini-DV cassettes that store 60 minutes of data. Although these files can also be transferred directly to the computer as they are recorded and saved to disk, we prefer offline downloads to the computer to control file sizes. The FireWire connection allows the host computer monitor to display the video recordings on a large screen during the experiment, even if the data are not immediately stored by its hard disk.

Firewire provides two-way data transfers of both audio and video, allowing the host computer to control operation of the camcorder playback functions, or those of any DV cassette player (such as the Sony DV Video Walkman) with an i.LINK (1394) connector. Device control is implemented with digital editing software such as Premiere (Adobe Systems, Inc.) or Final Cut Pro (Apple Computer). These applications include plug-ins for reading the DV format and for FireWire control of the recorder. Data transfer is implemented by entering the time code of the beginning and end frames of the DV file (in- and out-frames) and then initiating capture. Selected frames are displayed as capture progresses, providing visual feedback to the user. Batch capture of multiple clips from the same source DV cassette automates and speeds up the file transfer protocols.

We found that long sequences of interesting data were best broken up into smaller clips for ease of analysis of the spike train files. Even with compression, 2.5 minutes of DV data yields movie files over 550 MB. It is best to screen the original DV movies on a video monitor before movie capture in order to designate the time code of in- and out-frames for the clip to be analyzed. The complete sequence can be recreated after the spike data are analyzed using the editing software, or selected portions of data from several clips can be linked to illustrate particular response features. Similarly, spike trains from multiple clips can be pooled with the analysis instruments described below.

Detailed records of the time code of specific events logged during the experiment facilitate selection of sequences containing relevant information for off-line capture. Good record keeping is particularly important if multiple cameras are used simul-

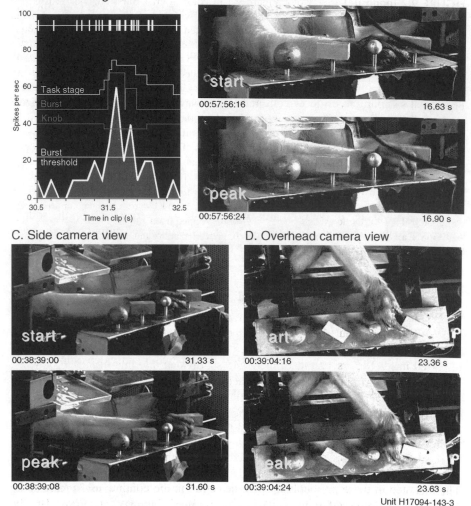

FIGURE 9.2

taneously to provide different views of the subject. One can allow the cameras to roll continuously as electrodes are repositioned to record what might be unexpected, but interesting, behaviors as the session proceeds.

9.4 HOW ARE THE NEUROPHYSIOLOGICAL DATA ANALYZED?

The digitized spike trains or other electrophysiological signals can be analyzed using standard techniques after conversion from QuickTime to the appropriate format used by the analysis software. In our implementation, we take advantage of the fact that neurophysiological data analysis programs such as Superscope II (GW Instruments, Inc.) and Igor Pro (WaveMetrics, Inc.) are able to read AIFF (Audio Interchange

FIGURE 9.2 (opposite) (See Color Figure 9.2 in color insert.) Burst analysis is used for objective evaluation of behaviors recorded with digital video. **A**. Spike train (top trace) and instantaneous firing rate (bottom, blue-and-white filled graph) during 2 s of continuous recording excerpted from a 2.5 minute duration video clip. **White threshold trace**: Burst threshold set 1 SD above the mean rate for the entire 2.5 min clip. **Green burst trace**: mean firing rate during the period of suprathreshold firing; bursts were offset by 48 spikes/s to aid visualization of the raw data. **Yellow task stage trace**: Upward deflections indicate the start of stages 1–4 (approach, contact, grasp, and lift); downward deflections mark the onset of stages 5–8 (hold, lower, relax, and release). **Red knob trace**: downward pulses span the contact through relax stages when the hand interacts with the object; the amplitude is proportional to the object's distance from the medial edge of the shape box. **B-D**. Video frames captured by three DV camcorders at the start of the burst, and at the moment of peak firing; images were cropped to highlight actions of the hand. Data from the three recorders were synchronized by matching the firing rates recorded on their audio channels. Spike trains in A were processed from recordings by the side camera. B and D show images captured with Sony DCR-TRV900 camcorders; images in C were cropped from a wider angle view recorded with the Canon XL1 instrument. Note the improved clarity and detail in all six panels compared to the Hi-8 images in Figure 9.1.

File Format) files. We therefore make a duplicate copy of the QuickTime audio track(s) using the File Export command in the editing software, selecting audio only and AIFF format.* Our protocols use Superscope II, as its built-in functions automatically implement many of the desired analyses. AIFF files from each clip are imported into a Superscope instrument and subdivided into one-minute fragments, which are analyzed independently to facilitate interactive spike separation. This improves processing efficiency when setting spike thresholds or cluster boundaries, as a single channel sampled at 32 kHz yields nearly 4 MB of data per minute. The entire data trace is reconstructed afterwards by linking sequential data fragments.

9.4.1 SPIKE SEPARATION

The first step in these protocols is spike detection in the continuous 32 kHz audio signal. Algorithms used for spike recognition are detailed in Ro et al. (1998). Briefly, positive and negative going thresholds are applied to the electrophysiological data trace, and it is scanned using the Pulse Analysis function in Superscope. An interactive clustering method segregates spikes of individual neurons from each other, and from noise transients, based on the action potential peak-to-peak amplitude and rise time. Spike time stamps for each isolated cell are represented in a continuous wave as 1V, 200 µs pulses occurring at the corresponding time in the data sample; separate pulse waves are synthesized for each spike detected. This standardization of the real data facilitates quantitative analyses in which the time base is linked to a behavioral event, such as a stimulus, rather than to real time in the recording session. It also reduces the file size, in our case by a factor of 4.

* Adobe Premiere (v. 5) only supports AIFF file export in Mac OS; Final Cut Pro (v. 2) supports both Mac OS and Windows for audio file export.

A list of spike time stamps is also exported to a Superscope journal and saved as a text file, which can be opened in spreadsheets such as Excel. This list is used to display continuous spike records from the clip (top traces in Figures 9.2A and 9.3), and to compute instantaneous firing rates from the reciprocal interspike interval. The latter, or the binned frequency plots compiled by burst analysis, can be matched to a duplicate copy of the spike train acquired online by standard A/D converters. This permits temporal linkage of the behavioral responses captured in video images to sensor data from stimulators, the manipulandum, EMG recordings, or spike trains from multiple electrode arrays.

9.4.2 BURST ANALYSIS

Objective data analysis is facilitated using a technique we call "burst analysis" (Babu et al., 2000; Debowy et al., 2000). Burst analysis provides an objective method for correlating firing patterns with activity as it relies upon the behavior of the cell under analysis as an alignment metric rather than subjective standardization of the animal's behavior. It is implemented by parsing the entire spike train of a clip into 100 ms bins to calculate instantaneous firing rates. The algorithm uses the Superscope Time Histogram function [TimeHisto (d1, threshold, bins)], where $d1$ specifies the name assigned to the standardized pulse wave depicting the spike time stamps, *threshold* is a value between 0 and 1V for detecting spike pulses, and *bins* specifies the computed number of bins to be returned (the total duration of the pulse wave divided by the desired bin width). The TimeHisto function returns a wave containing the total threshold crossings per bin, which are then converted to instantaneous firing rates by dividing each value by the bin width measured in seconds.

Blue-and-white graphs in Figures 9.2A and 9.3 display the measured instantaneous firing rates as a function of time in the clip. Bursts of high activity in these graphs are detected using the Pulse Analysis function in Superscope (Debowy et al. 2000). Mean firing rates for the entire 2–3 minute clip, and the standard deviation, are computed from the binned firing rate values and serve as thresholds (white trace, Figure 9.2A) for computerized scans of the firing rate graph. The total spike count during each burst is stored together with markers designating its onset, peak, and end times; spike counts per burst are converted to mean firing rates by dividing them by the burst duration. The green trace in Figures 9.2A and 9.3 designates the burst amplitude and time course when the threshold was set 1 SD above the mean rate, and provides a simplified representation of firing patterns superimposed on the average firing rate graphs.

Spike bursts 1 or 2 SD above the mean are selected for further analysis of the video records. Time markers designating the onset, peak, and end times of the burst are translated into video time code to capture the matching digitized images of the animal's behavior. In the example shown in Figure 9.2, the burst onset coincided with hand contact with the left rectangular knob when the animal performed a prehension task; activity was maximum 267 ms (8 frames) later when the object was fully grasped and lift was initiated. Views from the three different cameras were synchronized by recording responses from the same microelectrode on all three sets of audio channels, and aligning the bursts in the spike trains. Once the time codes

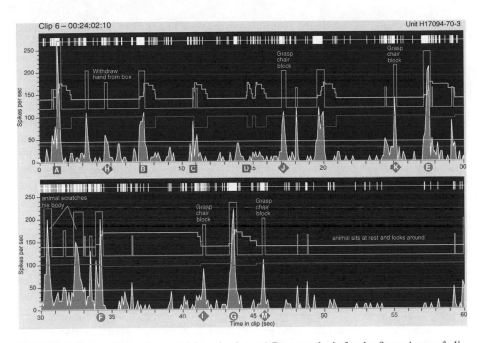

FIGURE 9.3 (See Color Figure 9.3 in color insert.) Burst analysis for the first minute of clip 6 from a neuron recorded in area 5 of PPC; same format as Figure 9.2A. Firing rates in the burst trace were offset by 120 spikes/s to aid visualization of the raw data. Symbols below the abscissa refer to the matching images in Figure 9.1 captured at the peak of these bursts. Bursts are correlated with grasping movements when the fingers flexed. **A-D.** Prehension of the right rectangular knob; note the difference in grip styles used in A-C, which evoked vigorous responses, and in D when no burst was detected. **E-G.** Prehension of the small round knob. **H.** Withdrawal of the hand from the box evoked a weak burst. **J-M.** Spontaneous grasp of a structural block on the chair during rest intervals evoked smaller amplitude, shorter duration bursts than grasp of the knobs tested in the task.

of continuous recordings were synchronized, analyses performed for one set of clips could be matched to views obtained from the other cameras.

The usefulness of the burst analysis technique for comparing responses to different behaviors is illustrated in Figure 9.3. These continuous records show the first minute of recording from another clip derived from our studies of prehension behaviors in monkeys. In order to study prehension in a reproducible manner, we trained the animals to perform a task in which objects are grasped and manipulated in response to visual cues. The objects tested were rectangular, round, and cylindrical knobs that protruded from a box placed directly before the animal ("shape box," Figures 9.1 and 9.2). The animal was reinforced for grasping a visually-cued knob, and lifting it vertically, 5–8 mm, until an upper stop was contacted; the knob had to be returned to the lowest position before the next trial was initiated. We divided the task into 8 stages — 1: Approach. 2: Contact. 3: Grasp. 4: Lift. 5: Hold. 6: Lower. 7: Relax. 8: Release. Stages 1–3 occurred during object acquisition; stages 4 and 5, during manipulation; and stages 6–8, upon relaxation and release of grasp. Start times of the 8 task stages were determined by visual inspection of the digitized movies; the frame time code for each stage was stored in a spreadsheet and converted to real time in the clip. The onset of stages 1–4 is represented in Figures 9.2A and 9.3 by upward deflections of the yellow trace; downward deflections mark the start of stages 5–8.

Statistical parsing of the spike train in Figure 9.3 identified consistent patterns of hand action, which are shown in Figure 9.1. This neuron appeared to sense the application of grasp by the hand to a variety of objects. The largest bursts occurred when the animal grasped one of the knobs while performing the prehension task (A, B, C, E, F, and G in Figures 9.1 and 9.3). In each case, the animal enclosed the test object tightly between the fingers and palm; note the similarity of hand posture used to grasp each object. The superimposed graphic representations of the task stages (yellow trace) and knob location (red trace) indicate that peak activity on these trials occurred during the grasp and/or lift stages. Firing rates were higher if the knob was approached from another location than if it was regrasped. In addition, we noted trials in which no significant change in activity occurred (e.g., Figure 9.3D). Visual inspection of the corresponding video images showed that the animal had in fact satisfied the task requirement to lift the specified object, but instead of grasping it, had deflected the knob upward from below, pushing it with extended fingertips (Figure 9.1D).

A set of narrower bursts characterized periods when the animal spontaneously grasped a structural block on the chair assembly while resting between trials (J, K, L, and M), or scratched its body. These behaviors were characterized by active flexion of the fingers. In contrast, the cell fired weakly or was silent during the hold stage of the task, when the hand remained immobile while maintaining grip force (first downward deflection in the yellow trace), or when the animal sat quietly with the hand at rest while looking around the laboratory (interval 46–59 s).

9.4.3 SPIKE RASTERS AND PSTHS

Another way to evaluate neuronal firing patterns is to select a particular behavior, or set of behaviors, and use it to align the spike trains from repeated trials in order to construct spike rasters and PSTHs. We supplement the voltage transients normally used as trigger signals for such analyses with temporal markers created in software from the time code of behaviors logged during visual inspection of the video images. For example, when analyzing responses recorded during the prehension task, we used the time code at the initial hand contact with the object, or the stabilization of grasp, to set markers on the spike trace bracketing a 2 or 4 s interval. Time stamps of firing during this period were measured as detailed in Ro et al., (1998) to construct raster displays; responses were also binned for statistical analyses and PSTHs.

Figure 9.4 illustrates an example of a raster from the prehension task compiled from the spike trains shown in Figure 9.3 (Trials 1–9), plus the subsequent 15 responses recorded later in the same clip. Here, we display activity during a 4 s interval bracketing hand contact with the knobs; the colored markers indicate the timing of the task stages. As noted in the burst analysis, the firing rate of the cell increased abruptly as approach began (golden trace), regardless of the knob tested. Activity peaked following contact, as the grasp was secured (magenta) and lift started

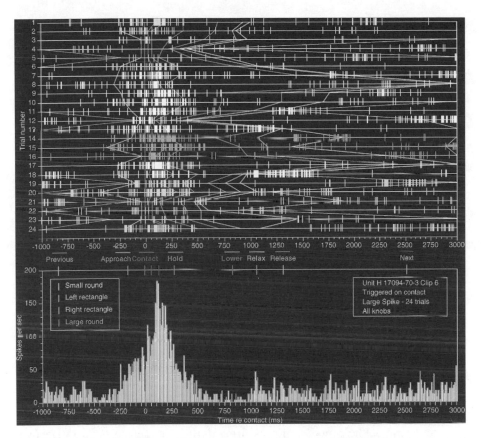

FIGURE 9.4 (See Color Figure 9.4 in color insert.) Rasters and PSTHs aligned to contact for an area 5 neuron. Trials 1–9 occurred during the first minute of the clip, and are shown in Figures 9.1 and 9.3; the remainder occurred in the subsequent 1.5 min. Spikes have been color-coded by the knob tested (see key); responses were similar for the four knobs tested. Color-coded lines spanning the raster mark the onset times of the task stages on each trial; the unlabelled magenta and dark blue lines indicate the grasp and lift stages, respectively. The PSTH averaged responses from all of these trials (binwidth = 10 ms); colored vertical bars mark the mean onset times of the task stages relative to contact in this clip.

(dark blue). Firing dropped precipitously at the onset of holding (light blue), and did not resume until after grasp was released (cyan). Significant activity was observed immediately upon release only when it was followed by approach toward another object.

Although this neuron responded to tactile stimulation of the hand, the onset of activity 250 ms before contact suggests a role in motor planning of grasping. A further increase in firing rates during grasping indicates that tactile cues relevant to task performance were enhanced during active acquisition of objects, while irrelevant stimuli may be suppressed by hand movements (gating studies reviewed in Chapman et al., 1996). The sensory responses of cortical neurons illustrated in Figure 9.4 may match expected and actual sensations to guide subsequent hand actions.

Raster displays provide a higher resolution view of the spike train than that obtained by burst analysis, because the data are spread out over a longer period. However, they shed little light on the activity in the intertrial interval. In addition, because the time base is fixed in duration, one often sees responses to the preceding or subsequent trials at the margins of the displays (previous and next markers, respectively, in Figure 9.4). Thus, there are tradeoffs between the two methods of data presentation that should be balanced by examining firing patterns with both techniques.

9.4.4 STATISTICAL ANALYSES OF FIRING RATES

The original raster-PSTH instrument described in our earlier report (Ro et al., 1998) was modified to provide binned representations of individual trials, using the Superscope Append function. PSTH bins between the task stage markers were summed on each trial and divided by the stage duration to permit statistical analyses of average firing rates per task stage. Repeated-measures multivariate ANOVA protocols were applied to these rates and to behavioral data as described in Debowy et al., (2001), allowing us to subdivide the population of neurons studied in primary somatosensory (SI) and posterior parietal cortex (PPC) into functional classes. Prehension task-facilitated neurons (65%) showed statistically significant increases in average firing rates compared to baseline during object acquisition and manipulation (Gardner et al., 1999); task-inhibited cells (35%) showed significant reductions in firing during the contact through lift stages (Ro et al., 2000). These populations were further subdivided by statistical analysis into groups whose spike trains were tuned to single task stages, spanned two successive stages, or were multiaction.

We analyzed response profiles of mean firing rate in the hand area of SI and PPC to determine the period(s) of peak firing during the prehension task (Debowy et al., 2001). The application of grip force on objects was found to be a more effective stimulus for these cortical neurons than its maintenance or release. Four times as many neurons fired at peak rates during acquisition stages (approach, contact, grasp) than upon relaxation of grasp, and their firing rates were higher. Hand positioning at contact excited the largest number of cells, but grasping evoked the highest firing rates in the population. The grasp stage also coincided with maximal inhibition of task inhibited cells. Holding evoked the lowest mean rates and had the fewest tuned cells.

Neurons in anterior SI, posterior SI, and PPC differed in the emphasis placed upon particular behaviors. These findings were manifest by the proportion of cells firing at peak rates during specific task stages, as well as by the total fraction of the population exhibiting significant excitation or inhibition coincident with these behaviors. The data support models in which PPC plans hand movements during prehension, rather than guiding their execution. PPC responses during the task preceded those in SI, peaking at or before hand contact with objects. PPC had the highest proportions firing at peak rates during approach, as the hand was preshaped for grasp, and the most facilitated responses. PPC firing rates were significantly depressed when the hand interacted directly with objects during grasping and manipulation; only 10% fired at peak rates during grasping while inhibition rose to 43%.

PPC firing patterns therefore appear predictive of hand behavior rather than reactive to tactile stimulation.

Sensory monitoring of hand–object interactions occurred primarily in SI, where cells responded to specific hand behaviors. Anterior SI neurons were most sensitive to hand positioning upon objects, while posterior SI neurons signaled application of grip force. 60% of neurons recorded in anterior SI increased firing rates significantly over baseline at contact; 38% fired at peak rates, and 10% were inhibited. The release of grasp evoked peak firing in only 5% of these neurons. Application of grip force by the hand was monitored most closely in posterior SI, as 80% of these neurons were facilitated or inhibited during grasping, and 31% fired at peak rates. Inhibitory responses were more prevalent in posterior than in anterior areas, providing a quiet background for detection of slippage when the object was held or transported. Our data suggests that SI neurons signal the onset and completion of specific actions and may enable error correction through feedback loops to primary motor areas. Anatomical connections from SI to PPC may supply information necessary to update grasp programs that seem to be formulated in PPC.

9.4.5 RECEPTIVE FIELD MAPPING

These techniques for analyzing neural correlates of trained behaviors can also be used to study natural hand movements that are difficult to instrument, such as grooming or feeding behaviors (Babu et al., 2000). They are also suitable for mapping receptive fields with hand-held probes. One of the problems encountered, when probes such as swabs or brushes are applied to the hand, is defining the exact skin positions touched, and quantifying the evoked neural responses. Here is an application where DV analyses are particularly helpful. Figure 9.5A shows a 6-sec segment of the spike train recorded as the receptive field of a cortical neuron was mapped; the nine images were cropped to show only the animal's hand and the actions of the investigator while probing the hand. Stroking the palm began at 40.7 s, just prior to image C, and was repeated three times, ending at 43.6 s. Peak responses occurred when the interdigital pad below digit 5 was contacted (D, E). Localized stroking (F, G) and taps (I, J) on this palm pad evoked strong activity. Removing the swab from the hand silenced the cell (H, and blank intervals between bursts I, J, and subsequent ones). This neuron therefore responded preferentially to stroking or tapping the most medial interdigital palm pad, and weakly to touch along the ulnar margin.

9.5 HOW ARE HAND BEHAVIORS QUANTIFIED USING DIGITAL VIDEO?

Images of the hand in Figures 9.1, 9.2, and 9.5 were obtained by exporting individual video frames from the editing software, saving them as 720×480 pixel TIFF files, and cropping them to highlight the most relevant information. The hand kinematics can also be abstracted from such images and quantified by placing the images in separate layers of an Adobe Illustrator project, and tracing outlines of the hand, arm, test objects on the shape box, or other features of interest. Stacking a series of such images in separate layers allowed us to animate the drawings, showing the dynamics

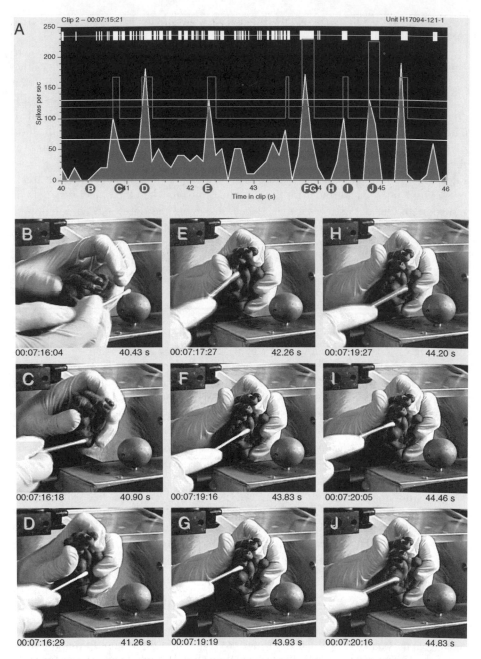

FIGURE 9.5 (See Color Figure 9.5 in color insert.) Mapping receptive fields on the palm with a cotton swab for a neuron recorded in anterior SI cortex. **A.** Burst analysis of firing patterns during a 6 s interval; mean firing rates during bursts are offset by 100 spikes/s. B-J. Cropped images of the hand coincident with the markers on the abscissa in A. The strongest responses occurred to touch applied to the interdigital pad below digit 5. DV images captured with Sony DCR-TRV900 camcorder.

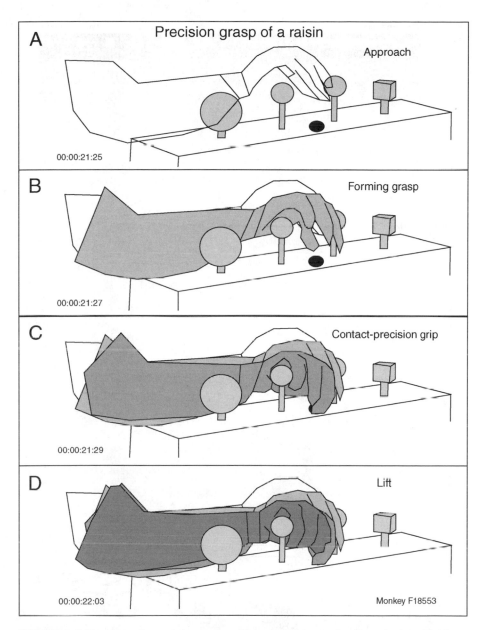

FIGURE 9.6 Layering of successive tracings illustrates how an animal acquired a raisin from the top of the shape box. Note the preshaping of the fingers in B just prior to contact. Time code indicates that 66 ms elapsed between A and B, and B and C, and 133 ms between C and D.

of movement of different parts of the hand as the action progressed. In the example shown in Figure 9.6, the animal picked up a raisin from the top of the shape box. The preshaping of the fingertips for precision grip is clearly visualized, and the grip

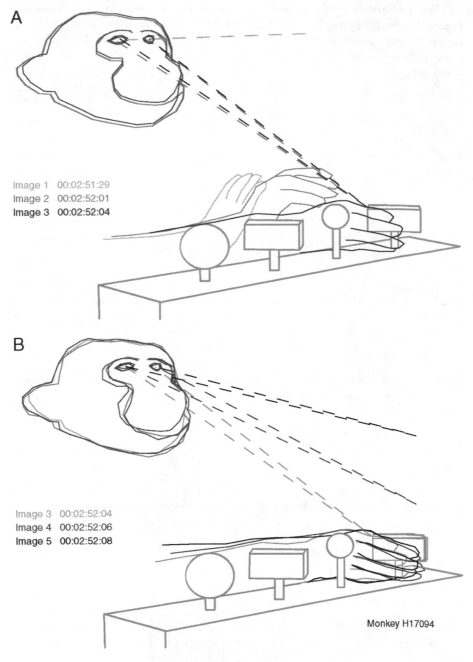

A

Image 1   00:02:51:29
Image 2   00:02:52:01
Image 3   00:02:52:04

B

Image 3   00:02:52:04
Image 4   00:02:52:06
Image 5   00:02:52:08

Monkey H17094

FIGURE 9.7

aperture can be measured. Forearm movements were stabilized as the fingers scooped up the food morsel and enclosed it in a fist; finger movements were clearly resolved at 30 fps because they are relatively slow compared to reaches to distant targets.

   The kinematics of the prehension task during a typical trial are shown in Figure 9.7. In this example, the hand released the right rectangular knob just before image 1, and moved medially toward the left rectangular knob. The wrist was pronated midway during the reach (image 2), and then supinated to clear the space between objects, as the target was acquired (image 3). After contact, the hand was slid forward along the knob's vertical surface until the hidden edge closest to the palm reached the base of the fingers, when grasp was tightened (image 4), and lift began (image 5). In parallel with the hand movement, the animal shifted its gaze (dashed line) from the monitor indicating the cue location (image 1), toward the object to be grasped in image 2, and maintained gaze at the grasp site until the hand was placed there (image 3). The animal's gaze was lifted back toward the monitor for feedback, as the knob was rotated and lift began (image 5).

   As these objects are rigid bodies, we can project their outline through the fingers, and estimate where they might rest in the hand. The drawings in Figure 9.7 indicate that the hand was positioned for grasp to avoid pressing the edge of the object into the soft tissue of the palm. Hand movement ceased when the inner edge was placed at the metacarpal-phalangeal (MCP) joint.

   Another of the animals studied used a different grasp strategy for performing the task, but achieved the same goals. That monkey used an overhand grasp, placing the hand on the top surface of the knob, and pushed it upward with the heel of the hand (see e.g., Figure 9.1 in Gardner et al., 1999). A similar kinematic analysis revealed that the animal used different tactics when grasping rectangular and round objects. During approach toward the rectangular knob, the animal positioned the hand above the knob, with the proximal phalanges aligned to the flat upper surface. She then lowered the hand so that contact was made on the finger shafts, with the inner edge falling into the groove formed by the MCP joints, and the outer one at the proximal interphalangeal (PIP) joints. Grip between the digit tips and proximal palm pads was tightened with little lateral or sliding motion. In this manner, neither edge contacted the fleshy parts of the hand. When approaching the large round knob, the animal aimed the center of the palm toward the forward, curved surface. Upon contact, the hand was slid upward to the top surface while the grip between the thumb and digit tips was secured. The forward sliding movement was combined with the upward pressure from the wrist to lift the object.

FIGURE 9.7 (opposite) Kinematics of the prehension task traced from a series of sequential video frames. A. Approach trajectory. Drawings of three frames spaced 66 ms apart outline the hand posture during movements between the right and left rectangular knobs. The dashed lines project the center of gaze, which moved from the computer monitor that provided location cues (image 1) to the selected grasp site (image 2); gaze was maintained at the grasp site until the hand was placed there (image 3). B. Grasp and manipulation. After contact, the hand was slid forward along the knob surface as grasp was initiated, until the inner edge reached the base of the fingers (image 4). Gaze was lifted back toward the screen as grasp was tightened, and the object lifted by the hand (image 5). The hand posture in these drawings is nearly identical to that shown in Figure 9.2 from another session, demonstrating the stereotyped behavior adapted by each animal when it performed the task.

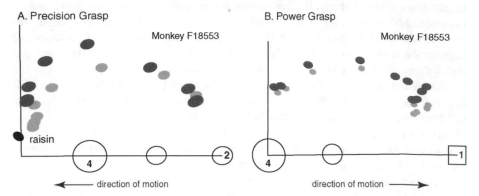

FIGURE 9.8 Oval markers of the wrist position on successive frames illustrate the hand trajectory during acquisition of a raisin at the lateral edge of the shape box (precision grasp, left panel), and during reaches between knobs 4 and 1 during the prehension task (power grasp, right panel). The greatest distance moved between frames occurred midway between the start and end points. Dark and light colored ovals indicate the first and second trials of each type.

Motion trajectories can also be constructed from the images. For example, in Figure 9.8, we drew oval-shaped markers at the wrist in a series of aligned frames, and then placed all of the markers in one layer. This allowed us to visualize the entire path of hand movement between knobs 4 and 1 (right panel), and from knob 2 to a raisin placed near the right edge of the shape box (left panel); two trials are illustrated for each movement. As in previous kinematic studies using expensive infrared (IRED) monitoring systems (Jeannerod, 1984; Jeannerod et al., 1995; Roy et al., 2000), these reach-to-grasp movements were executed in a smooth trajectory in which the peak velocity of the wrist occurred midway between start and end locations. Furthermore, although the 30 fps sample rate of DV is an order of magnitude slower than that of IRED systems, this was sufficient to outline the reach path. The DV frame rate is adequate for measuring manipulatory movements of the fingers, especially when the proper view angle is selected (Figures 9.2, 9.6, and 9.7).

9.6 WHAT INSTRUMENTATION IS RECOMMENDED?

9.6.1 VIDEO EQUIPMENT

The applications described in this chapter require only consumer DV equipment with external audio input jacks, remote controlled operation, and time code. Two such camcorders are the moderately priced Sony DCR-TRV900, which we use to provide frontal and overhead views of the hand (Figures 9.2B and 9.2D), and the high-end Canon XL1 we position for wide-angle lateral views of the animal's upper body and head (Figure 9.2C). Both of these 3-CCD cameras yield excellent picture quality, but are optimized for different features. The Sony model is compact, and small enough to be held in one hand. This is advantageous, particularly for over head views when light weight is desirable. In addition, the swing-out, swivel-mounted 3.5" color LCD screen is useful for remote viewing when framing the image and adjusting zoom depth. The Canon is a more robust instrument, and should be positioned on a heavy-duty tripod. It has interchangeable lenses, which makes it particularly well suited for microscopic applications.

Some camcorders feature progressive scan CCDs, in which horizontal lines are scanned sequentially, rather than with the interlaced mode typical of standard video, in which half-frames containing the odd or even lines are scanned alternately in the two fields comprising a video frame. Progressive scan yields sharper still images, but because the picture is refreshed 30 times per second, the video appears jumpy. In interlaced mode, a new image is constructed at twice the frame rate (60 fields per second for NTSC video); it is preferable for analyses of movement. The Sony camcorder allows the user to select either interlaced or progressive scan recording modes.

Audio is the major advantage of the Canon XL1 recorder. It has four independent audio channels, each equipped with shielded RCA input jacks and independent level control knobs; the camcorder can operate in 16-bit 48 kHz mode for 2-channel recording, or 12-bit 32 kHz for 2- or 4-channel operation. The latter is particularly useful for multiple electrode applications, or for simultaneous direct recording of signals from tactile stimulators or from EMG amplifiers. The major disadvantage of the Sony DV camcorder is crosstalk between audio channels through the stereo minijack audio input connector, making this unit inappropriate for multiple electrode or other applications where two channels must be recorded with minimal crosstalk. However, it provides excellent fidelity for single electrode recordings, particularly when used in combination with an external adapter providing gain control and shielded inputs (Beachtek DXA 4S dual XLR adapter).

9.6.2   Computer Hardware and Software

All data processing in our laboratory is done on Macintosh G3 or G4 computers, from the initial digital video file acquisition, through data analysis, to preparation of figures for publication. Macs are particularly well suited for all types of multimedia applications, especially video products using QuickTime. Many of the commercial software applications described in this chapter have been adapted for Windows platforms, but we have no direct experience using that hardware for this purpose. QuickTime comes installed on all new Macintosh computers, and most have built-in FireWire ports for direct connection to DV camcorders. QuickTime can also be downloaded for free from the Apple website.

As DV entails very large data files acquired in real time, a G3/300 or faster computer is required, along with at least 128 MB of RAM memory, a 16 GB A/V (audio/video rated) hard drive, and a removable storage device, such as a 2-GB Jaz drive, for archiving video files. An external Ultra Wide A/V hard disk drive, dedicated to storage of video and audio, is recommended for avoiding fragmentation when combining small clips into a larger data file. DV camcorders are connected to the computer's FireWire port, using a 4-pin to 6-pin IEEE 1394 cable.

Two major classes of software are required for these protocols: digital editing software and data analysis software that can read AIFF files. Both Adobe Premiere and Apple's Final Cut Pro are easy to use, excellent editing programs. They are necessary for downloading the video and audio files into the computer, for viewing the movies in real time or frame-by-frame, for logging time codes, for converting the electrophysiological data in the audio signal to AIFF files for quantitative analysis, and for exporting single frame images of the behavior. They can also be used to make movies that link together different behaviors or different neurons, for special effects, titles, and fancy transitions between clips.

Superscope II and Igor Pro are useful neurophysiological packages for data acquisition, analysis, and display. All of our custom written software was created using Superscope II, which is available only for Macintosh computers. It should be possible to create similar data analysis software using Igor Pro, using the algorithms described here and in our earlier publications (Debowy et al., 2001; Ro et al., 1998). Both of these programs make extensive use of spreadsheets, using text files to transfer data in and out of the analysis programs, and for further processing. Statistical analyses are performed with Statview (SAS Institute, Inc.). We typically use Delta-Graph 4.0 (SPSS, Inc.) for displaying rasters, histograms, and other graphic data, and assemble multiple data panels with Canvas (Deneba Systems, Inc.).

Two software products from Adobe Systems, Inc. are used for image processing: Photoshop for processing TIFF images exported from the video files, and Illustrator for tracing the hand kinematics from the single frame images. Multiple image figures are assembled for publication using QuarkXPress.

9.7   WHAT CAN WE LEARN FROM IMAGING ACTIONS OF THE HAND?

The hardware and software tools described in this chapter have allowed us to study hand behaviors that are not directly instrumented. Although other modes have allowed investigators to analyze grasping behaviors in both humans and monkeys, they either did not visualize the kinematics of prehension (Johansson, 1996; Muir and Lemon, 1983; Picard and Smith, 1992; Sakata et al., 1995; Salimi et al., 1999; Wannier et al., 1991; Westling and Johansson, 1987), or else did not correlate the behaviors with the simultaneously occurring electrophysiological processes (Jean-nerod et al., 1995; Roy et al., 2000). The DV recording methodology eliminates the problem of synchronization of action with neuronal signals characteristic of the IRED system, because the images and spike recordings are digitized together, and therefore share a common time base.

The 30 fps image capture rate appears to be sufficient for analysis of the neural correlates of hand actions, as demonstrated in our published work (Debowy et al., 2001; Gardner et al., 1999; Ro et al., 1998, 2000). Video images of hand behaviors allowed us to categorize the relative timing of firing patterns of neurons studied with a prehension task. Our analyses suggest that somatosensory areas of parietal cortex do not function in a simple hierarchical fashion. Instead, because PPC neurons were found to fire before hand contact with the test objects, this region appears to anticipate and plan grasping behaviors rather than elaborate features detected at earlier stages of cortical processing. On average, neurons in SI fired after PPC, providing feedback to the motor programs planned there or elsewhere.

The possibility of quantifying the hand posture from several different viewing angles will also provide useful information about coordinated actions of the hand, that supplement data from force transducers and other sensors. Kinematic analyses of the sort described in Section 9.4 will be particularly useful for determination of whether cortical neurons differentiate the geometric shape of round and rectangular objects, or simply encode the behavioral strategies used to grasp them.

DV technology opens up the possibility of quantitative analyses of spontaneous hand behaviors that are not trained and overlearned. The natural hand kinematics of grooming and feeding behaviors can be dissected into simpler components with these techniques, and their neurophysiological correlates measured. Many of the hand behaviors studied with other techniques have been practiced to the extent that they have become robotic, and perhaps unphysiological. It is therefore important to be able to compare the trained behaviors to those used by animals for purposes they consider useful, and whose kinematics are self-generated and self-paced. DV imaging is sufficiently flexible to permit both types of analyses with a standardized, relatively simple set of tools.

The DV methodology described in this chapter has potential for wider applicability than simple correlation of spike trains with hand movements. Any type of behavioral activity that can be imaged through the camcorder's viewfinder could, in principle, be analyzed with these methods. Obvious experimental uses include studies of locomotor behavior, pathfinding in mazes, or tactile exploratory movements (Lederman et al., 1988). Possible clinical applications include studies of motor disorders or stereognostic tests of cognitive function in which recordings of EMG or EEG would replace the spike trains described in this report. The brain's own body image would thus be complemented by objective measurements of the body's actual behavior.

9.8   USEFUL WEBSITES (LISTED ALPHABETICALLY)

- Apple Computer: *http://www.apple.com*. Follow links to QuickTime, FireWire, and Digital Video for further technical information.
- Bruxton Instruments: *http://www.bruxton.com*. Macintosh data acquisition and analysis software with links to both Superscope II and Igor Pro. Nice monographs on data acquisition for neurophysiological recording.
- GW Instruments: *http://www.gwinst.com*. Information about Superscope II software used for data analysis in our application. User manuals and a demonstration package are available for download at this site.
- WaveMetrics, Inc.: *http://www.wavemetrics.com*. Information about Igor Pro software, another structured programming language that can read AIFF files, and be used for analysis of DV data. It is available in both Macintosh and Windows versions.

ACKNOWLEDGMENTS

Major funding for this investigation has been provided by Research Grant R01-NS11862 from the National Institute of Neurological Disorders and Stroke. Additional support has been supplied by the MD–PhD Program (Daniel J. Debowy, Edward H. Hu) and Honors Program for medical students (Shari Reitzen) at New York University School of Medicine, and by fellowships to summer interns (Maria Chu and Jill Sakai) sponsored by the Sackler Institute of Graduate Biomedical Studies of New York University. We thank Dr. Daniel Gardner for many helpful comments and criticisms of earlier versions of this chapter.

REFERENCES

Babu, K. S., Debowy, D. J., Ghosh, S., Hu, E. H., Harris, A., Natiello, M., and Gardner, E. P., Spike burst analysis: a tool for analysing natural prehension behaviors recorded with digital video. *Soc. Neurosci. Abstr.* 26, 2202, 2000.

Bovik, A., Ed. *Handbook of Image and Video Processing.* San Diego, CA: Academic Press, 2000.

Chapman, C. E., Tremblay, F., and Ageranioti-Bélanger, S. A., Role of primary somatosensory cortex in active and passive touch. In: Wing, A. M., Haggard, P., and Flanagan, J.R., Eds. *Hand and Brain.* Academic Press, San Diego, CA, pp. 329, 1996.

Debowy, D. J., Babu, K. S., Hu, E. H., Ghosh, S., Harris, A., Natiello, M., and Gardner, E. P., Spike burst analysis: a tool for analysing trained prehension behaviors recorded with digital video. *Soc. Neurosci. Abstr.* 26, 2202, 2000.

Debowy, D., Ghosh, S., Ro, J. Y., and Gardner E. P., Comparison of neuronal firing rates in somatosensory and posterior parietal cortex during prehension. *Exp. Brain Res.* 137, 269, 2001.

Edin, B. B. and Abbs, J. H., Finger movement responses of cutaneous mechanoreceptors in the dorsal skin of the human hand. *J. Neurophysiol.,* 65, 657, 1991.

Edin, B. B. and Johansson, R. S., Skin strain patterns provide kinaesthetic information to the human central nervous system. *J. Physiol. Lond.,* 487, 243, 1995.

Fenton, A. A. and Muller, R. U., Using digital video techniques to identify correlations between behavior and the activity of single neurons. *J. Neurosci. Meth.,*70, 211, 1996.

Gardner, E. P., Ro, J. Y., Debowy, D., and Ghosh, S., Facilitation of neuronal activity in somatosensory and posterior parietal cortex during prehension. *Exp. Brain Res.,* 127, 329, 1999.

Hulliger, M., Nordh, E., Thelin, A. E., and Vallbo, Å.B., The responses of afferent fibers from the glabrous skin of the hand during voluntary finger movements in man. *J. Physiol. Lond.,* 291, 233, 1979.

Jeannerod, M., The timing of natural prehension movements. *J. Mot. Behav.,* 16, 235, 1984.

Jeannerod, M., Arbib, M. A., Rizzolatti, G., and Sakata, H., Grasping objects: the cortical mechanisms of visuomotor transformation. *TINS,* 18, 314, 1995.

Johansson, R. S., Sensory control of dexterous manipulation in humans. In: Wing, A. M., Haggard, P., and Flanagan, J. R., Eds., *Hand and Brain.* Academic Press, San Diego, CA, pp. 381, 1996.

Lederman, S. J., Browse, R. A., and Klatzky, R. L., Haptic processing of spatially distributed information. *Percept. Psychophys.,* 44, 222, 1988.

Muir, R. B. and Lemon, R. N., Corticospinal neurons with a special role in precision grip. *Brain Res.,* 261, 312, 1983.

Picard, N. and Smith, A. M., Primary motor cortical activity related to the weight and texture of grasped objects in the monkey. *J. Neurophysiol.,* 68, 1867, 1992.

Ro, J. Y., Debowy, D., Ghosh, S., and Gardner, E. P., Depression of neuronal firing rates in somatosensory and posterior parietal cortex during object acquisition in a prehension task. *Exp. Brain Res.,* 135, 1, 2000.

Ro, J. Y., Debowy, D., Lu, S., Ghosh, S., and Gardner, E. P., Digital video: a tool for correlating neuronal firing patterns with hand motor behavior. *J. Neurosci. Meth.,* 82, 215, 1998.

Roy, A. C., Paulignan, Y., Farne, A., Jouffrais, C., and Boussaoud, D., Hand kinematics during reaching and grasping in the macaque monkey. *Behav. Brain Res.,* 117, 75, 2000.

Sakata, H., Taira, M., Murata, A., and Mine, S., Neural mechanisms of visual guidance of hand action in the parietal cortex of the monkey. *Cereb. Cortex,* 5, 429, 1995.

Salimi, I., Brochier, T., and Smith, A. M., Neuronal activity in somatosensory cortex of monkeys using a precision grip. I. Receptive fields and discharge patterns. *J. Neurophysiol.,* 81, 825, 1999.

Wannier, T. M. H., Maier, M. A., and Hepp-Reymond, M. C., Contrasting properties of monkey somatosensory and motor cortex neurons activating during the control of force in precision grip. *J. Neurophysiol.,* 65, 572, 1991.

Westling, G. and Johansson, R. S., Responses in glabrous skin mechanoreceptors during precision grip in humans. *Exp. Brain Res.,* 66, 128, 1987.

10 Deciphering the Code — Dynamic Modulation of Neural Activity During Tactile Behavior

Randall J. Nelson, Mikhail A. Lebedev, and Yu Liu

CONTENTS

10.1 INTRODUCTION

Sir Charles Sherrington, in the Hughlings Jackson Lecture given on January 29, 1931, paid homage to his predecessor by stating a number of comments that are no less attractive today. He suggested that Jackson would have been pleased with the thought that "The nervous system is indeed both a form and a series of events. These have to be confronted together even for inquiry which concerns itself with function as its chief aim."* He goes on to suggest that Jackson appreciated the inextricable link between sensation, perception, intention and behavior. "Hughlings Jackson, in his writings, turns back and forth between muscular co-ordination and mental experience, as if for him, they were but aspects of a single theme. It may be that to decipher how nerve manages muscle is to decipher how nerve manages itself."** Thus, for many years, deciphering the signals from afferents and directed toward effectors has proven to be both a challenge and a problem for those seeking to understand the working nervous system as the nervous system is working.

One way to approach the challenges and problems inherent in studying the somatosensory system is to consider what might be called "tactile fidelity." Fidelity in tactile sensation is not unlike that in music, and the mode of operation of peripheral tactile sensors very closely resembles the operation of a violinist described in a citation from Chris Dobrian. "In playing a single brief note, a violinist combines bow angle, bow speed, bow pressure, bow placement, bow attack, finger placement, finger movement, finger pressure, in addition to whatever totally involuntary muscular movements may be caused by nervousness, coffee consumption, humidity, unknown electrical discharges in the brain, etc. — and all of these factors are changing from millisecond to millisecond."***

The purpose of this chapter is to outline some of the approaches that have been taken as experimenters attempt to decipher the fidelity of the representation of tactile information by the cerebral cortex. By nature, this chapter cannot review all the methods for quantifying the central representations of peripheral inputs. We will focus on the representation by the primary somatosensory cortex of the temporal properties of peripheral inputs. Temporal signals play an essential role in the representation of dynamic external events, such as vibration.[1] Moreover, because external objects are typically scanned by actively moving skin surfaces along them, their static characteristics, such as texture and shape, are temporally encoded by neuronal discharges, as well (e.g., Reference 2). Several examples, both historical and contemporary, will be discussed. These will serve as illustrations of what can be gleaned from certain types of experiments and what has yet to be defined.

As late as the mid-1950s, there was still relative disagreement about the anatomical parcellation of the cytoarchitectural fields that comprise the cortical recipient zones for inputs from peripheral body parts (see Reference 3). Significant progress has been made in defining and characterizing the spatial representations of exteroceptive and proprioceptive inputs to the cortex.[4] As well, the responses of single cortical neurons or groups of neurons to controlled peripheral inputs are being defined and analyzed in increasing detail. For a number of important questions being posed, it has been, and continues to be, important to ask how well the signals transduced in the periphery are represented centrally. Said in another way, is fidelity of peripheral inputs maintained and under what conditions does it vary as a function of the behavioral state of the animal? To address these questions, we turn to studies

---

* C.S. Sherrington, Quantitative management of contraction in low-level co-ordination. *Brain* 54: p. 1.
** Ibid., p. 27.
*** Chris Dobrian, Music and Language (1992; http://www.arts.uci.edu/dobrian/CD.music.lang.htm).

of the sense of flutter-vibration, since the stimuli that result in activation of this sense have easily quantifiable temporal signatures. Seminal work several decades ago established how mechanoreceptive afferents respond to sinusoidal skin stimulation superimposed on indentations.[5] Following step indentation, most afferents respond with brief, sometimes high-frequency bursts of impulses. After these initial bursts, there are at least three different response patterns, depending on the type of afferent fibers being examined. Afferents from slowly adapting mechanoreceptors discharge throughout indentation, albeit at rates lower than the initial burst, whereas those from rapidly adapting mechanoreceptors do not (see Reference 6 for review). The rapidly adapting afferents, and their cortical counterparts, the quickly adapting and pacinian-receptive cortical units, are of most interest for this discussion since they can be entrained to fire at the frequency of periphery stimuli.[7] Quickly adapting units respond best to lower frequency sinusoids (10–60Hz) while pacinian units encode stimuli of higher frequencies (60–400Hz). Quickly adapting units are rather abundant whereas units with pacinian-like properties make up less than 10% of the sampled neurons in virtually all studies. If the goals of an experiment are to test the fidelity with which peripheral stimuli are represented within the cortex, to determine if there are regional differences in responsiveness, and/or to determine if changes in behavioral conditions cause changes in responsiveness, then the cortical responses to time-varying stimuli serve as a way to probe the system. In short, given that the experimentally produced peripheral stimuli have a temporal pattern, the degree to which cortical neurons preserve that pattern can be measured and the conditions that result in a breakdown of faithful stimulus representation can be addressed.

10.2 WAYS OF REPRESENTING THE CODE

Before examining the responses of cortical neurons for changes in representational fidelity, it is important to have some idea of what constitutes fidelity in afferent fiber activity patterns and how these patterns change with stimulus frequency and intensity. Returning to the work of Talbot, Mountcastle, and colleagues, two points should be noted. First, afferent fibers responding to flutter/vibration are sensitive to stimulus amplitude.[5] Below a certain stimulus strength, fibers do not respond with an impulse for every cycle of the stimulus. Rather, they may skip beats, resulting in intervals between the impulses at multiples of the stimulus period. These intervals can be detected as peaks in interval histograms. At high amplitudes, fibers may respond with multiple impulses per stimulus cycle, resulting in interval histograms with one or more peaks at intervals shorter than the stimulus period. Often, a peak in the interval histogram will be present, under these conditions, at an interval corresponding to one-half of the stimulus period. This effect reflects the tendency for the afferents to discharge at opposite phases of the stimulus cycle corresponding to stimulus velocity peaks. Between high and low stimulus amplitudes exists a range of stimulus amplitude for which fibers, if being stimulated at appropriate frequencies, are capable of issuing, on average, one impulse per stimulus cycle. Second, as the stimulus frequency diverges from the frequency to which the afferent unit responds most sensitively, the increase in amplitude necessary to shift the impulse pattern from less than one, to the one impulse per cycle pattern increases. The same is true for the decrease in amplitude needed to shift from a more than one pattern to a one-for-one pattern. The range of amplitudes over which one impulse is elicited per stimulus cycle narrows. Thus, afferents are tuned, both to frequency and, in a sense, to the amplitude of peripheral stimuli. At least for rapidly adapting afferents, it has been shown recently that impulse entrainment remains stable within and across trials.[8] This is true even in the face of increases in response intensity and the fact that the capacity to discriminate peripheral stimuli can and often does increase with time.

10.2.1 RECENT CONSIDERATIONS OF THE TEMPORAL ASPECTS OF INPUT REPRESENTATIONS

Considerable effort by others also has been devoted to understanding firing patterns of neurons within the sensorimotor cortices and relating the time-variant patterns of CNS activity to the actual pattern of the stimuli being delivered and the behaviors being produced (some recent studies include References 9–19). Mechanoreceptive afferents respond relatively consistently over time.[8,20] Experience-dependent response plasticity in sensorimotor cortical neurons, however, may be continually modulated by mechanisms that probably involve changes in the threshold that allow calcium entry, are NMDA receptor dependent, and are GABA regulated.[21-23] A recent report has suggested that experience-dependent cortical plasticity may result from at least two mechanisms.[24] The first appears to be a change in the excitatory connectivity of cortical elements; the second may be a release from inhibition by cortical interneurons. The structure of spike trains that represent peripheral stimulus features may degrade as the information ascends in the CNS. By the time that it reaches the cortex, the fidelity with which the temporal pattern of activity is synchronized with peripheral events may have diminished.[2] The activity patterns of single cortical neurons may not exhibit as faithful a representation of stimuli as ensembles of neurons. Single neuron changes in firing rate may not convey as much information about stimulus characteristics as populations of neurons, either by singular or correlated activity (see References 25–27).

10.2.2 SPIKE TRAIN QUANTIFICATION

There have been several metrics designed to permit the quantification of spike train entrainment. Each has strengths and weaknesses. As examinations of neural data collected during experiments occurs, further refinements in data analysis techniques also occur. To evaluate some of these, we have constructed an artificial dataset, with known characteristics. The rationale for doing this has been that certain effects can be seen during visual inspection of data. The goal has been, and continues to be, to develop some objective means of expressing the goodness-of-fit between peripheral inputs and central representations thereof. Given this, it should be possible to examine changes in representational fidelity as a function of experimentally manipulated environmental conditions. In many experiments, these variables are behavioral-context linked.

Figure 10.1 shows a portion of the aforementioned dataset. Above is the depiction of a peripheral input, in the form of a sine wave, which we can imagine being

A.   Stimulus

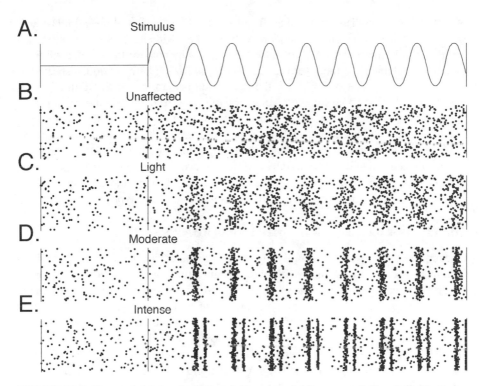

B.   Unaffected

C.   Light

D.   Moderate

E.   Intense

FIGURE 10.1 The model "stimulus" (Panel A) and model "responses" (Panels B–E) shown in raster plots. Raster plot (Panels B–E) are constructed such that each dot represents an action potential and each row represents an experimental trial. The data mimic what might be seen if a cortical neuron was unaffected (B) by a peripheral stimulus, or if that neuron responded to light (C), moderate (D), or intense (E) stimuli. Tight vertical distributions of dots indicate spikes that occur in each trial at the same time relative to the stimulus. The width of these distributions is indicative of the degree of entrainment of the response to the stimulus.

delivered to a location on the skin at time zero. Arrayed below are four possible responses, shown in the form of rasters, which differ in the extent to which they represent entrainment to the stimulus. From above, the first represents several trials in which, although stimulation evoked an increase in firing rate, this response was not visibly affected by stimulus periodicity. The second shows what might happen if a neuron received a light stimulus, and consequently was lightly entrained to it. With moderately intense peripheral stimulation, visible neural entrainment would be expected in certain neurons. However, if the stimulus is intense, as previously observed, neurons may fire more than one spike per stimulus cycle and therefore exhibit a second, relatively consistent vertical band in raster displays.

    The datasets used to construct the panels were derived based on the following assumptions. First, there is a fixed probability of spike occurrence in a given stimulus cycle, equivalent to mean firing rate. Second, each spike is phase-locked to the stimulus with a certain precision. Accordingly, a spike was assigned to every 3 out of 4 stimulus cycles. The timing of each spike relative to the start of the stimulus cycle was set to have a uniform distribution within a time window of a fixed width and position within the cycle. The widths of these time windows for the unaffected, light, moderate, and intense stimuli were 1, $1/6^{th}$, $1/18^{th}$, and $1/36^{th}$ of the period, respectively. It should be noted that, although for the unaffected pattern, the distribution of spikes was uniform for the whole stimulus cycle, even this pattern contained information about the stimulus periodicity because one-fourth of the stimulus cycles was devoid of spikes. For the intense stimulus, independent of the presence of the first impulse, a second impulse was generated three-fourths of the time within the time window of the same width, but its position was advanced by one-quarter cycle. These data can be used, then, to evaluate relative strengths and weaknesses of data analysis methods for quantifying stimulus-related responses and qualitatively assessing response types.

10.2.3 AVERAGE FIRING RATE AFTER INITIAL RESPONSES

One conventional way to quantify the response of cortex neurons to peripheral stimuli is to calculate average firing rate during a time interval following stimulus onset. We have reported elsewhere[28] that some SI neurons exhibit very similar firing rates for structurally different spike patterns. Our simulated data for the first three cases (Figure 10.1) was also characterized by the same firing rate for different temporal patterns of spikes. Figure 10.2 shows post-stimulus time histograms (PETHs) constructed from the data represented in Figure 10.1. It can be seen that firing rate calculated for the time window 0–150 ms relative to stimulus onset does not provide a clear indication of certain features readily detectable by visual inspection. Despite the differences in apparent entrainment appreciated upon visual inspection of the rasters (Figure 10.1) or PETHs (Figure 10.2), the upper three panels have no significant difference in MFR. Certainly, one could argue that the lower panel is both qualitatively and quantitatively different from the upper three. However, the goal is to detect and demonstrate subtle changes in fidelity with which peripheral stimuli are represented. As such, it would be essentially impossible to compare what might be significant changes in temporal entrainment in three of the four panels solely using MFR as the metric. A representative example of actual somatosensory cortical neural activity in an awake, behaving monkey during vibratory-triggered wrist movements is shown in Figure 10.4A. It is clear from visual inspection of this plot that the major effect of vibration is in the temporal structure of spikes, not in MFR. The presence of vibratory-entrained activity can be also seen in panel 4B, in which spike occurrences are represented using a raster display. Thus, MFR may indicate general changes in responsiveness under some experimental conditions, but no information about the temporal structure of the neuronal signal may be gleaned from this type of analysis.

10.2.4 PERCENTAGE ENTRAINMENT OF THE RESPONSE
TO THE STIMULUS

Another approach to describing the fidelity of the central representation of peripheral inputs with temporal patterns is to construct a metric that evaluates spike occurrences

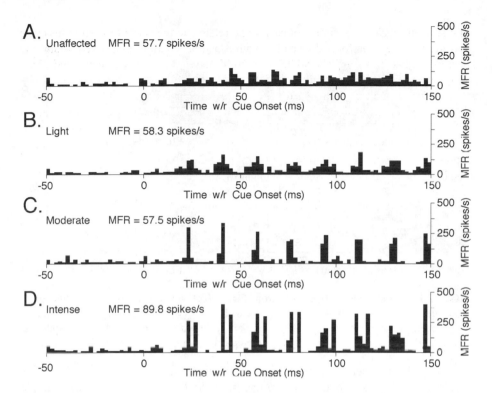

FIGURE 10.2 Histograms constructed from the data presented in Figure 10.1. Note that although changes in the apparent degree of entrainment can be appreciated by visual inspection, the mean firing rate (MFR) in the upper three panels remains about the same. Only when the response becomes saturated (lower panel) is there an appreciable change in firing rate.

in light of expectations derived from the characteristics of the stimulus itself. One of the great benefits of using flutter vibration as a peripheral stimulus is that it and neuronal activity associated with it can be described as a function of stimulus phase. Provided that stimulus phase is consistent as a function of presentation time, then spike occurrences translate directly to the phase of the stimulus at that time. If then, the response latency (i.e., the conduction time from periphery to the central location) remains relatively constant, shifts in phase plots may indicate changes in representational fidelity that accompany changes in the sensory environment. Figure 10.3 shows the phase plots that were derived from the datasets previously described above. Several measures can be calculated using phase plots.

Two decades ago, Ferrington and Rowe[29] used an entrainment index they called Percentage Entrainment (PE) to quantitatively describe the distribution of spike occurrences relative to the phase angle of the sinusoidal stimuli delivered to the sensory periphery. PE is calculated by plotting the distribution of spike occurrences as a function of stimulus phase, binning the results in some reasonable way. By then, calculating the number of spikes in each half-cycle and comparing that ratio with the total, the PE, which ranges between 0.5 and 1.0 can be expressed. Figure 10.3 illustrates what happens if the 40 trials from the sample datasets are

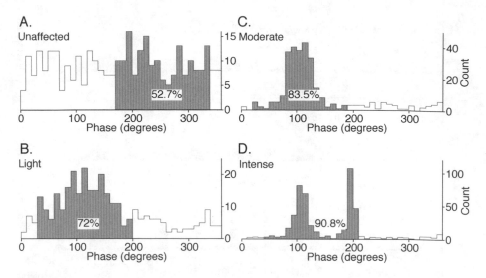

FIGURE 10.3 Phase plots of the data from Figure 10.1 where each bin of the histogram represents spikes occurring in a 10° phase bin with respect to the frequency of the stimulus. The percentage entrainment (PE) is for the half-cycle with containing the most spikes is indicated in each panel. The technique for the construction of these plots is in the text.

plotted in this manner. It is easy to see that PE correlates fairly well with increases in entrainment. It is difficult to determine from the PE values if this rather simple approach really describes the spike train characteristics that one can appreciate from visual inspection. There is really rather little change in the numerical values of PE for panels B, C, and D. However, visual inspect of Figure 10.1 would lead one to expect greater differences in the measures since the spike trains are clearly different. In addition, PE is rather insensitive at describing multimodality in the phase distributions of spike occurrences. While PE can serve as a rough indicator of entrainment since it is assumed that perfect entrainment would be characterized by all the spikes occurring in one half-cycle, many subtle characteristics of the response are observed. For example, it is not hard to imagine that shifts in spike times of several milliseconds, correlated with shifts in the mean phase of the cycle distribution of tens of degrees might not result in any change in PE at all. Such shifts could occur if cortical neurons recipient of peripheral responses fired at shorter latencies indicating increased sensitivity or decreased suppression of their activity. In addition, PE alone makes it difficult to describe the degree of variance of spike occurrences distributed around the mean phase in the cycle distribution histogram. Thus, if trial-by-trial variations occur, they would tend not to influence PE greatly. A plot showing the phase relationship of a monkey somatosensory cortical neuron to a vibratory stimulus that triggered self-paced ballistic movements of the hand at the wrist joint is shown in Figure 10.4E. The relatively narrow distribution of the histogram is a good indicator of the tight temporal coupling between the stimulus and the response to it.

At this point, a central assumption should be discussed. That assumption is that in the primary somatosensory cortex there is a subpopulation of neurons that represents

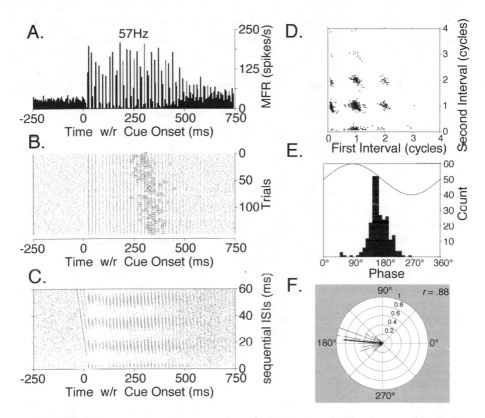

FIGURE 10.4 The responses of an area 1 cortical neuron to 57 Hz vibratory stimuli that trigger wrist flexions. **A.** A histogram of the neuron's response to 160 stimulus presentations. **B.** A raster display of spike occurrences as described in Figure 10.1. Open squares indicate the onset of wrist flexion movements **C.** Sequential interspike intervals (ISIs) plotted with respect to vibration onset which was the animals cue to move. For a given spike occurrence plotted on the abscissa, the values on the ordinate correspond to all subsequent spikes occurring within 60 ms of that spike. **D.** A joint interval scatterplot where the first and second ISIs are plotted on the abscissa and ordinate, respectively, for spikes occurring in the first 125 ms after the onset of the vibratory stimulus. Clusters on the diagonal indicate spiking at times equal to multiples of the period of the stimulus. **E.** A phase plot for spikes occurring in the epoch mentioned above. Bins are 10°. **F.** The values for the phase plot show in E, replotted in polar coordinates and showing the length of the mean vector (r) calculated from the equation included in the text. The length of the rays representing phase bins have been scaled so that the largest reaches the outer diameter of the polar plot; the mean vector length is plotted in the radial scale.

characteristics of peripheral stimuli by temporal patterns of neuronal discharges, e.g., vibratory entrainment, rather than merely by firing rates. Following this line of reasoning, changes in the degree of variability in spike occurrences may indicate changes in fidelity for that subpopulation. Shifts in the distribution of spikes without significant changes in variability may indicate changes in threshold for response, assuming that neurons spike once per stimulus cycle. These shifts could be associated with stimulus intensity, as well as in central processes. The former might cause peripheral receptors to activate at a different time whereas the latter might result from the release or application of centrally generated inhibition. We will not dispute that other cortical neurons may actually impart their information by changes in firing rates.[30]

10.2.5 SYNCHRONICITY OF THE RESPONSE WITH THE STIMULUS

A few years ago, we began experimenting with ways of expressing how well neuronal activity patterns were entrained with peripheral stimuli, and whether changes in the fidelity of the cortical representation of vibratory stimuli varied as a function of cortical location, receptive field type, and behavioral requirements.[28] The reasons for doing this will be discussed below. Turning to the ways this was done, we took advantage of the fact that if sinusoidal stimuli were presented at known times, we could measure each post-stimulus spike occurrence relative to the stimulus phase at that time. This required that a continuous representation of the data be constructed so that temporally adjacent points maintained their adjacency in the representation despite the incrementing of the number of cycles since stimulus onset. Luckily, this problem had been previously solved for data describing directional biases through the use of circular statistics.[31,32] As we discussed earlier, because vibratory stimuli are harmonic, a unit vector representing the stimulus phase can be thought of as rotating with a constant angular velocity around the center of a circular diagram. Spike occurrences can be represented as vectors depicting instantaneous phase of the stimulus at the time of a spike. To calculate time-varying representations of the mean phase angles, we used a sliding time window with the width equal to one to four stimulus periods (depending on the stimulus frequency) and calculated the mean and standard deviation of the response phase within this window. From the standard deviation, we derived a parameter we called "Synchronicity" by determining the difference between the floating angular standard deviation and that deviation expected from a uniform (completely random) distribution over the vibratory cycle, divided by the latter. This yielded a parameter which ranged between zero (uniform distribution) and one (constant phase relationship). Details of the derivation of this metric and its application to data from cortical recordings can be found in the previously mentioned article.[28] The benefit of this way of measuring fidelity of stimulus representation in the cortex is that it is, by nature, time varying. Thus, changes in fidelity as a function of different phases of a single behavioral task, or across similar epochs of different tasks could be compared relatively objectively. It was found that some, but not all, cortical neurons that represented vibratory stimuli did so rather well during approximately 100 ms following the onset of a vibratory cue signaling that a wrist movement could begin. However, just prior to the onset of movement, but generally before the onset of activity in muscles that will generate the upcoming movement, the fidelity with which somatosensory cortical neurons represented vibratory stimuli degenerated. Interestingly, this was at about the same time that tactile discrimination threshold in humans increases and the ability to sense peripheral stimulation degrades.[33,34]

10.2.6 PHASE AND MEAN PHASE OF THE RESPONSE

Many of the measures needed to provide an accurate description of the phase relationship of spikes occurrences relative to periodic peripheral stimuli can be derived from circular statistics[31,32] and visualized in phase plots constructed in polar coordinate systems. This type of representation of the occurrence of spike on a phase plane probably stems initially from a study in the late 1960s, which quantified the phase relationship of spikes associated with auditory stimuli.[35] Briefly, if each spike during a period of interest is thought of as a unit length vector at a phase angle between 0° and 360° then the distribution of a spike train can be represented by the distribution of a number of these vectors in the circle of a polar plot. Given that this is done, then the mean vector of the distribution can be calculated where its angle is the mean phase of the distribution and its vector strength, in the form of its vector length from 0 to 1, is a direct representation of the degree to which the spikes are phase-locked with the stimulus. The general formula of the aforementioned authors[31,32] can be used to calculate the mean vector of the spike occurrences:

$$r = (1/n) \bullet [(\Sigma \sin (2\pi f\ t_i))2 + (\Sigma \cos (2\pi f\ t_i))^2]^{1/2}$$

where r is the mean vector length, f is the frequency of vibration, t_i represents the time of spike occurrences, and n is the number of spikes. The value of r ranges from 0.0 to 1.0, where 0.0 corresponds to the uniform cycle distribution, and 1.0 corresponds to discharges at a constant phase relative to the stimulus cycle. It is this phase-locking or entrainment that we said earlier was an indication of the fidelity of the central representation of peripheral sensory events. In addition to the mean vector length, the angular deviation, the polar version of the standard deviation, can be calculated and several determinations can be made from that. One of the most important calculations that then must be done is to determine if the phase distribution of neuronal discharges is non-uniform (i.e., perhaps indicating vibratory entrainment), by using Rayleigh Test (Reference 31; pp. 54-55). In this test, a z-score is calculated according to the formula:

$$z = nr^2$$

where r is the mean vector length and n is the number of elements in the parent distribution. The values of z greater than 4.6 were taken as satisfying the criterion of statistical significance, which corresponded to $p <= 0.01$. Another is to re-represent the vector distribution, by doubling the phase angle, if the initial distribution appears bimodal.[31] This is sometimes necessary if a spike train contains spikes centered around both a phase and its antiphase. The resultant mean vector angle would tend to misrepresent the actual distribution since it would be at approximate right angles to both arms of the distribution and the mean vector length would be uncharacteristically short (see below). Data from an area 1 somatosensory cortical neuron, plotted in polar coordinates, is displayed in Figure 10.4F. The length of the mean vector, 0.88, indicates that the neuron is tightly coupled to the sinusoidal stimulus presented to its peripheral receptive field.

We sought ways to represent the fidelity with which neuronal discharges are entrained to the vibratory stimulus as a function of behavioral time. In our initial approach, with time relative to some important event represented on the abscissa, the phase of each spike relative to a sinusoidal stimulus was plotted on the ordinate (i.e., Reference 28; Figure 10.2E and 10.2F). One benefit of this sort of plot is that the relationship of response to the periodic stimulus can be easily visualized. Another is that if individual entries are replotted relative to some other behaviorally significant time, the same values of response phase can be used, although shifted in time. A difficulty with this sort of plotting procedure is that the values of the phase represented on the ordinate are discontinuous. This is because phase distributions that span the boundaries between the end of one cycle and the start of another may appear on opposite ends of the plot although they are actually continuous if the phase plane is represented from $-180°$ to $180°$ rather than from $0°$ to $360°$. To cope with this problem, we selected individually for each illustrated neuron whether to use the range $-180°$ to $180°$ or $0°$ to $360°$. The mean phase was also plotted for the selected range. Continuous shifts in the mean phase as the function of response phase indicated that responses to the stimuli started to occur earlier or later. These shifts are related to the changes in the responsiveness of neurons to the periodic inputs. Plots of mean phase also convey visually the information about the width of cycle distribution, and the plots of synchronicity quantify it.

10.2.7 INTERSPIKE INTERVALS, JOINT INTERVAL PLOTS, AND AUTOCORRELOGRAMS

The previously mentioned expression of Synchronicity served as a simple means of visualizing entrainment and the consistency of spike occurrences as a function of stimulus phase. However, it quantified only the distribution of discharges over the vibratory cycle, and left other spike-train features out. Additional information can be obtained from examining interspike intervals (ISIs). If, as previously done for response phase, the interspike intervals are plotted as a function of behavioral time, a time-dependent representation of the spike train periodicity can be constructed. This form of visualization has several benefits. As with the phase plots described above, shifting trial centering points so that they align with behaviorally significant events does little to degrade the image one sees. If the spike train contains ISIs of about the same duration, these will be viewed as a horizontal cluster of dots. The width of the cluster, then, is an indication of the consistency of the ISIs around a central tendency. If the stimulus eliciting the spikes is itself periodic, as are flutter and/or vibratory stimuli, then values on the ordinate can also be scaled in stimulus cycles. Another benefit of this form of data representation is that it does not necessarily make assumptions about the presence of periodicities at a given frequency in the spike trains. Moreover, it can reveal intrinsic periodicities present in spike trains even without an external periodic input. For example, plotting ISIs of a neuron that is intrinsically rhythmic or is driven by a secure driver at a frequency not previously known will result in a horizontal band of dots representing those ISIs. Thus, this form of data representation can allow for a crude but sometimes adequate assessment of the spike train's frequency spectrum.

Some basic variants of ISI plots can provide additional information about the temporal characteristics of spike trains. Joint interval scattergrams[26-40] provide a means of analyzing the serial dependencies of the ISIs. Simply, in these plots for a given spike occurrence (n), the plotted point represents the ISI between it and the next spike (n+1) on the abscissa. The ISI involving spike n+1 and spike n+2 forms the value of the ordinate. Clusters of dots at regular intervals on both axes are often seen.[40] Those equidistant along both axes represent ISIs that have occurred at the primary frequency component of the spike train. Clusters represented asymmetrically along the axes represent either multiple spikes per cycle or missing spikes (instances where a beat has been missed). Diagonal lines indicate regions of unentrained spikes (see Reference 40 for discussion). By this method, unentrained spike trains or those without intrinsic rhythmicity are represented without clusters and with dense groups of points near either axis.

Another way to visualize the consistency of spike occurrences is to construct expectation density (ED) histograms that illustrate the autocorrelation function.[1,7,39,41-43] Conventionally, an interval is chosen over which all spike occurrences from a reference point are compiled into a histogram with binwidths that may or may not have any relationship to the expected periodicity of the spikes. For example, one may choose to plot the time of occurrence of all spikes for several hundred milliseconds after the occurrence of each spike or after the occurrence of some other event (see Reference 39 for our implementation). If the ISIs are relatively consistent, the ED histogram will contain multiple peaks at intervals that thus give an indication of the firing pattern. The difference in the heights of the peaks in the ED histogram as a function of time gives an indication of this consistency as well. Variations in ISIs are correlated with progressively decreasing heights of peaks further from the point of reference. If the ISIs are quite consistent, the heights of the peaks will be approximately the same. Renewal density (RD) histograms can be used to determine the serial dependency in spike trains. One may think of a renewal process as if there is a clock whose countdown time is fixed and that is constantly reset by the occurrence of a spike. If the ISIs are shuffled or randomized, thus destroying their serial order, nonrenewal processes will have quite different ED and RD histograms. If, on the other hand, the occurrence of spikes takes on the properties of a renewal process, there may be little difference in the ED and RD histograms. This comparison can be very important since it has been argued that, in the case of externally driven activity, a peak in the RD histogram is smaller than the corresponding peak in the ED histogram.[1,7] However, neurons that are entrained to peripheral stimuli, that are themselves periodic, often exhibit little difference when ISIs are shuffled to construct RD histograms.[39] Histogram plots of ED and RD have a disadvantage since they represent large intervals and thus do not give much information about instantaneous changes in firing pattern consistency.

Returning to basic ISI plots, if, instead of simply plotting the first ISI on the ordinate for any given point on the abscissa, one plots all ISIs subsequent to a given point, for the extent of the range of the axis, a sort of running autocorrelogram is constructed. When these plots are examined closely, neuronal activity across several trials is compressed into a single representation. Rhythmic or entrained activity is represented as a series of horizontal bands. The width of these gives an indication of the degree of entrainment; the narrower the band, the better the entrainment. The consistency of the entrainment, in part, is indicated by the number of horizontal bands apparent in the plot. The number of bands is dependent on the extent of the axis and the frequency to which the spike train is entrained. This form of data representation, however, does little to convey the strength of the entrainment other than by the width of the bands. Figure 10.4C presents an example of this type of plot for the cortical neuron that has been used to illustrate the other types of data representation. A sinusoid which set the manipulandum in motion at time zero caused periodic spike occurrences in virtually all trials, as seen in Figure 10.4B. The multiple horizontal bands result from the security with which the stimulus drives the neural activity. Despite the continuing presence of the stimulus, the periodic response pattern is disrupted at about 325–500 ms after stimulus onset. This is about the time at which the monkey initiated a hand movement in response to the stimulus. The disruption in the temporal properties of the response pattern was probably due to receptive field surface shear as the movement began. Despite this, remnants of the temporal pattern are still visible in Panel C until about 500–650 ms post-stimulus. At this point, on average, the stimulus was turned off.

One way to achieve a representation of entrainment strength is to bin the data and encode the count in each bin using either gray or color scales. When this is done the resultant plot, in essence, retains those features of a running autocorrelogram described above, but also adds some information about the relative strength of the correlation between the driving stimulus and the spike trains associated with it. As such, these plots contribute additional information over and above the traditional expectation density histograms. To our knowledge, we are the only ones to have employed joint expectation density histograms of a spike train plotted against themselves to show preservation of rhythmic activity but alteration of intrinsic frequency during the course of a behavioral trial.[44] Extrapolating from this, it is not hard to imagine shuffling the ISIs as in a renewal density histogram (RD; References 1,7, and 41–42) to determine if the rhythmicity observed in the ED is preserved in the RD. If so, the driving force is probably intrinsic or extremely secure and extrinsic. If the periodicity breaks down with shuffling, the driving force is most likely from an extrinsic source (see Reference 39 for conceptual review).

10.3 WHAT THE CODE CAN REVEAL

Given that there are many ways of describing sensory responsiveness and quantifying the fidelity with which neuronal responses represent actual stimuli, the question becomes, what questions can be answered using the techniques outlined above? Our laboratory's long-term goal has been to gain a clearer understanding of how somatosensory information is processed dynamically during purposeful hand movements. It is important to understand when and under what behavioral circumstances the responsiveness of primate sensorimotor cortical neurons to peripheral inputs is altered. By understanding how, where, and when somatosensory responsiveness is modified during behavior, sensory processing will not only be better understood; deficits in processing can be more readily assessed and localized.

The somatosensory system must convey information to the central nervous system (CNS) about both external and internal sensory environments. Objects coming in contact with skin surfaces or causing deflection in limbs (exteroceptive inputs) must be represented accurately in time and space so that appropriate behaviors can be planned and executed. These behaviors may either maintain that contact, as in tactile exploration, or disengage from that contact, as in avoidance. Moreover, the CNS must also maintain spatially and temporally accurate representations of inputs from the body's own effectors, such as those encoding muscle tension and joint angle (proprioceptive inputs). Finally, to operate efficiently, the CNS probably maintains internal representations of expected sensory feedback from movements and anticipated results of the actions toward which the movements themselves are directed.[45-51] It is thought that central representations of inputs, intentions, and actions are continually updated and that the need for updating is closely linked to behavioral outcome, especially during goal-directed movements.[51-52]

10.3.1 CHANGES IN SENSORY RESPONSIVENESS

Before we examine behavior-dependent modulation of responsiveness, however, we need to ask a more straightforward question. That is, how do responses vary as a function of changes in the frequency of vibratory stimuli? There are several possibilities. Previous work (e.g., see Reference 7) would suggest that when less than optimal stimuli are delivered to a cortical neuron's peripheral receptive field, that neuron responds either poorly or not all. The tuning curves for afferent fibers and for cortical neurons that are the recipients of the ultimate projections of those fibers, illustrate that thresholds for activation of cortical neurons are higher when stimulus frequency is either higher or lower than that for which the neuron is tuned. These observations, done under rather static conditions, do not address what happens when the peripheral input is pertinent to some subsequent behavior. Over the past decade, we have conducted a number of experiments in which simple hand movements about the wrist joint were triggered by vibratory stimuli delivered to the monkey's palm, by way of a smooth aluminum plate. In different blocks of trials, we, at times, vary the frequency of the vibratory stimulus used to trigger the movements that the monkeys had previously been taught to make. Each monkey held a steady wrist position by pressing downward on the plate because the handle, if not actively held in a neutral position, had a tendency to elevate.

Figure 10.5 shows the results for one area 1 somatosensory cortical neuron, recorded while a monkey made several wrist flexions. The illustration consists of three columns, marked with the frequency of the vibratory go-cue that was delivered at time zero. These stimuli remained on until well after the monkey began the appropriate movement (open squares). In the first column, the apparent entrainment of the cortical neuron's firing to the frequency of the peripheral stimulus can be seen in the post-stimulus time histogram (Panel A), the raster plot (Panel B), and in the ISI plot (Panel C). In Panel D, the spike occurrences for the first 125 ms of the reaction time interval are plotted in polar coordinates relative to the phase of the stimulus at that time. In the column representing the 27 Hz stimulus trials, a slight bimodality in the phase distribution can be seen (Panel D). When the phase angles

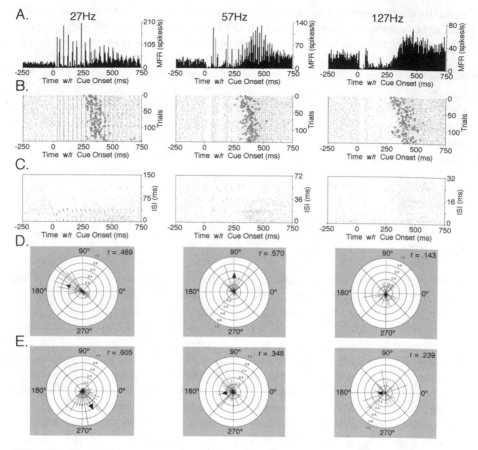

FIGURE 10.5 Histograms (A), rasters (B), ISI scattergrams (C) and, in D and E, polar plots of phase relationships of the spikes occurring in the first -125 ms after the onset of 27 Hz (left column), 57 Hz (middle column), and 127 Hz (right column) stimuli that triggered wrist flexions. Unlike Figure 10.4, scattergrams in C show only the single ISI for a spike occurring at the position on the abscissa and the next spike. Horizontal bands at twice the stimulus period, therefore, would indicate a "skipped cycle." In D, the polar plots are of actual phase relationships. In E, the phase angles have been doubled because of the apparent bimodality of some of the phase distributions. See text and references for conditions under which this manipulation of data the representation is appropriate. Arrows in both D and E indicate the length of the mean vector (r). The rays representing the binned phases of spike occurrences are normalized as described in Figure 10.4.

are doubled and replotted (Panel E), the length of the mean vector (r) improves somewhat. That is, the value is greater on a scale from 0 to 1. The center column, that for the 57 Hz trials, illustrates both the strengths and the weaknesses of each of these ways of visualizing and quantifying the data. It is clear from the histogram that the post-stimulus firing rate is lower than that for the 27 Hz trials. It is also evident that there are gaps in which there is little or no spiking. These occur immediately after stimulus onset and just prior to movement onset (open squares).

Thus, this neuron may have had its activity suppressed from background levels (while the monkey held a neutral position and awaited the vibratory go-cue). Notably, during the two brief phasic episodes of firing that occur between cue- and movement-onset, the neuron's activity is loosely entrained to the stimulus, as seen in the ISI plot (Panel C, center column). By examining this plot, it is also possible to see that the activity of this neuron becomes entrained at about the same time as wrist movements begin. The polar plots in Panels D and E must be interpreted with some caution. While the value representing the length of the mean vector in Panel D of this column is indeed greater than for the length of the mean vector represented in the left column, the number of spikes contributing to the calculation of this value was much smaller. Although both plots show distributions that are statistically different from a uniform distribution, there is no doubt that this neuron is better entrained to the lower of these stimulus frequencies. In the right column, the firing rate immediately after the onset of the 127 Hz go-cue is noticeably suppressed. There appears to be a brief return to the background firing rate at about 100 ms after stimulus onset. The activity of this neuron then appears to have been suppressed until the onset of wrist movements. After movements were begun, the activity does not appear to have been entrained to this highest of the three frequencies.

From observations like these, we could make several conclusions. First, somatosensory cortical neurons that are vibratory-responsive are not always entrained to periodic stimuli presented in their peripheral receptive fields. This certainly is to be expected, especially if the amplitude of the stimulus remains constant but its frequency is varied. The stimulus may have been at a frequency for which the neurons were not tuned. Second, neurons that are entrained to the frequency of a behaviorally important vibratory stimulus do not necessarily maintain background firing rates if the stimulus frequency is not optimal; their activity can be suppressed. Finally, no single method, at least of those illustrated, is sufficient to fully describe the stimulus-related activity of neurons of the type shown in Figure 10.5. However, several methods, taken together, can give a rather complete description of a neuron's behavior as the animal is behaving.

The somatosensory system is distinctive in that peripheral discharges occur not only as the result of interaction with external objects being sensed but also during any movement of the part of the body sensing them. Hence, the central nervous system must be able to distinguish peripheral signals carrying information from external objects from signals that are by-products of movements. It has been suggested that the central nervous system anticipates peripheral inputs in advance of movements, and generates so-called efference copy that is then compared with the peripheral discharge.[53] According to this theory, modulation should occur at about the time of movement initiation, interact with redundant or competitive signals, vary with the properties of the sensory inputs, and lead to some sensations being suppressed, while others are facilitated. However, we still know very little about how and which behaviors modify the responsiveness of single units or groups of sensorimotor cortical neurons that process inputs during hand movements. It has become increasingly clear that central responses to peripheral stimuli may be context-dependent and that the CNS has the ability to multiplex, that is, select the information channel(s) at any given time that are most relevant for current behaviors. In humans, gating of somatosensory-evoked potentials before the onset of muscle activity (measured by electromyography, EMG) is thought to result from cortically generated premotor events and is not due to corticofugal influences.[54-58] This gating may represent corollary discharge[59] or efference copy[60-61] by which motor centers inform the sensory cortices that subsequent movements are centrally generated.[62-66] It has been suggested that "the cutaneous input may be inhibited [before and] during movement to facilitate transmission of information from other receptors more important for the control of movement"[67] (such as those from joint afferents and muscle receptors). This modulation "…has been interpreted as [resulting in] an improvement in signal to noise ratio"[33] and may also be important in reducing both cutaneous and deep inputs that might interfere with motor control.[67-71] The representation of motor commands in sensorimotor cortex has been reviewed recently.[53] It has been difficult to prove when and where motor commands modify sensory responses and movement-related neuronal activity. To better understand the behavior-dependent modulation of responsiveness, we have examined this modulation during a couple of common behaviors that have given us insights into sensory processing in general.

10.3.2 SENSORY RESPONSIVENESS CHANGES WITH CHANGES IN EXPECTATION

First, we altered reward schedules to determine if sensory responses and movement-related single-unit activity of sensorimotor cortical neurons changes as a function of the outcome of the previous behavioral trial. We hypothesized that the activity of sensorimotor cortical neurons would be modulated with changes in reward predictability and Knowledge of Result (KR). We thought that sensory responses would be better related to vibratory stimuli and movement-related activity would be more tightly coupled to movement kinematics immediately following unsuccessful behavioral outcomes.

Three monkeys made vibration-cued or visually-cued ballistic wrist flexion and extension movements (Figure 10.6). They received rewards on average 75% of the correctly performed trials. Fruit juice rewards were withheld pseudo-randomly and for incorrect behavioral responses (movements in the non-cued direction). A total of 147/356 sensorimotor cortical neurons (Figures 10.6B and 10.6C) from two monkeys were analyzed in detail because these neurons showed vibratory go-cue-related activity changes and had been held long enough to record at least 25 trials for each movement direction. Data were separated into five types, based on if a trial 1) followed a rewarded trial, 2) was unrewarded, 3) followed an experimentally withheld reward, 4) contained a movement error, or 5) followed an unrewarded trial due to a movement error.

Several neuronal activity measures were calculated and compared. Trial-dependent differences in mean firing rates as a function of trial type were calculated while the monkeys waited for go-cues (BKG), during the initial response to vibration (RAMFR) and following the initial vibratory (SAMFR) by conventional means (Figure 10.6D). Data were represented as histograms and rasters depicting single-trial spike occurrences with corresponding wrist position traces for comparison (Figure 10.6E and 10.6F). In some instances, scatterplots of interspike intervals

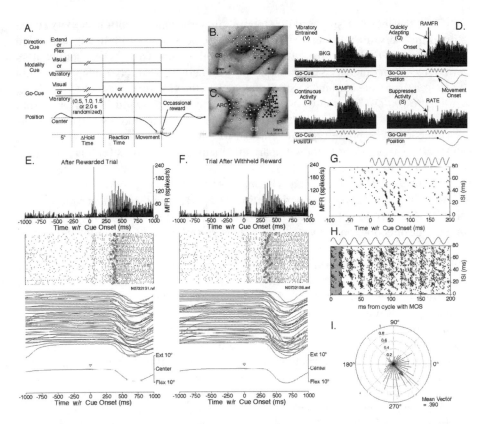

FIGURE 10.6 A. Schematic of the Withheld Reward Task. Each animal held a steady position for a random time, detected the go-cue, and then made ballistic wrist flexions or extensions. Requested movement directions and the probability of reward were variable. B and C. Dorsolateral views of the recorded regions in two monkeys (N and E). Filled circles indicate surface location of penetrations containing cue-related neurons. ARC = Arcuate, CS = Central, and IP = Intraparietal Sulci. D. Population histograms of area 1 neurons depicting four types of cue responses. Epochs for measuring firing rates are shown. E and F. Top: Histograms showing mean firing rates (bin width 5 ms). Rasters in which each dot represents a spike and each row represents a single trial. Dark marks indicate movement onset. Bottom: Single and average wrist position traces. G and H. Interspike intervals (ISI) constructed by plotting all the ISIs for 80 ms following a spike in behavioral time. Rhythmic firing appears as clusters. MOS = movement onset. I. Polar phase plot of spike occurrences in H. Rays represent 10° bins. Arrow indicates length of mean vector, indicating the degree of entrainment ranging from 0 to 1.

(ISIs) as a function of time before or after behaviorally significant events were examined for periodicities indicative of vibratory-stimulus entrainment (Figures 10.6G and 10.6H). For neurons with vibratory go-cue-related activity increases, mean vector lengths resulting from polar plots of spike occurrences were calculated as a function of the stimulus phase at which theses spikes occurred (Figure 10.6I).

First, the types of single-unit responses in sensorimotor cortex to vibrotactile stimuli that served as go-cues for ballistic movement of the stimulated hand were characterized. Four distinctly different types were found. Three types were characterized by activity increases that were either continuous (C neurons), adapted quickly (Q neurons), or were entrained to the frequency of the vibratory stimulus (V neurons; see Figure 10.6D). The fourth type had their activity suppressed (S neurons) from BKG levels at short latencies after stimulus onset. The distributions of 147 neurons as a function of response type and cortical location (area 4 motor cortex, areas 3a, 3b, 1, 2/5, and posterior to the intraparietal sulcus parietal cortex [P]) can be found in Table 1 of Reference 72.

The behavioral patterns of each animal were analyzed to determine if the onsets of cue-related activity changes and reaction times (RTs) and movement times (MTs) differed as a function of receptive field (RF) type, cortical location, response class, movement direction, and reward history. Onsets varied as a function of the former two, but not the latter three characteristics. When rewards were withheld experimentally, but not because of a behavioral error, RTs and MTs decreased. RTs were shorter than baseline in trials immediately after withheld rewards, independent of movement direction. For flexion movements, RTs were also shorter than baseline immediately following trials in which the animal failed to receive a reward because of movement in the wrong direction. In addition, also for flexion movements, RTs were longer in those trials where directional failures occurred. MTs immediately following directional failures were not different than baseline, regardless of movement direction. MTs immediately following trials in which reward for correct performance were withheld were significantly shorter than baseline, independent of movement direction.

After examining changes in behavior, firing rates and stimulus-entrainment were analyzed to determine if they varied with reward history. After experimentally withheld rewards, but not after behavioral errors, activity increased significantly in areas 3a, 3b, and 1 during the instructed delay period (IDP; after instructions but before go-cues). The initial activity bursts (QAMFR) in area 1 and the parietal cortex increased. The degree of entrainment of neurons in area 3a, as determined by the mean vector length, decreased significantly just before movements and, during movements, the same was true for both areas 3a and 3b.

Finally, the data from neurons with similar response characteristics from a single cortical area (area 1) were combined. This population analysis revealed additional characteristics of neurons belonging to relatively homogeneous response classes that were not obvious when records from single units were examined individually. The population of neurons that fired continuously in response to vibratory go-cues (C neurons) showed what seemed to be entrainment to the vibratory stimulus during, and in the trial after, directional mistakes (Figure 10.7, upper row, panels 3 and 4 from left). The population of neurons whose cue-related bursts of activity quickly returned to background rates (Q neurons) showed some entrainment to the vibratory stimulus frequency during the initiation and execution of movements. The population of neurons with suppressed activity in response to vibratory go-cues (S neurons) appeared to fire somewhat periodically at about 30–40 Hz before the vibration onset. The vibratory-entrained neurons (V neurons) shown increased "burstiness" at stimulus onset in instances where the animal made a directional mistake and in trials immediately after those mistakes (Figure 10.7, bottom row, panels 3 and 4 from left). These properties just described were not as clearly visible in records of single units as they were for the populations of area 1 neurons.

The results of this population analysis lead to several suggestions. The underlying assumption is that, by grouping neurons from a single cortical area that have similar response patterns, information can be obtained about how these neurons might work together during behavior. When viewing displays of these compiled spike patterns, certain temporal characteristics emerge. Continuously spiking (C) neurons seem to become weakly entrained to peripheral stimuli during and immediately after mistakes made by the monkeys. As is seen clearly in some single records (Figures 10.6G and 10.6H) neurons that adapt quickly to peripheral stimuli also may be weakly entrained, but in this case, after movements have begun. The neurons whose firing rates are suppressed at stimulus onset appear to fire rhythmically, while the animals hold against a light load and await vibratory go-cue onset. It seems unlikely that these neurons are the same type as the one illustrated in Figure 10.5; however, multiple vibratory frequencies were not tested in this study. Thus, without further experimentation, their relationship to neurons entrained to low frequency vibration but suppressed by higher frequencies remains unclear. The neurons that were clearly entrained to the vibratory go-cues (V) were somewhat more responsive to the onset of the stimulus before directional mistakes and in the trials immediately following them. Entrainment following the initial burst, however, did not seem qualitatively different as a function of whether or not previous trials were rewarded or mistakes had been made. Taken together, these findings suggest that sensory responsiveness can and does change on a trial-by-trial basis and that these changes can be visualized either for single or composite records by some of the techniques described above.

10.3.3 SENSORY RESPONSIVENESS DURING GUIDED MOVEMENTS

A second set of experiments was done to determine if sensory responses of sensorimotor cortical neurons are best related to stimuli when movements are guided by vibratory stimuli and if selective enhancement of cortical activity is specific to cells representing the peripheral stimulus location. As noted before, previous studies in our laboratory have lead to the suggestions that central inputs may modulate sensorimotor cortical responses to peripheral stimuli, but that these responses themselves in sensorimotor cortical areas may be gated when the peripheral inputs specify wrist movement endpoints.

Monkeys were trained to make wrist extensions and flexions after holding steady positions for 0.5–2.0 seconds (Figure 10.8A). They held against a load opposing flexion movements. Vibration (Vb) to the monkey's palm, appearance of visual (Vs) targets, or both Vb and Vs combined served as go-cues for identical movements in corresponding paradigms (VIB-only, VIS-only, and COM). Movement end points were specified by increases in vibration to the monkey's palm, superimposition of the wrist movement cursor with visual targets, or both. Trials were run in same-cue blocks where movement direction was alternated every tenth correctly executed trial. Wrist movement was pseudorandomly chosen to be 5° or 10° flexions or extension

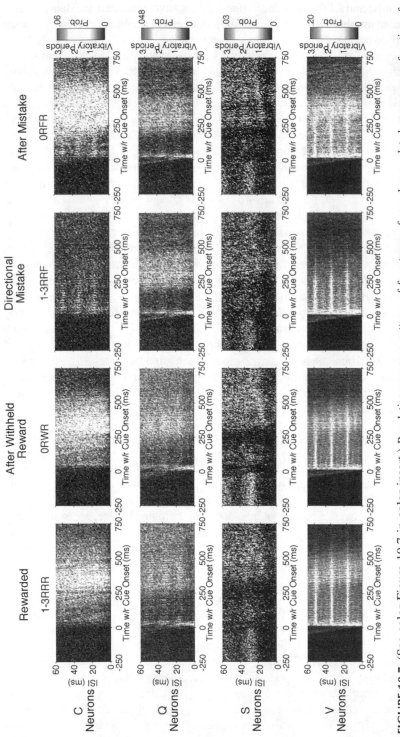

FIGURE 10.7 (See color Figure 10.7 in color insert.) Population response patterns of four types of area 1 cue-related neurons as a function of trial type and behavioral time. Scatterplots constructed by binning all ISIs (5 ms × 1 ms) and normalizing by spikes/trial to give probability of spike occurrences in a bin. Probability indicated by attributed color. Tick marks to the right indicate time corresponding to the period of the 57 Hz vibratory stimulus. Response types as in Figure 10.6. Nomenclature for trial types as in text.

in the VIS-only, and COM paradigms. To date, we have examined the data for only 10° movement amplitudes from one monkey. All correct movements were rewarded in these self-paced tasks. The animal was not penalized for variations in RTs, MTs, or intertrial intervals. Complete records (all three paradigms) were obtained for 403/691 task-related neurons; 156 were located in pre-central and 247 in post-central cortical areas. As shown in Figure 10.8B, reconstruction of the electrode penetrations indicated that the recordings in the post-central cortex were mainly in areas 3a, 3b, and 1.

The general behavior of the monkey was quantified by examining records from the post-central cortical neurons. RTs, during the VIS-only task, were greater than those for VIB-only and COM for both wrist extension and flexion movements (p < 0.01). MTs for wrist flexion were longer than that of wrist extension. However, there is no significant difference for MTs across the three paradigms as a function of direction.

Analyses were conducted for 74/247 (~30%) post-central cue-related neurons using analytical procedures developed in the previously described study. These neurons have been grouped, by visual inspection, into C (n = 25), Q (n = 17), S (n = 16), and V (n = 16) neurons, based on their responses to vibratory stimuli. Figure 10.8C shows the responses of a V neuron which exhibited no cue-related discharges in VIS-only trials, but was entrained to the vibratory stimulus in both VIB-only and COM-cue paradigms. These paradigms will subsequently be mentioned collectively as the vibratory paradigms (or trials). Activity onsets, BKG, QAMFR, SAMFR were measured. Mean vector length was calculated for all neurons with cue-related firing rate increases during the vibratory paradigms.

The animal behaved similarly during the vibratory paradigms, but there were some notable differences in response patterns. During post-central neuron recording, RTs for the vibratory trials averaged 275± 11 ms and were significantly shorter than those for VIS-only trials (297± 15; p<.0001; repeated measures ANOVA). BKG was not different across the three paradigms. Comparing the firing rates across the vibratory paradigms yielded no statistically significant differences in QAMFR and SAMFR, although firing rates during these two epochs, as well as during BKG, tended to be higher in VIB-only trials. However, across these paradigms, there were significant differences in mean vector length for the V neurons but not for C and Q neurons. V neurons were better entrained to the premovement part of the 57 Hz vibratory cue in the COM-cue trials (mean 0.393± .24) than in the VIB-only trials (mean 0.275± .17; p<.0001; repeated measures ANOVA). Quantifying the entrainment during movement to the targets was somewhat difficult because the frequency of the vibration is time-variant (by experimental design) and the MTs, while relatively uniform, vary by times equal to several periods of the final 127 Hz vibratory stimulus. These variations can be seen in the population plots presented in Figure 10.9.

From this figure, several initial observations, later confirmed statistically, were noted. It appeared that in the vibratory paradigms, both C and Q neurons encoded the vibratory stimuli to some extent. While the general probability of spike occurrence in any given bin was higher in the the VIB-only paradigm for these classes of neurons, the spread of the distribution was also wider, suggesting that entrainment may not be as good as in the COM-cue paradigm. It appeared that V neurons may

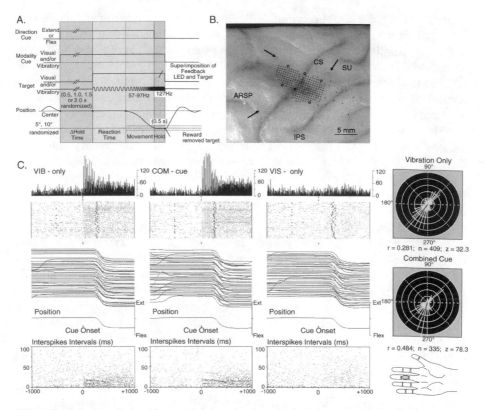

FIGURE 10.8 (See Color Figure 10.8 in color insert.) **A**. Schematic of the Guidance Task with the three variations tested. Animals held a steady position for a random time, detected the target, and then made wrist flexion or extension movements and held at the target zone. Targets in the visual-only trials (VIS-only) were LEDs signaling the position to which the animal moved. Approach to the target in the vibratory-only trial (VIB-only) was signaled by increasing the vibratory frequency as the animal neared the target and decreasing frequency if he moved away from it. At the target, the vibratory frequency abruptly increased by 30 Hz. The third condition combined both visual and vibratory targets (COM-cue). All correct trials were rewarded. **B**. Dorsolateral view of the cortical surface of the most extensively studied monkey showing the locations of penetrations in the pre- and post-central cortices. Anterior (left); Medial (up). ARSP = Arcuate Spur; IPS = Intraparietal Sulcus; CS = Central Sulcus; SU = Superior Parietal Dimple. Open circles depict locations of electrolytic lesions. Arrows show postmortem pin marks placed relative to recording chamber coordinates. **C**. (below). A vibration-responsive neuron from primary somatosensory (SI) cortex with an RF on the third digit. The cycle distribution of the vibratory response became more bimodal in vibration-only trials and the mean vector length (r) was shorter. Increased bimodality or cycle distribution broadening occurred for most SI vibration-responsive neurons.

be more poorly entrained to vibratory stimuli prior to movement but better entrained during movement and once the target was achieved and the new position held. This suggestion is based on the intensity of the spike occurrences at multiple periods of the final (127 Hz) stimulus frequency at times (~400–900 ms) when the monkeys held the handle still at the new position. In several cases, activity patterns immedi-

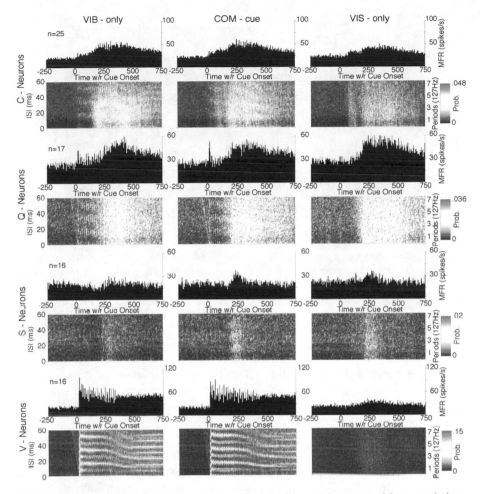

FIGURE 10.9 (See Color Figure 10.9 in color insert.) Response patterns of four populations of postcentral cortical cue-related neurons as a function of trial type and behavioral time. Scatterplots as described in Figure 10.7. Response types as in Figure 10.6. Initial vibratory cue frequency was 57 Hz, but increased linearly as animals moved the handle toward a target 10° from center. When the target was reached, the final stimulation frequency was 127 Hz. Thus, periods shown at the right of each scatterplot are for that frequency. Entrainment is evident in V neurons; weaker entrainment appears present in Q and C neurons.

ately after go-cue onset were qualitatively different when changes in vibratory frequency specified target location. In these instances, cessations in activity were seen that did not occur in either VIS-only or COM trials.

Appropriate quantification methods of the entrainment during and after movements are still being developed. Conceptually, the latter seems easier to accomplish since the records could be aligned in a time corresponding to the beginning of a vibratory cycle in which the end of the movement occurs. Thus, the records would maintain their temporal association with the 127 Hz stimulus, independent of movement

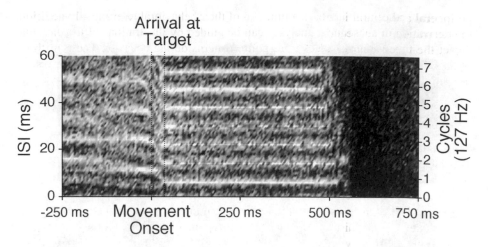

FIGURE 10.10 (See Color Figure 10.10 in color insert.) Scatterplot of the lower left panel of Figure 10.9 with data centered on wrist movement onset. When the target was reached, the final stimulation frequency was 127 Hz. Periods shown at the right of the scatterplot are for that frequency, which was presented while the animal held the handle 10° from the neutral position. Entrainment of this population of neurons is clearly seen for both the vibratory stimulus present before the animal moved and once he attained the target zone.

time. To initially see if this might be the case, data can be plotted using one of the methods described above. Figure 10.10 shows the re-plotting of one of the population scattergrams from Figure 10.9, only this time with the data centered on the onset of wrist movement. In this case, it is the lower left panel which depicts the ISIs for the area 1 V neurons. Although mean vector analyses have not been conducted as yet for these data, several things are clear. First, these neurons initially show entrainment to the 57 Hz vibratory stimulus that acted as the go-cue to trigger the movements. Second, during the movement (from movement onset to arrival at target), there is apparently a smooth transition in the ISIs from the initial to the new (127 Hz) vibratory frequency that reflects the experimental design. Finally, once the target, and hence the new handle position is attained, these neurons appear to be entrained to the high frequency and not suppressed by it as was the neuron whose records are illustrated in Figure 10.5. The distribution of ISIs, once the target is reached, is quite narrow, suggesting that these neurons closely follow the frequency of the vibration of the handle at the target zone.

These observations suggest that when movement endpoints are specified by somatosensory cues alone, initial increases in firing rates may be accompanied by degradations in the fidelity of responses. Prior to movement, ongoing responses to vibration are suppressed for some types of neurons. Some neurons are more poorly entrained by vibratory stimuli in trials without explicit visual targeting cues. It appears from population plots that this entrainment may improve significantly once movements, guided by vibration alone, begin. If so, it would be consistent with the idea that suppression of activity (or correlated firing) occurs before instances of highest activity/movement coupling to improve signal-to-noise ratios. Certainly, more studies are needed to examine suppression and facilitation of responses to peripheral and central inputs as a function of their behavioral relevance. In addition, the derivation of subsequent analyses can be guided by illustrations of the data that depict the time-variant changes in response patterns during behavior. These methods of data illustration can be used to directly compare response patterns which, for behavioral trials, vary in duration, with respect to behaviorally significant events.

10.4 SUMMARY

This chapter began by asking how one might go about determining the fidelity with which peripheral stimuli are represented by sensorimotor cortical neurons. To start to answer this question, the responses to vibratory stimuli were considered because these stimuli, by nature, have a temporal signature. Neurons that spike with relatively consistent ISIs corresponding to the frequency of known peripheral stimuli are well documented.[7] What historically has been difficult to do is to find the means of quantifying the fidelity of signal representation in the responses of neurons to those signals. Several ways of doing so have been presented here. Perhaps none is sufficient alone. Taken together, however, these analytical tools provide a means of suggesting answers to several questions about how cortical representations work when animals are working.

As stated earlier, what is presented here is by no means an exhaustive review of all the ways in which data may be represented to address the question of response fidelity. That was not the intent. The intent was to describe some forms of illustrating pertinent data in forms that are readily tractable to those not necessarily in the subdiscipline and measurements that can accompany these forms of data representation. Some changes in sensory responsiveness are far from subtle. For those, differences in mean firing rates may be sufficient to test the hypotheses under consideration. Certainly, stimulus intensity is well correlated with firing rates over much of the dynamic range of sensorimotor cortical neurons. Stimulus location in this system is represented in a number of topographic, though distorted, representations of the sensory periphery.[4] However, some influences on sensation of peripheral events and on behaviors that may accompany these sensations are subtle, such as attention, intention to move, expectation of goal attainment, and Knowledge of Result (KR). It is therefore reasonable to think that firing pattern changes associated with the alterations just mentioned may themselves be subtle. These changes may be in the finer aspects that represent fidelity and not necessarily as mean firing rate changes. Some of the data analyses and data representations discussed above are useful in deciphering the code that the periphery sends the central nervous system as the brain deals with the sensory environment and its own body image.

ACKNOWLEDGMENTS

The original work described herein and the formulation of this manuscript was funded by research grant R01-NS36860 from the National Institute of Neurological Disorders and Stroke to Randall J. Nelson. We thank John M. Denton for his assistance in drafts of the manuscript.

REFERENCES

1. Mountcastle, V. B., Steinmetz, M. A., and Romo, R., Frequency discrimination in the sense of flutter: psychophysical measurements correlated with postcentral events in behaving monkeys, *J. Neurosci.*, 10, 3032, 1990.
2. Ahissar, E., Sosnik, R., and Haidarliu, S., Transforming from temporal to rate coding in a somatosensory thalamocortical pathway, *Nature*, 406, 302, 2000.
3. Woolsey, C. N., Organization of somatic sensory and motor areas of the cerebral cortex. In: *Biological and Biochemical Bases of Behavior*, The University of Wisconsin Press, Madison, 63, 1958.
4. Kaas, J. H., Jain, N., and Qi, H. X., The organization of the somatosensory system in primates. In: *The Somatosensory System: Deciphering the Brain's Own Body Image*, Nelson, R. J., Ed., CRC Press, Boca Raton, 2001.
5. Talbot, W. H., Darian-Smith, I., Kornhuber, H. H., and Mountcastle, V. B., The sense of flutter-vibration: comparison of the human capacity with response patterns of mechanorecptive afferents from the monkey hand, *J. Neurophysiol.*, 31, 301, 1968.
6. Johansson, R. S., Tactile afferent units with small and well documented receptive fields in the glaborous skin area of the human hand. In: *Sensory Functions of the Skin of Humans*, Kenshalo, D. R., Ed., Plenum, New York, 129, 1978.
7. Mountcastle, V. B., Talbot, W. H., Sakata, H., and Hyvarinen, J., Cortical neuronal mechanisms in flutter-vibration studied in unanethetized monkeys. Neuronal periodicity and frequency discrimination, *J. Neurophysiol.*, 32, 452, 1969.
8. Whitsel, B. L., Kelly, E. F., Delemos, K. A., Xu, M., and Quiberra, P. M., Stability of rapidly adapting afferent entrainments responsivity, *Somatosens. Motor Res.*, 17, 13, 2000.
9. DiCarlo, J. J. and Johnson K. O., Velocity invariance of receptive field structure in somatosensory cortical area 3b of the alert monkey, *J. Neurosci.*, 19, 401, 1999.
10. Gardner, E. P., Ro, J. Y., Debowy, D., and Ghosh, S., Facilitation of neuronal activity in somatosensory and posterior parietal cortex during prehension, *Exp. Brain Res.*, 127, 329, 1999.
11. Nelson, R. J., Modulation of somatosensory cortical responsiveness following unexpected behavioral outcomes. In: *Somatosensory Processing: From Single Neuron to Brain Imaging*, Rowe, M. J., Ed., Gordon and Breach Science Publishers of Harwood, North Ryde, Australia, 209, 2000.
12. Nicolelis, M. A. L., Ghazanfar, A. A., Stambaugh, C. R., Olivera, L. M. O., Laubach, M., Chapin, J. K., Nelson, R. J., and Kaas, J. H., Simultaneous encoding of tactile information by three primate cortical areas, *Nat. Neurosc.*, 1, 621, 1998.
13. Pruett, J. R., Sinclair, R.J, and Burton, H., Response patterns in second somatosensory cortex (SII) of awake monkeys to passively applied tactile gratings, *J. Neurophysiol.*, 84, 780, 2000.
14. Ro, J. Y., Debowy, D., Ghosh, S., and Gardner, E. P., Depression of neuronal firing rates in somatosensory and posteriorparietal cortex during object acquisition in a prehension task. *Exp. Brain Res.*, DOI 10.1007/s002210000496, URL:http://dx.doi.org/10.1
15. Salinas, E. and Romo, R., Conversion of sensory signals into motor commands in primary motor cortex, *J. Neurosci.*, 18, 499, 1998.
16. Sergio, L. E. and Kalaska, J. F., Changes in the temporal pattern of primary motor cortex activity in a directional isometric force vs. limb movement task, *J. Neurophysiol.*, 80, 1577, 1998.

17. Tommerdahl, M., Delemos, K. A., Favorov, O. V., Metz, C. B., Vierck, C. J., and Whitsel, B. L., Response of anterior parietal cortex to different models of same-site skin stimulation, *J. Neurosci.*, 80, 3272, 1998.

18. Tommerdahl, M., Delemos, K. A., Whitsel, B. L., Favorov, O. V., and Metz, C. B., Response of anterior parietal cortex to cutaneous flutter vs. vibration, *J. Neurosci.*, 82, 16, 1999.

19. Zhang, J., Riehle, A., Requin, J., and Kornblum, S., Dynamics of single neuron activity in monkey primary somatosensory motor cortex related to sensorimotor transformation, *J. Neurosci.*, 17, 2227, 1997.

20. Vega-Bermudez, F. and Johnson, K. O., Surround suppression in the responses of primate SA1 and RA mechanoreceptive afferents mapped with a probe array, *J. Neurophysiol.*, 81, 2711, 1999.

21. Alloway, K. D. and Burton, H., Differential effects of GABA and bicuculline on rapidly- and slowly-adapting neurons in primary somatosensory cortex of primates, *Exp. Brain Res.*, 85, 598, 1991.

22. Meyers, W. A., Churchill, J. D., Muja, N., and Garraghty, P. E., Role of NMDA receptors in adult primate cortical somatosensory plasticity, *J. Comp. Neurol.*, 418, 373, 2000.

23. Whitsel, B. L., Favorov, O., Delemos, K. A., Lee, C. J., Tommerdahl, M., Essick, G. K., and Nakhle, B., SI neuron response variability is stimulus tuned and NMDA receptor dependent, *J. Neurosci.*, 81, 2988, 1999.

24. Lebedev, M. A., Mirabella, G., Erchova, I., and Diamond, M. E., Experience-dependent plasticity of rat barrel cortex: redistribution of activity across barrel-columns, *Cereb. Cortex*, 10, 23, 2000.

25. Ageranioti-Belanger, S. A. and Chapman, C. E., Discharge properties of neurons in the hand area of primary somatosensory cortex in monkeys in relation to the performance of an active tactile discrimination task. II. Area 2 as compared to areas 3b and 1, *Exp. Brain Res.*, 91, 207, 1992.

26. Roy S. and Alloway K. D., Synchronization of local neural networks in the somatosensory cortex: a comparison of stationary and moving stimuli, *J Neurophysiol.*, 81, 999, 1999.

27. Vaadia, E., Haalman, I., Abeles, M., Bergman, H., Prut, Y., Slovin, H., and Aersten, A., Dynamics of neural interactions in monkey cortex in relation to behavioral events, *Nature*, 373, 515, 1995.

28. Lebedev, M. A., Denton, J. M., and Nelson, R. J., Vibration entrained and premovement activity in monkey primary somatosensory cortex, *J. Neurophysiol.*, 72, 1654, 1994.

29. Ferrington, D. G. and Rowe, M., Differential contributions to coding of cutaneous vibratory information by cortical somatosensory areas I and II, *J. Neurophysiol.*, 43, 310, 1980.

30. Romo, R. and Salinas, E., Touch and go: decision-making mechanisms in somatosensation, *Annu. Rev. Neurosci.*, 24, 107, 2001.

31. Batschelet, E., *Circular Statistics in Biology*, Academic Press, New York, 1981.

32. Zar, J. H., *Biostatistic Analysis*, Prentice-Hall, Englewood Cliffs, 1974.

33. Coquery, J. M., Role of active movement in control of afferent input from skin in cat and man, In: *Active Touch: The Mechanisms of Object Manipulation: A Multidisciplinary Approach*, Gordon, G., Ed., Pergamon, Oxford, 1978.

34. Williams, S. R., Shenasa, J., and Chapman, C. E., Time course and magnitude of movement-related gating of tactile detection in humans. I. Importance of stimulus location, *J. Neurophysiol.*, 79, 947, 1998.

35. Goldberg, J. M. and Brown, P. B., Response of binaural neurons of dog superior olivary complex to dichotic tonal stimuli: Some physiological mechanims of sound localization, *J. Neurophysiol.*, 32, 613, 1969.

36. Rodieck, R. W., Kiang, Ny-S., and Gerstein, G. L., Some quantitative methods for study of spontaneous actvity of single neurons, *Biophys. J.*, 2, 351, 1962.

37. Surmeier, D. J. and Towe, A. L., Properties of proprioceptive neurons in the cuneate nucleus of the cat, *J. Neurophysiol.*, 57, 938, 1987a.

38. Surmeier, D. J. and Towe, A. L., Intrinsic features contributing to spike trains patterning in proprioceptive cuneate neurons, *J. Neurophysiol.*, 57, 962, 1987b.

39. Lebedev, M. A. and Nelson, R. J., Rhythmically firing (20–50 Hz) neurons in monkey primary somatosensory cortex: Activity patterns during initiation of vibratory-cued hand movements, *J. Comp. Neurosci.*, 2, 313, 1995.

40. Lebedev, M. A. and Nelson, R. J., High-frequency vibratory sensitive neurons in monkey primary somatosensory cortex: entrained and nonentrained responses to vibration during the performance of vibratory-cued hand movements, *Exp. Brain Res.*, 111, 313, 1996.

41. Perkel, D. H., Gerstein, G., and Moore, G., Neuronal spike trains and stochastic point processes. I. The single spike train, *Biophys.*, 7, 391, 1967.

42. Poggio, G. F. and Viernstein, L. J., Times series analysis of impulse sequences of thalamic somatic sensory neurons, *J. Neurophysiol.*, 27, 517, 1964.

43. Zadeh, L. A., Signal-flow graphs and random signals, *Proc. I. R. E.*, 45, 1413, 1957.

44. Lebedev, M. A. and Nelson, R. J., Rhythmically firing neostriatal neurons in monkey: Activity patterns during reaction-time hand movements, *J. Neurophysiol.*, 82, 1832, 1999.

45. Blakemore, S. J., Goodbody, S. J., and Wolpert, D. M., Predicting the consequences of our own actions: the role of sensorimotor context estimation, *J. Neurosci.*, 18, 7511, 1998.

46. Gerdes, V. G. and Happee, R., The use of internal representation in fast goal directed movements: a modeling approach, *Biol. Cybern.*, 70, 513, 1994.

47. Jenmalm, P. and Johansson R. S., Visual and somatosensory information about object shape control manipulative fingertip forces, *J. Neurosci.*, 17, 4486, 1997.

48. Kalaska, J. F. and Crammond, D. J., Deciding not to go: neuronal correlates of response selection in a go/no-go task in primate premotor and parietal cortex, *Cereb. Cortex*, 5, 410, 1995.

49. Kalaska, J. F., Scott, S. H., Cisek, P., and Sergio, L. E., Cortical control of reaching movements, *Curr. Opin. Neurobiol.*, 7, 849, 1997.

50. Ruiz, S., Crespo, P., and Romo, R., Representation of moving tactile stimuli in the somatic sensory cortex of awake monkeys, *J. Neurophysiol.*, 73, 525, 1995.

51. Vetter, P. and Wolpert, D. M., Context estimation for sensorimotor contol, *J. Neurophysiol.*, 84, 1026, 2000.

52. Bard, C., Fleury, M., Teasdale, N., Paillard, J., and Nougier, V., Contribution of proprioception for calibrating and updating the motor space, *Can. J. Physiol. Pharmacol.*, 73, 246, 1995.

53. Nelson, R. J., Interactions between motor cortex commands and somatic perception in sensorimotor cortex, *Curr. Opin. Neurobiol.*, 6, 801, 1996.

54. Cohen, L. G. and Starr, A., Localization, timing, and specificity of gating of somatosensory evoked potentials during active movement in man, *Brain*, 110, 451, 1987.

55. Papa, S. M., Artieda, J., and Obeso, J. A., Cortical activity preceding self-initiated and externally triggered voluntary movement, *Mov. Disord.*, 6, 217, 1991.

56. Rushton, D. N., Rothwell, J. C., and Craggs, M. D., Gating of somatosensory evoked potentials during different kinds of movement in man, *Brain*, 104, 465, 1981.

57. Schmidt, R. F., Schady, W. J. L., and Torebjork, H. E., Gating of tactile input from the hand. I. Effect of finger movement, *Exp. Brain Res.*, 79, 97, 1990a.

58. Schmidt, R. F., Torebjork, H. E., and Schady, W. J. L., Gating of tactile input from the hand. II. Effect of remote movements and anaesthesia, *Exp. Brain Res.*, 79, 103, 1990b.

59. Sperry, R. W., Neural basis of the spontaneous optokinetic response produced by visual inversion, *Exp. Zool.*, 92, 263, 1950.

60. von Holst, E. and Mittelstaedt, H., Das reafferenzprinzip. Wechselwirkungen zwischen zentralnervensystem und peripherie, *Natuwissenschaften*, 37, 464, 1950.

61. von Holst, E., Relations between the central nervous system and the peripheral organs, *Br. J. Anim. Behav.*, 89, 1954.

62. Chapman, C. E., Jiang, W., and Lamarre, Y., Modulation of lemniscal input during conditioned arm movements in the monkey, *Exp. Brain Res.*, 72, 316, 1988.

63. Miles, F. A. and Evarts, E. V., Concepts of motor organization, *Ann. Rev. Psychol.* 30, 327, 1979.

64. Nelson, R. J., Activity of monkey primary somatosensory cortical neurons changes prior to active movement, *Brain Res.*, 406, 402, 1987.

65. Pertovaara, A., Kemppainen, P., and Leppanen, H., Lowered cutaneous sensitivity to nonpainful electrical stimulation during isometric exercise in humans, *Exp. Brain Res.*, 89, 447, 1992.

66. Seal T. and Commenges, D., A quantitative analysis of stimulus- and movement-related responses in the posterior parietal cortex of monkey, *Exp. Brain Res.*, 58, 144, 1985.

67. Dyhre-Poulsen, P., Perception of tactile stimuli before ballistic and during tracking movements. In: *Active Touch: The Mechanisms of Object Manipulation: A Multidisciplinary Approach*, Gordon, G., Ed., Oxford, Pergamon, 171, 1978.

68. Dyhre-Poulsen, P., Increased vibration threshold before movements in human subjects, *Exp. Neurol.*, 47, 516, 1975.

69. Jiang, W., Chapman, C. E., and Lamarre, Y., Modulation of somatosensory evoked responses in the primary somatosensory cortex produced by intracortical microstimulation of the motor cortex in the monkey, *Exp. Brain Res.*, 80, 333, 1990a.

70. Jiang, W., Lamarre, Y., and Chapman, C. E., Modulation of cutaneous cortical evoked potentials during isometric and isotonic contractions in the monkey, *Brain Res.*, 536, 69, 1990b.

71. Wiesendanger, M. and Miles, T. S., Ascending pathway of low-threshold muscle afferents to the cerebral cortex and its possible role in motor control, *Physiol. Rev.*, 62, 1234, 1982.

72. Nelson, R. J. and Lebedev M. A., Activity of monkey sensorimotor neurons varies with reward history: I. Vibrotactile cue responses after withheld rewards, *J. Neurophysiol.*, submitted, 2001.

11 A Computational Perspective on Proprioception and Movement Guidance in Parietal Cortex

Paul E. Cisek

CONTENTS

11.1 FUNCTIONAL AND DESCRIPTIVE MODELS OF CORTICAL MOVEMENT CONTROL

Computational models can offer much to our attempts to understand biological motor control. Even a naïve model can at least act as a mnemonic device, helping to organize one's knowledge of a body of data. Of course, most models aspire to much more. One would expect that a good model not only provides a coherent and unifying explanation of data, but also generates testable hypotheses, which can influence future experimental work. Ultimately, what is most desirable is a complementary relationship between models and experiments, where each serves to refine the other.

For early neurophysiological work on the cortical control of movement, a highly influential model was the transcortical servo (Phillips, 1969). The transcortical servo hypothesis, inspired by cybernetic theories of control systems (Wiener, 1958), proposed that the well-known feedback organization of spinal reflexes was repeated again at the cortical level. Support for this notion came from analyses of reflex latencies, where short reflex components (12 ms, in monkeys) were considered spinally-mediated, while longer reflexes (30–40 ms) were considered to involve a feedback loop through cortical regions, a feedback loop which can exhibit a high degree of context-dependent plasticity (Evarts, 1973). Neurophysiological recordings showed that, consistent with the notion of a transcortical servo, many pyramidal tract neurons (PTNs) preferentially respond to opposite directions of active and passive movements (Conrad, Meyer-Lohmann, Matsunami, & Brooks, 1975; Evarts & Fromm, 1981) and are potentiated by perturbations opposing a voluntary movement (Evarts & Tanji, 1974). The transcortical servo hypothesis seemed to fulfill the highest expectations for a model — it explained and unified data, and made predictions that were often experimentally confirmed.

When researchers began investigating more complex multi-joint arm reaching movements in the early 1980s, the theoretical emphasis shifted somewhat. It was observed that many cells in the motor regions of the cerebral cortex were broadly tuned to the direction of a movement. This was first reported for cells in the primary motor cortex (Georgopoulos, Kalaska, Caminiti, & Massey, 1982) and later for parietal (Kalaska, Caminiti, & Georgopoulos, 1983) and premotor (Caminiti, Johnson, Galli, Ferraina, & Burnod, 1991) cells as well. Moreover, new analyses were able to reconstruct the direction of a movement from the firing pattern of a population of cells (Georgopoulos, Caminiti, Kalaska, & Massey, 1983), even during curved movements (Schwartz, 1993). The central interest of these new studies shifted away from questions of the control circuitry toward questions of the variables represented by the cortical cells and the sensorimotor coordinate transformations that compute these representations. Many recent experiments exemplify this attitude by focusing upon the issue of how the activity of particular neural populations covaries with some relevant sensory or motor variables, such as target position, intended direction, movement extent, movement speed, direction of load, etc. (Ashe & Georgopoulos, 1994; Fu, Suarez, & Ebner, 1993; Georgopoulos, Ashe, Smyrnis, & Taira, 1992; Georgopoulos, Caminiti, & Kalaska, 1984; Georgopoulos, Kalaska, Caminiti, & Massey, 1982; Georgopoulos, Kettner, & Schwartz, 1988; Kalaska, Caminiti, & Georgopoulos, 1983; Kalaska, Cohen, Hyde, & Prud'homme, 1989; Kalaska, Cohen, Prud'homme, & Hyde, 1990; Kettner, Schwartz, & Georgopoulos, 1988; Lacquaniti, Guigon, Bianchi, Ferraina, & Caminiti, 1995; Messier & Kalaska, 2000; Moran & Schwartz, 1999; Schwartz, 1993; Scott & Kalaska, 1997; Sergio & Kalaska, 1998). In parallel, experimental procedures often do not address issues such as whether a cell is a PTN or not, but instead emphasize sophisticated multi-variate regressions of its activity.

One may say that while the classic transcortical servo was a model of the *functional* circuitry involved in generating movement, many recent models are an account of how internal activities *describe* relevant variables. It may be useful to consider this shift in the broader context of 20th Century brain science. During the 1940s and 1950s, control theory and cybernetics emphasized feedback (Ashby, 1956; Wiener, 1958), which later came to influence theories of behavior (Ashby, 1965; Powers, 1973) and functional interpretations of motor systems (Phillips, 1969). At that time, circuit models were in vogue. However, by the mid-1960s, the "cognitive revolution" focused the attention of theorists onto the representational abilities of the brain, toward questions of how the brain builds internal representations of the external world (Marr, 1982), and how it generates representations of movement plans (Miller, Galanter, & Pribram, 1960). With this shift of attention, models eventually became more descriptive, addressing questions of how internal brain states correlate with sensory or motor variables. Over time, this new attitude came to influence motor control theory and experimental design, emphasizing representations over circuit models.

Of course, both functional and descriptive models are useful, but both are also subject to certain pitfalls. Functional circuit models tend to be very simplified. Cortical anatomy is much more complex than the simple "circles and lines" diagrams in which distinct elements perform distinct functions. In the real cortex, functionality is smeared, with individual cells belonging, in different degrees, to various commingled populations (Johnson, Ferraina, Bianchi, & Caminiti, 1996; Wise, Boussaoud, Johnson, & Caminiti, 1997) and even appearing to change their relation to motor output over the course of learning (Gandolfo, Li, Benda, Schioppa, & Bizzi, 2000; Wise, Moody, Blomstrom, & Mitz, 1998). Descriptive models are better suited to capture these trends.* On the other hand, a purely descriptive model often only restates the data and does not propose a mechanistic explanation of how the cells actually work together to produce movement. Descriptive models are also vulnerable to relating cortical activity to the wrong variables. Unless one has a functional circuit in mind *a priori*, one is left with little else but attempts to infer functional roles by correlating activity to only the measured kinematic or kinetic output variables, none of which may actually be of interest to cells embedded within a network controlling nonlinear muscles and joints (Fetz, 1992).

Several years ago, Dan Bullock, Steve Grossberg, and I attempted to make some sense of cortical data using the approach of functional modeling. To do so, we developed a mathematical model of the neural circuit involving cortical areas 4 and 5, basal ganglia, and spinal cord, controlling a single agonist-antagonist joint (Bullock, Cisek, & Grossberg, 1998; Cisek, Grossberg, & Bullock, 1998). This model, illustrated in Figure 11.1, proposes to explain how neurophysiologically–identified cell types, each with characteristic temporal patterns, work together to implement voluntary movements with observed psychophysical features. Here, I will refer to this model as FAVITE, or Feedback-Assisted Vector-Integration-to-Endpoint, because it is founded upon the Vector-Integration-to-Endpoint scheme of Bullock and Grossberg (1988) and expanded to include various sources of peripheral feedback.

In my admittedly biased opinion, the FAVITE model is a good starting point for interpreting neural data from a functional perspective. Like the transcortical servo hypothesis, the model frames neurophysiological data in a pragmatic context, in which neural responses can be understood as contributing to specific aspects of a mechanism for movement control. For example, phasic bursts in area 4 during reaction time (Kalaska, Cohen, Hyde, & Prud'homme, 1989) are interpreted as generating the force pulse needed to overcome the resting inertia of the arm. In

---

* However, a model being developed by Burnod and colleagues (1999) illustrates that functional models can also capture these trends while providing some pragmatic rationale for them.

contrast, phasic activities of posterior area 5 cells (Crammond & Kalaska, 1989; Kalaska, Cohen, Prud'homme, & Hyde, 1990) are interpreted as a "Difference Vector," used to guide the limb in the direction of a target. Thus, although these two signals may look similar, they perform very different functional roles in the context of the model. The FAVITE model is a set of such hypotheses expressed through a mathematical formalism, which enables simulation of various movement scenarios, including voluntary and passive movements, exertion of forces against obstacles, deafferented operation, movement in various perturbing force-fields, and proprioceptive illusions induced by muscle tendon vibration.

In this chapter, I re-examine some of the basic assumptions of the FAVITE model in light of some recent work on movement control, particularly studies of the parietal cortex. This includes work which supports the core features of the model and confirms its predictions, as well as studies which force some modifications. I describe some of these necessary modifications and propose further testable predictions.

11.2 OVERVIEW OF THE FAVITE MODEL

I will not burden the reader with the mathematical equations available in Bullock et al. (1998) and Cisek et al. (1998). I will instead sketch out the general ideas of the model with rather broad strokes. In essence, the FAVITE model is a circuit which generates a kinematic movement command online through internal feedback and superimposes dynamics compensation upon it (see Figure 11.1). Unlike many traditional control schemes from robotics, no "desired trajectory" is prepared and optimized before movement begins. Instead, only a desired target is specified and then the arm moved toward it by gradually reducing a Difference Vector (DV), computed as the difference between the current and desired limb positions (what may be called a "motor error"). In this sense, FAVITE operates like a kinematic transcortical servo, except that the reduction of the DV can be voluntarily controlled and its speed independently gated. During DV reduction, a descending positional command controlling the resting length of muscles is shifted from an initial to a final position. However, unlike equilibrium point models (Bizzi, Accornero, Chapple, & Hogan, 1984; Bizzi, Hogan, Mussa-Ivaldi, & Giszter, 1992; Feldman, 1974; Feldman, 1986), the descending command is not purely positional, but also combines dynamics compensation for inertial and static forces.

The model, in essence, consists of seven hypotheses regarding the organization of movement control. These are:

1. A representation of current limb position (called the Perceived Position Vector, or **PPV**) exists in anterior area 5 and is computed on the basis of efference copy from area 4 and feedback from primary muscle spindles routed through area 2.

2. A voluntary arm movement involves the reduction of a Difference Vector (**DV**), which is computed in posterior area 5 as the difference between the PPV and a Target Position Vector (**TPV**), which represents the target position. The DV may be activated, or "primed," before the movement is released.

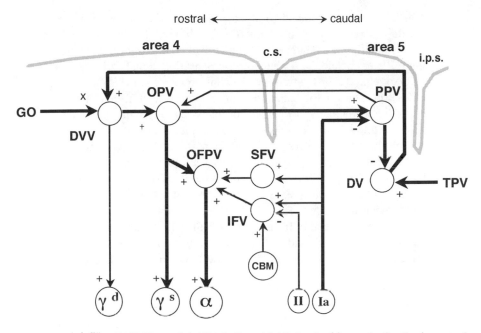

FIGURE 11.1 The FAVITE model. Thick lines highlight the kinematic feedback control aspects of the model, with thin lines representing additional compensatory circuitry. GO — scaleable gating signal; DVV — Desired Velocity Vector; OPV — Outflow Position Vector; OFPV — Outflow Force + Position Vector; SFV — Static Force Vector; IFV — Inertial Force Vector; PPV — Perceived Position Vector; DV — Difference Vector; TPV — Target Position Vector; CBM-cerebellum γ^d — gamma-dynamic motoneuron; γ^s — gamma-static motorneuron; α — alpha motoneuron; Ia — type-Ia afferent fiber; II — type-II afferent fiber; c.s. — central sulcus; i.p.s. — intraparietal sulcus. The symbol + represents excitation, - represents inhibition, $\times$ represents multiplicative gating, and $+\int$ represents integration over time. (From Bullock, D., Cisek, P., and Grossberg, S., *Cerebral Cortex*, 8, 48, 1998. With permission.)

3. The DV projects to a Desired Velocity Vector (*DVV*) in area 4 "phasic movement-time" cells,[*] where it is scaled by a GO signal arriving from the basal ganglia. The GO signal controls the onset time and the speed of movement. The DVV serves as a velocity command which controls changes in the descending positional command from area 4 (see hypothesis 4 below) and also projects to spinal gamma-dynamic motor neurons.

4. In area 4, an Outflow Position Vector (*OPV*) projects a positional command to alpha- and gamma-static motor neurons. The OPV also serves at the source of the efference copy signal to the PPV (see hypothesis 1 above). During voluntary movement, the OPV gradually integrates the DVV command over time, shifting the hand toward the target. As the hand approaches the target, the DV in area 5 and the DVV in area 4 are both reduced to baseline, at which point the OPV integration ends and movement ceases.

---

[*] The classification of area 4 cells used here is based on Kalaska et al. (1989).

5. The descending command to alpha motor neurons is augmented by an Outflow Force + Position Vector (*OFPV*) in area 4 "phasic-tonic" cells, which superimposes static and dynamic load compensation upon the shifting OPV command. Two kinds of compensation signals are described in Bullock et al. (1998): a transient Inertial Force Vector (*IFV*), which provides force pulses for launching and braking the limb and a sustained Static Force Vector (*SFV*), which compensates for gravity and other static loads. Although in the mathematical implementation of the FAVITE model, these signals were generated via peripheral feedback (for the sake of mathematical simplicity), we actually prefer to conceive of them as being supplied by a forward dynamics model, possibly implemented in the cerebellum (Bullock & Grossberg, 1991; Ito, 1984; Kawato & Gomi, 1992; Miall & Wolpert, 1996; Vilis & Hore, 1980).

6. A reciprocal connection from the area 5 PPV to the area 4 OPV enables the descending command to track any movements imposed by external forces and keeps muscle spindles loaded and in their range of optimal sensitivity.

7. In different movement contexts, the strength of some of the circuit connections may be varied to modify the operating characteristics of the system. For example, during fast movements, peripheral sensitivity can be reduced to shift the system from a feedback controller toward a feedforward controller.

The first six of these hypotheses are described in detail by Bullock et al. (1998), where evidence for them is reviewed. The seventh hypothesis is discussed by Cisek et al. (1998) in the context of various proprioceptive illusions induced by muscle tendon vibration, as well as movements in the presence of elastic loads and Coriolis forces. In this chapter, I will focus primarily on the first two hypotheses, those concerned with the computations proposed to occur in the posterior parietal cortex.

Below, I attempt to bring together diverse sources of experimental data on the functional role of the parietal cortex. Perhaps some of this data is being inappropriately forced into my particular perspective. The parietal cortex is a complex brain region, with many functional roles. Its study is made difficult by a lack of consensus on what regions in the human are homologous with those in the monkey, and even by a lack of a consistent nomenclature for specific recording sites. Therefore, the following discussion should merely be taken as well-intentioned speculation.

11.3 REPRESENTATIONS OF CURRENT POSITION AND INTENDED DIRECTION IN PARIETAL CORTEX

Several lines of evidence led us to propose that the representation of current limb position (PPV) exists in anterior area 5 and the difference vector (DV) in posterior area 5. Activity in parietal area 5 has long been considered to reflect postural information (Sakata, Takaoka, Kawarasaki, & Shibutani, 1973) and the sustained activity of area 5 cells shows a clear relationship to the position of the hand in space (Georgopoulos, Caminiti, & Kalaska, 1984; Lacquaniti, Guigon, Bianchi, Ferraina,

& Caminiti, 1995). For some cells, this relationship is linear (Georgopoulos, Caminiti, & Kalaska, 1984). Thus, it is often suggested that a representation of current body posture exists in parietal cortex and possibly involves area 5.

However, it has long been acknowledged that the parietal cortex is not simply a sensory region as originally conceived (Mountcastle, Lynch, Georgopoulos, Sakata, & Acuna, 1975). During voluntary movements, many area 5 cells respond before the onset of movement (Burbaud, Doegle, Gross, & Bioulac, 1991) and this response has been shown to be central in origin (Bioulac & Lamarre, 1979; Burbaud, Gross, & Bioulac, 1985). Several recent neurophysiological studies suggest that activity in the parietal cortex is strongly related to the intended movement (Bracewell, Mazzoni, Barash, & Andersen, 1996; Mazzoni, Bracewell, Barash, & Andersen, 1996; Snyder, Batista, & Andersen, 1997; Snyder, Batista, & Andersen, 2000). Neglect studies show that the parietal lobe plays a role in some aspects of motor preparation (Driver & Mattingley, 1998; Husain, Mattingley, Rorden, Kennard, & Driver, 2000). The dorsal visual stream, originally conceived as specialized for spatial vision (Ungerleider & Mishkin, 1982), is being reconsidered as a system for visuomotor guidance (Goodale & Milner, 1992; Milner & Goodale, 1995). All of these results are consistent with the proposal that early movement planning occurs in the parietal lobes, and some of this may involve the construction of representations functionally equivalent to the PPV and DV of the FAVITE model.

While the activity of many cells in area 5 reflects position, the movement-time activity of many "variational" neurons reflects the difference between the final and initial positions, like a DV (Lacquaniti, Guigon, Bianchi, Ferraina, & Caminiti, 1995). Further supporting the hypothesis of a DV, many cells in area 5 predict the intended movement direction during an instructed-delay period (Crammond & Kalaska, 1989), even after a monkey is instructed to make no movement at all (Kalaska & Crammond, 1995). Furthermore, as a population, area 5 cells are insensitive to external loads (Kalaska, Cohen, Prud'homme, & Hyde, 1990; Kalaska & Hyde, 1985), just as one would expect of kinematic position (PPV) and intended direction (DV) representations.

On both anatomical and physiological grounds, several authors have suggested that area 5 should be divided into anterior and posterior sub-regions (Burbaud, Doegle, Gross, & Bioulac, 1991; Johnson, Ferraina, Bianchi, & Caminiti, 1996). Anterior area 5 projects to the more caudal pre-central areas, while posterior area 5 cells project to more rostral regions such as rostral area 4 and area 6 (Johnson, Ferraina, Bianchi, & Caminiti, 1996; Johnson, Ferraina, & Caminiti, 1993; Jones, Coulter, & Hendry, 1978; Strick & Kim, 1978; Zarzecki, Strick, & Asanuma, 1978). The basic anatomical organization is a nested loop structure, in which cells on opposite sides of the central sulcus have similar response properties and are reciprocally connected (Johnson, Ferraina, Bianchi, & Caminiti, 1996; Pandya & Kuypers, 1969). The cells closer to the central sulcus (area 4 and anterior area 5) tend to be more tonic and more related to limb position, while cells further out from the central sulcus (rostral area 4 and posterior area 5) tend to be more phasic and related to direction (Johnson, Ferraina, Bianchi, & Caminiti, 1996). Thus, we proposed that the PPV resides among tonic, position-related cells in anterior area 5, which are reciprocally connected (perhaps polysynaptically) with tonic cells in area 4 (OPV). We also proposed that the DV resides among phasic, direction-related, primable cells in posterior area 5, which are connected with phasic movement-time cells in area 4 (DVV) (see Figure 11.1).

These proposals predict that during voluntary movements, posterior area 5 cells should become active first (because the DV is the first to receive the new target information), followed by area 4 cells (which implement the shift of the descending command toward the target), followed by anterior area 5 cells (which reflect the changing position of the limb). This order of events has been observed by Burbaud et al. (1991). Total deafferentation of the forelimb does not eliminate early changes in area 5 occurring before movement onset (possibly related to DV computation), but suppresses late neural changes in area 5 (Bioulac & Lamarre, 1979).

If a movement is obstructed, the activities of both PPV and DV populations should maintain a nearly constant level of activity, while the obstruction keeps the limb stationary. This was demonstrated by Evarts and Fromm (1981) in a paradigm in which a voluntary pronation or supination movement was transiently obstructed. As predicted for the DV, post-central cell activity remained constant during the obstruction, and decayed to baseline when the movement was released. That is, at all times, it signaled the remaining distance between the current hand position and an intended target. (It should be noted, however, that the example cell from Figure 8 of Evarts and Fromm [1981] was found in the region between areas 2 and 5, and it is not known whether similar cells exist in the more posterior part of area 5, where we would expect to find a DV representation).

In summary, there is good evidence in support of PPV and DV representations in parietal area 5. Recent studies are, for the most part, consistent with this proposal and do not force a reformulation of the core themes of the FAVITE model. However, we should never expect a model to be impervious to conflicting data for too long — that would be unrealistic. Below, I describe data which already reveals important modifications which must be made to the model. I particularly emphasize a recent study by Desmurget et al. (1999) involving transcranial magnetic stimulation.

11.4   SHIFTING TARGETS AND THE FAVITE MODEL

Desmurget et al. (1999) suggested that the posterior parietal cortex is involved in representing current limb position (PPV) and the difference between that and the target position (DV). Their conclusion was based on the use of Transcranial Magnetic Stimulation (TMS) to disrupt a subject's ability to update their control of a reaching movement to a target that unpredictably jumps during the movement. It was shown that without TMS stimulation, subjects could reliably and smoothly update their hand trajectory so as to reach a target which jumped. This suggests the operation of an online feedback system, similar to the one in the FAVITE model. However, with stimulation over the medial intraparietal sulcus, subjects did not update their trajectories and instead reached to the *original* location of the target. This led Desmurget et al. to propose that the posterior parietal cortex "functions as a 'neural comparator' to compute the current motor error and allow updating of the muscle activation pattern" (Desmurget, Epstein, Turner, Prablanc, Alexander, & Grafton, 1999, p. 565). This is equivalent to the hypothesis of a DV which continuously represents the remaining distance and direction to the target, and is used to shift the descending pre-central command to the muscles, such that the hand is brought toward the target (Bullock, Cisek, & Grossberg, 1998). Furthermore, Desmurget et al. also propose that the posterior parietal cortex "can evaluate the current location of the hand by integrating proprioceptive signals from the somatosensory areas and efferent copy signals from the motor regions" (Desmurget, Epstein, Turner, Prablanc, Alexander, & Grafton, 1999, p. 565). This is equivalent to our suggestion of a PPV computed from muscle-spindle signals routed through area 2 and efference copy from area 4 (Bullock, Cisek, & Grossberg, 1998).

However, in addition to providing more indirect evidence for PPV-like and DV-like computations in posterior parietal cortex, the Desmurget et al. study also points out that there must be more to the story. In their study, subjects *did* manage to reach toward the targets even when TMS was applied to their parietal cortex.[*] They did not adjust for target jumps and were quite inaccurate, but the trajectories of their movements were still oriented in the right general direction and did not suddenly halt, reverse, or fly off at random angles. If the DV of a FAVITE circuit were suddenly suppressed or injected with random values, the effect on the trajectory would be devastating. Can the model be reconciled with these observations?

I believe that there are two possible ways of explaining the data of Desmurget et al. in the context of the FAVITE model. I describe both of these explanations below and propose some experiments which can distinguish between them.

11.4.1 DIFFERENCE VECTOR COMPUTATION USING A FORWARD MODEL

One way of explaining the Desmurget et al. results involves the concept of an internal "forward model" (Miall & Wolpert, 1996). It has often been suggested that well-practiced movements are guided primarily by predictive internal estimates (a forward model) of the state of effectors, and rely less on online feedback (Bhushan & Shadmehr, 1999; Flanagan & Wing, 1997; Miall & Wolpert, 1996; Wolpert, Ghahramani, & Jordan, 1995). While a reaching movement may initially be slowly guided through feedback control, over time the proprioceptive signals can be anticipated and internally generated in advance, turning the system into a feed-forward controller. This is the simplest way in which a control system can avoid the instabilities associated with long conduction delays, and there is strong evidence that such a strategy is used by the nervous system (reviewed in Kalaska, Scott, Cisek, & Sergio, 1997). Forward models are often proposed as useful for dealing with the dynamics of movement, and the cerebellum is often implicated as their substrate (Bullock & Grossberg, 1991; Ito, 1984; Kawato & Gomi, 1992; Miall & Wolpert, 1996; Vilis & Hore, 1980). In the FAVITE model, we proposed that the launching and braking pulses evident in area 4 phasic–tonic cells are provided by a cerebellar forward model (Bullock, Cisek, & Grossberg, 1998, p. 52). In addition to that, a forward model may also be useful for providing a DV for the modulation of the pre-central descending command, as described below.

---

[*] This very important observation led Iacoboni (1999) to conclude that "reaching movements are fully planned before movement onset." However, I believe that this takes the data too far. I question that conclusion below, and present several alternative hypotheses which are consistent with available data.

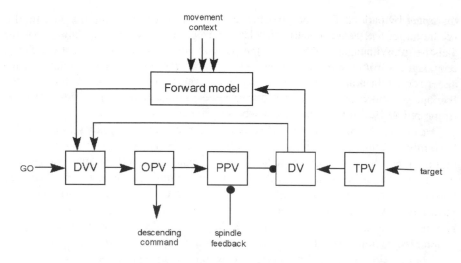

FIGURE 11.2 The kinematic sub-circuit of the FAVITE model, augmented with a forward model for predicting the DV profile. Lines with arrowheads indicate excitatory connections. Lines ending in circles indicate inhibitory connections.

In the original implementation of the FAVITE model (Bullock, Cisek, & Grossberg, 1998), the only source of directional information for the precentral DVV was the area 5 DV, generated online using the area 5 PPV. However, suppose that a second source of directional information comes from a forward model, which has learned to anticipate the temporal profile of the DV for some reasonably familiar movement context (Figure 11.2). During performance, the input to the pre-central DVV may come solely from the forward model, and the parietal DV computation may only be used when sudden unexpected perturbations (such as a change in target location) disrupt the course of movement.

Suppose that the transcranial magnetic stimulation used by Desmurget et al. (1999) disrupted the PPV and DV computations occurring in the parietal cortex. This means that the forward model has become the only source for guiding movement, and all perturbations, including changes in target location, are ignored. The accuracy of movement is compromised, but the hand does at least hit the right ballpark. In other words, one can hypothesize that in addition to an "online DV" in area 5, there are other sources of directional information available to update the pre-central descending command, including perhaps a forward model of the kinematics.

A number of recent studies support the hypothesis that the posterior parietal cortex is primarily involved in the online guidance of movements using currently available (and attended) sensory information. During tasks requiring the recalibration of sensorimotor transformations, whether adapting to displacing prisms (Clower, Hoffman, Votaw, Faber, Woods, & Alexander, 1996) or to a rotated reference frame (Ghilardi, Ghez, Dhawan, et al., 2000), positron emission tomography reveals activity changes in the posterior parietal cortex. Lacquaniti et al. (1997) found increased activation in the inferior parietal lobule in an immediate-pointing task, as compared with a task in which pointing movements were made to a previously memorized location. Consistent with these results, the accuracy of memory-guided saccades is disrupted by parietal TMS only when the stimulation occurs during the presentation of the target and not when it occurs later during the delay (Brandt, Ploner, Meyer, Leistner, & Villringer, 1998), as if the parietal cortex is primarily used to convert *currently available* sensory information into a form useful for movement. A similar conclusion was reached by Goodale et al. (1994), who found that a patient with damage in her ventral visual stream was able to scale her grip size to the size of a target object in real time, but not when a delay was imposed between vision of the object and onset of the reach. That is, her intact parietal cortex could only use currently available information for scaling the grip. Rushworth et al. (1997b) found that a lesion of area 5, area 7b, and medial intraparietal area (MIP) did not strongly affect well-practiced reaches with the same start and end position (which could be memorized), but severely impaired reaches made from different starting positions (which had to be planned each time *de novo*). All of these results implicate the posterior parietal cortex in online guidance of movements by current sensory information, and suggest that if a forward model is used to control well-practiced movements in a feed-forward manner, then it must reside elsewhere in the brain.

11.4.2 SEPARATE CONTROL FOR VISUAL VS. PROPRIOCEPTIVE GUIDANCE

However, still another possibility exists. Suppose that there are separate sub-regions of the parietal cortex responsible for guiding movement based on visual information vs. movement based on proprioceptive information. Such regional specialization has been demonstrated in a lesion study by Rushworth, Nixon, and Passingham (1997a). These investigators removed areas 5, 7b, and MIP in three monkeys previously trained to reach for visually presented targets and to reach in the dark for targets defined by arm position. While the lesion did not disrupt reaching in the light, it caused a marked inaccuracy when reaching in the dark. Conversely, lesions of areas 7a, 7ab, and LIP (lateral intraparietal area) in three other monkeys disrupted visually-guided reaching, but not proprioceptively guided reaching in the dark. This led Rushworth et al. (1997a) to conclude that areas 5/7b/MIP are involved in proprioceptively guided reaching, while areas 7a/7ab/LIP are involved in visually guided reaching. In terms of the FAVITE model, this translates to separating the TPV, PPV, and DV of Figure 11.1. into two parallel sub-circuits, each operating in a different coordinate system, as shown in Figure 11.3.

In Figure 11.3b, the top row of PPV, DV, and TPV populations operate in an "extrinsic" coordinate system based on the visual input (indicated by an "e" suffix), while the bottom row of PPV, DV, and TPV populations operate in an "intrinsic" arm-based coordinate system (indicated by an "i" suffix) defined by efference copy and proprioceptive feedback. The upper row might therefore involve the 7a/7ab/LIP region of Rushworth et al. (1997a), while the lower row may involve area 5/7b/MIP. Learned coordinate transformations link corresponding representations across the two levels. The DIRECT model of Bullock et al. (1993) discusses PPV and DV computations in different reference frames and how transformations between them can be learned during development. The circuit described here shares many central elements with that model, the only difference being the addition of the computation of a DV in intrinsic coordinates based on PPVi and TPVi elements.

a) Original model   **b)** Unlumped model

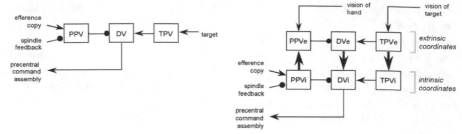

FIGURE 11.3 Unlumping of the PPV, DV, and TPV stages of the FAVITE model into separate populations operating in different coordinates. (a) Original model, without reference to coordinate systems. (b) Unlumped model. The top row of populations operate in "extrinsic" spatial coordinates defined by visual input and the bottom row in "intrinsic" arm coordinates defined by efference copy and proprioceptive feedback. Thick arrows indicate learned coordinate transformations.

The extrinsic PPV (PPVe) can be computed in one of two ways: either using visual feedback of the hand, if that is available, or otherwise through a transformation of the PPV in intrinsic coordinates (PPVi). The DV in intrinsic coordinates (DVi) can also be computed in one of two ways: either through a transformation of the extrinsic DV (DVe), if available, or otherwise by subtracting the PPVi from a motor TPVi. In each case, the computation based on visual information takes precedence over the computation using proprioceptive intrinsic coordinate signals.

Visual presentation of a target activates the TPVe population and visual feedback of the position of the hand activates the PPVe. With these two sources of input, the DVe can be computed in extrinsic spatial coordinates and transformed into a DVi in intrinsic arm coordinates, which is then used to update the pre-central descending command. Note that the DVi is computed based on the available DVe, and not by subtracting the PPVi from the TPVi. Thus, during reaching movements to visual targets with full visual feedback of the arm, the PPVi and TPVi populations are not strongly involved, since the PPVe and TPVe are available. Assuming that the transformation from DVe to DVi is accurate (as it should be after years of practice), such conditions allow very accurate movement.

When vision of the arm is prevented, the PPVi comes into play. The PPVi is computed based on efference copy and spindle feedback (as in the original FAVITE model), and transformed into the PPVe through another well-practiced and accurate coordinate transformation. Thus, the difference vector is again computed in spatial coordinates (DVe) and then transformed into intrinsic arm coordinates (DVi). The reaching movement is less accurate under these conditions, because it relies on an imperfect estimate of the position computed using efference copy and proprioceptive feedback.

Under other conditions, the target may not be defined through the visual modality at all. Suppose instead that a somatosensory target is presented to the system, as in the case of scratching an itch. In this case, the TPVi is defined but the TPVe is not, and so the difference vector computation cannot occur in extrinsic spatial coordinates.

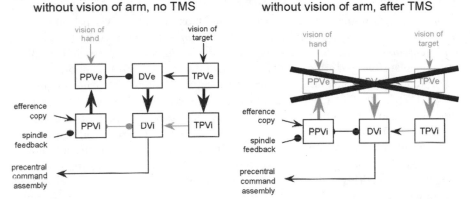

FIGURE 11.4 Hypothetical explanation of the Desmurget et al. (1999) results. Gray boxes and arrows are not involved in movement guidance. (a) Movements made without visual feedback and without TMS involve the computation of the extrinsic DVe using PPVe and TPVe. (b) Movements made with TMS involve the computation of the DVi directly using PPVi and TPVi in intrinsic arm coordinates.

Instead, it occurs directly in intrinsic coordinates (DVi) using the available TPVi and the somatosensory PPVi. In other words, all the activity during proprioceptively guided movement is confined to the bottom row of the Figure 11.3b circuit.

In summary, the circuit of Figure 11.3b allows very flexible use of various sources of information regarding current arm position and target location to allow reaching movements under many different conditions. I introduce it here because it also provides a functional context for interpreting the results of Desmurget et al., as discussed below.

In the Desmurget et al. (1999) experiment, there are essentially four cases: 1) movements to stationary targets without TMS; 2) jumping targets without TMS; 3) stationary targets with TMS; and 4) jumping targets with TMS. In all of these cases, the target itself is presented visually, but there is no visual feedback of the moving hand. Figure 11.4a illustrates what happens in the first two cases. Because the target is presented visually, it activates the TPVe population. However, since the hand is out of view, the only position information comes from efference copy and peripheral feedback, activating the PPVi population. Thanks to a learned transformation, the PPVi is converted into a PPVe, which then allows the computation of a DVe in extrinsic spatial coordinates. This is then coverted into a DVi, which controls the movement. During movement, the DVe is continuously recomputed based on the current PPVe and TPVe, and thus automatically corrects for target jumps. However, because the computation of the PPVi is not perfectly accurate, the movement trajectories exhibit some endpoint inaccuracies.

Consider now what happens in the cases with transcranial magnetic stimulation (Figure 11.4b). When the target is first visually presented at the beginning of a trial, it activates TPVe, which in turn generates TPVi activity through a learned coordinate transformation.* The initial DVe is computed in spatial coordinates based on the TPVe and PPVe generated from the PPVi, as above, and the movement begins. Shortly after movement onset, however, stimulation is applied over parietal cortex. *Suppose that the stimulation disrupts the operation of only the spatial representations — i.e., disrupting the upper row of Figure 11.4b* (such a localized effect is plausible given the results of Rushworth et al. (1997a) mentioned above, who produced just such an effect through localized lesions). That means that the PPVe, DVe, and TPVe representations are no longer available, and therefore the movement can now only be guided by the representations in arm coordinates. Because the TPVe was transformed into a TPVi before the stimulation started, a target position in intrinsic arm coordinates still exists. Together with the continuously computed PPVi, the TPVi is used to compute the DVi directly since the DVe is no longer available. This means that the arm will still be brought into the vicinity of the target, with some additional inaccuracies due to imperfections of the TPVi representation. However, because all sources of visual information have now been blocked, target jumps no longer affect the movement trajectory.

The intact Figure 11.3b circuit also explains some results on the automatic adjustments found when a target shifts during a reaching movement. Pisella et al. (2000) showed that normal subjects often could not help automatically adjusting their movement trajectory to a target jump even if explicitly instructed not to do so. It appeared that an on-line feedback mechanism automatically recomputed the current direction between the hand the target, and the subjects could not always inhibit this process voluntarily. A patient with a bilateral posterior parietal lesion did not show corrective movements when instructed not to correct, and showed only the slowest and deliberate corrections when instructed to correct. Thus, Pisella et al. concluded that an 'automatic pilot' for guiding reaching movements on-line exists in the posterior parietal cortex.

In the context of the Figure 11.3b circuit, this is equivalent to the continuous updating of the DVe population using visual information on the target location (TPVe), thus automatically adjusting the movement direction. For the parietal patient, this automatic updating of the extrinsic spatial DV is not available, and so movement guidance has to use other sources of directional information. However, for this same reason the parietal patient will not exhibit the automatic adjustments in movement trajectories seen for the control subjects.

Wolpert, Goodbody, and Husain (1998) describe a patient (P.J.) with a superior parietal lesion who is unable to maintain an internal estimate of the current position of her hand without visual feedback. As soon as she loses sight of her hand, she quickly perceives it to drift and even loses track of it altogether. P.J. is able to maintain a constant grip force with visual feedback of her performance, but without such feedback her grip force begins to decay within seconds. She is also more inaccurate during deliberately slowed movements than during self-paced movements.

---

* In a system with redundant degrees of freedom, this transformation is non-trivial. Because each extrinsic target position (TPVe) can be mapped onto many intrinsic target arm configurations (TPVi), the mapping is ill-posed. To address this difficulty, one might propose that the system simply has one preferred arm configuration for every extrinsic position.

From the perspective of the Figure 11.3b circuit, P.J.'s deficits may be explained if one postulates that her lesion impairs the computation and maintenance of the PPVi representation. Thus, her deficit is not evident when she is provided with continuous visual feedback that directly activates the spatial PPVe representation and allows her to estimate her arm's position. However, without feedback, her impaired PPVi will slowly corrupt the PPVe, leading to errors. This causes her to both misreport her current hand position and to decrease grip force, as all difference vector computations are compromised. During well-practiced self-paced movements, she can rely upon predictive forward models to guide the computation of the DVe, but slow movements without visual feedback are disrupted by online PPVi input which is compromised.

11.4.3 DISTINGUISHING BETWEEN THE TWO EXPLANATIONS

Above, I described two possible explanations for the results of the Desmurget et al. (1999) study in the context of the FAVITE model. I will call the first hypothesis the *forward model explanation* and the second hypothesis the *proprioceptive feedback explanation.* Since both candidate explanations are consistent with available data (at least to my knowledge), it is necessary to consider ways in which they could be experimentally distinguished.

One method is to introduce a physical perturbation into voluntary reaching movements. Suppose that a subject performs point-to-point reaching movements, without visual feedback of the hand, to stationary targets. During some of the movements, TMS is applied to the parietal cortex. As in the Desmurget et al. study, we expect that subjects will be able to complete the reach, albeit with low accuracy, even with TMS. Suppose, however, that during the movement, the subject's arm is perturbed in some way — perhaps using a temporary (~500 ms) obstruction, an off-axis load (Krakauer, Ghilardi, & Ghez, 1999), or a robotic manipulandum (Scott, 1999; Shadmehr & Mussa-Ivaldi, 1994). Because the arm is not visible, the only source of information on the perturbation comes from proprioceptive feedback. The question now is: does the subject compensate for the perturbation?

Under the assumptions of the forward model explanation, the TMS knocks out a parietal system for online control; and, therefore, one would predict that the subject *will not compensate* for the perturbation. Of course, spinal reflexes and biomechanical properties of muscles and joints will provide some low-level compensation, but not with the same properties as the entire intact system. That is, it should be possible to find specific differences between the compensatory actions in the non-TMS vs. TMS movements, especially with respect to the long-latency components, which presumably involve cortical control.

In contrast, under the assumptions of the proprioceptive feedback explanation, parietal TMS only knocks out *visually* guided online control, but spares the proprioceptively guided circuit. Therefore, one would predict that the subject *will compensate as before.* That is, there should be no differences in either the short- or long-latency components of the response to the perturbation between the non-TMS and TMS movements.

Note, however, that it is possible that both explanations are partly correct. The online system may indeed be separated into visually and proprioceptively guided sub-systems, but there may *also* exist a forward model. Therefore, an absence of compensation would not preclude the existence of a proprioceptive sub-circuit; however, it would confirm the existence of a forward model. Likewise, the presence of compensation would not preclude a forward model, but would confirm a proprioceptive subcircuit.

Another way to test the hypothesized circuit in Figure 11.3b is through an experiment in which visual feedback and proprioceptive feedback are dissociated. Suppose that a subject performs point-to-point reaching movements using a planar manipulandum, moving below an opaque surface, upon which their hand location is projected as a spot of light. During most movements, the spot of light veridically indicates hand location. During some of the movements, an unpredictable physical perturbation is applied to the manipulandum and the subject encouraged to compensate. During half of the perturbed trials, visual feedback of the hand location correctly reflects the direction of the perturbation. During the other half of perturbed trials, however, the visual feedback indicates the opposite direction of perturbation. For example, during movements away from the body, the actual perturbation physically deflects the limb toward the right, but the visual feedback displays a deflection toward the left. How does the subject compensate? In particular, what is the direction of the long-latency components of the subject's response?

According to the Figure 11.3b circuit, the compensation of a normal subject should be appropriate for the *visual* feedback of the perturbation. That is, the arm will actually deflect further off the correct path so as to bring the visible spot of light toward the target. The reason for this would be that the target for movement is defined in extrinsic visual space (TPVe), and the control is dominated by the visually-guided subcircuit in the top row of Figure 11.3b, where the representation of current end-effector position (PPVe) is determined through vision.

However, what would happen if transcranial magnetic stimulation were now applied to the posterior parietal cortex around the time of the perturbation? Suppose that, as proposed by the above proprioceptive feedback explanation of the Desmurget et al. (1999) data, the TMS disrupts only the visual portion of the Figure 11.3b circuit. In that case, control would now be dominated by the proprioceptively guided subcircuit in the bottom row of Figure 11.3b, dependent on a proprioceptively determined estimate of current limb position (PPVi). The resulting compensation would now be appropriate for the veridical direction of the physical perturbation, and the subject would correctly reach the target even while their visual feedback would indicate that they went in the wrong direction. In other words, one would find opposite directions of compensation for an identical physical perturbation during TMS and non-TMS trials.

The above proposals are, of course, very skeletal, and a great many details would need to be worked out to run such experiments. However, even these brief proposals illustrate some of the issues which arise when a functional model provides some context to experimental data.

In any case, the above considerations illustrate that it is somewhat premature of Iacoboni (1999) to conclude on the basis of the Desmurget et al. study that "reaching movements are fully planned before movement onset." The computation of a motor command may occur during the execution of movement, even if that computation is not guided by visual feedback, but by proprioceptive feedback or by a forward model. A novel visually guided movement may be guided online, using parietal cortex even if a well-practiced movement can be poorly executed without visual guidance and with disrupted parietal activity. Of course, these suggestions are quite general and do not depend upon the particulars of the FAVITE model.

11.5 CONCLUSIONS

As stated in the introduction, the usefulness of a functional model is that it provides a pragmatic context for interpreting data. It leads one beyond questions of how cellular activities co-vary with various movement-related information and toward questions of how that information is actually *used* in a mechanism for guiding movement. There are two reasons why this is important. First, the statement that some variable affects some pattern of activity is merely an observation and not an hypothesis. Testing that observation only confirms whether it is accurate, but does not leave us with an explanation. Second, by considering the pragmatic concerns facing a system such as the cortical circuits for movement control, one is led toward those issues which motivated the evolution of that system. Motor control is not about describing actions but about performing actions, and accuracy of a description does not always go hand in hand with accuracy of a movement. For example, a veridical representation of the current position of the limb is not always the most useful element for a successful control scheme. In the FAVITE model, the gain of peripheral feedback is deliberately reduced during fast movements in order to avoid the instabilities inherent in a feedback system with conduction delays (Cisek, Grossberg, & Bullock, 1998). This results in a system that does not represent position accurately, but performs better than a system which makes veridical position sense a priority.

There are other examples. The Difference Vector is fundamentally a variable useful for directing movement and determining its completion. In essence, all that a good DV has to do is point in the right direction and shut off when the hand reaches the target. The precise representation of distance is of lesser importance, as long as other (possibly pre-central) mechanisms take care of shaping the speed profile. This means that the DV can combine other sources of information without disrupting movement performance, as discussed in the example below.

Consider what happens when several possible targets for a movement are available. Based on data from the oculomotor system (Glimcher & Sparks, 1992; Platt & Glimcher, 1997) and the arm movement system (Cisek & Kalaska, 1999), one can speculate that the brain can begin to prepare several potential directional signals simultaneously before selecting one to guide overt execution. Kinematic trajectories of movements made in the presence of distractors support this proposal (Tipper, Howard, & Houghton, 1998; Tipper, Lortie, & Baylis, 1992). If that is the case, then how does the brain decide between different actions? One candidate mechanism is a competition between representations of different movement directions in the posterior parietal cortex. Suppose that different potential movements are represented in parietal cortex as different DVs, each of which represents direction but not distance.

If each of these DVs is scaled by various influences — such as salience (Colby & Goldberg, 1999), probability (Platt & Glimcher, 1999), or reward size (Platt & Glimcher, 1999) — then the competition may simply be determined by which DV has the highest level of activity. If so, then the activity of posterior parietal cells will not faithfully represent *any* particular movement-related variable at all, but will instead reflect a combination of many variables. In other words, parietal activity will not simply reflect attention (Colby & Goldberg, 1999) or intention (Snyder, Batista, & Andersen, 2000), but a particular functionally valuable mixture of sensory variables (i.e., salience), decision variables (i.e., payoff), and motor variables (i.e., intended direction).

Of course, functional models risk many pitfalls. First, every phenomenon has many possible explanations, and even when diverse data is viewed from a unified perspective there is always room for alternatives. The plausibility of one functional model does not preclude the plausibility of another, and neither does the absence of an alternative give support to any existing model. Second and more important, functional models are inevitably very simplified. Some data is always neglected, and deciding what to leave out and what to make central to model development is very precarious. The FAVITE model, for example, is fundamentally based upon a psychophysically inspired ancestor (Bullock & Grossberg, 1988) with particular emphasis placed upon cortical neurophysiology of areas 4 and 5 and studies of human proprioception. Perhaps lesion studies or functional imaging data would have been a better place to start. That cannot be known until other approaches are taken, models specified, predictions made, and experiments performed to distinguish alternatives. Hopefully, such a process will eventually converge toward some promising consensus.

REFERENCES

Ashby, W.R. *An Introduction to Cybernetics.* Chapman and Hall, 1956.

Ashby, W.R. *Design for a Brain: The Origin of Adaptive Behaviour.* London, Chapman and Hall, 1965

Ashe, J., and Georgopoulos, A.P. Movement parameters and neural activity in motor cortex and area 5. *Cerebral Cortex,* 4, 590, 1994

Bhushan, N., and Shadmehr, R. Computational nature of human adaptive control during learning of reaching movements in force fields. *Biological Cybernetics,* 81, 39, 1999.

Bioulac, B., and Lamarre, Y. Activity of post-central cortical neurons of the monkey during conditioned movements of a deafferented limb. *Brain Research,* 172, 427, 1979.

Bizzi, E., Accornero, N., Chapple, W., and Hogan, N. Posture control and trajectory formation during arm movement. *Journal of Neuroscience,* 4, 2738, 1984.

Bizzi, E., Hogan, N., Mussa-Ivaldi, F.A., and Giszter, S.F. Does the nervous system use equilibrium-point control to guide single and multiple joint movements? *Behavioral and Brain Sciences,* 15, 603, 1992.

Bracewell, R.M., Mazzoni, P., Barash, S., and Andersen, R.A. Motor intention activity in the macaque's lateral intraparietal area. II. Changes of motor plan. *Journal of Neurophysiology,* 76, 1457, 1996.

Brandt, S.A., Ploner, C.J., Meyer, B.U., Leistner, S., and Villringer, A. Effects of repetitive transcranial magnetic stimulation over dorsolateral prefrontal and posterior parietal cortex on memory-guided saccades. *Experimental Brain Research,* 118, 197, 1998.

Bullock, D., Cisek, P., and Grossberg, S. Cortical networks for control of voluntary arm movements under variable force conditions. *Cerebral Cortex*, 8, 48, 1998.

Bullock, D., and Grossberg, S. Neural dynamics of planned arm movements: Emergent invariants and speed-accuracy properties during trajectory formation. *Psychological Review*, 95, 49, 1988.

Bullock, D., and Grossberg, S. Adaptive neural networks for control of movement trajectories invariant under speed and force rescaling. *Human Movement Science*, 10, 1, 1991.

Bullock, D., Grossberg, S., and Guenther, F.H. A self-organizing neural model of motor equivalent reaching and tool use by a multijoint arm. *Journal of Cognitive Neuroscience*, 5, 408, 1993.

Burbaud, P., Doegle, C., Gross, C.G., and Bioulac, B. A quantitative study of neuronal discharge in areas 5, 2, and 4 of the monkey during fast arm movements. *Journal of Neurophysiology*, 66, 429, 1991.

Burbaud, P., Gross, C., and Bioulac, B. Peripheral inputs and early unit activity in area 5 of the monkey during a trained forelimb movement. *Brain Research*, 337, 341, 1985.

Burnod, Y., Baraduc, P., Battaglia-Mayer, A., Guigon, E., Koechlin, E., Ferraina, S., Lacquaniti, F., and Caminiti, R. Parieto-frontal coding of reaching: an integrated framework. *Experimental Brain Research*, 129, 325, 1999.

Caminiti, R., Johnson, P.B., Galli, C., Ferraina, S., and Burnod, Y. Making arm movements within different parts of space: The premotor and motor cortical representations of a coordinate system for reaching to visual targets. *Journal of Neuroscience*, 11, 1182, 1991.

Cisek, P., Grossberg, S., and Bullock, D. A cortico-spinal model of reaching and proprioception under multiple task constraints. *Journal of Cognitive Neuroscience*, 10, 425, 1998.

Cisek, P., and Kalaska, J.F. Neural correlates of multiple potential motor actions in primate premotor cortex. *Society for Neuroscience Abstracts*, 25, 381, 1999

Clower, D.M., Hoffman, J.M., Votaw, J.R., Faber, T.L., Woods, R.P., and Alexander, G.E. Role of the posterior parietal cortex in the recalibration of visually guided reaching. *Nature*, 383, 618, 1996.

Colby, C.L., and Goldberg, M.E. Space and attention in parietal cortex. *Annual Review of Neuroscience*, 22, 319, 1999.

Conrad, B., Meyer-Lohmann, J., Matsunami, K., and Brooks, V.B. Precentral unit activity following torque pulse injections into elbow movements. *Brain Research*, 94, 219, 1975.

Crammond, D.J., and Kalaska, J.F. Neuronal activity in primate parietal cortex area 5 varies with intended movement direction during an instructed-delay period. *Experimental Brain Research*, 76, 458, 1989.

Desmurget, M., Epstein, C.M., Turner, R.S., Prablanc, C., Alexander, G.E., and Grafton, S.T. Role of the posterior parietal cortex in updating reaching movements to a visual target. *Nature Neuroscience*, 2, 563, 1999.

Driver, J., and Mattingley, J.B. Parietal neglect and visual awareness. *Nature Neuroscience*, 1, 17, 1998.

Evarts, E.V. Motor cortex reflexes associated with learned movement. *Science*, 179, 501, 1973.

Evarts, E.V., and Fromm, C. Transcortical reflexes and servo control of movement. *Canadian Journal of Physiology and Pharmacology*, 59, 757, 1981.

Evarts, E.V., and Tanji, J. Gating of motor cortex reflexes by prior instruction. *Brain Research*, 71, 479, 1974.

Feldman, A.G. Change of muscle length as a consequence of a shift in an equilibrium of muscle load system. *Biophysics*, 19, 544, 1974.

Feldman, A.G. Once more on the equilibrium-point hypothesis lambda model for motor control. *Journal of Motor Behavior*, 18, 17, 1986.

Fetz, E.E. Are movement parameters recognizably coded in the activity of single neurons? *Behavioral and Brain Sciences*, 15, 679, 1992.

Flanagan, J.R., and Wing, A.M. The role of internal models in motion planning and control: Evidence from grip force adjustments during movements of hand-held loads. *Journal of Neuroscience*, 17, 1519, 1997.

Fu, Q.-G., Suarez, J.I., and Ebner, T.J. Neuronal specification of direction and distance during reaching movements in the superior precentral premotor area and primary motor cortex of monkeys. *Journal of Neurophysiology*, 70, 2097, 1993.

Gandolfo, F., Li, C., Benda, B.J., Schioppa, C.P., and Bizzi, E. Cortical correlates of learning in monkeys adapting to a new dynamical environment. *Proc. Natl. Acad. Sci. U.S.A.*, 97, 2259, 2000.

Georgopoulos, A.P., Ashe, J., Smyrnis, N., and Taira, M. The motor cortex and the coding of force. *Science*, 256, 1692, 1992.

Georgopoulos, A.P., Caminiti, R., and Kalaska, J.F. Static spatial effects in motor cortex and area 5: quantitative relations in a two-dimensional space. *Experimental Brain Research*, 54, 446, 1984.

Georgopoulos, A.P., Caminiti, R., Kalaska, J.F., and Massey, J.T. Spatial coding of movement: A hypothesis concerning the coding of movement direction by motor cortical populations. *Experimental Brain Research, Supplement* 7, 327, 1983.

Georgopoulos, A.P., Kalaska, J.F., Caminiti, R., and Massey, J.T. On the relations between the direction of two-dimensional arm movements and cell discharge in primate motor cortex. *Journal of Neuroscience*, 2, 1527, 1982.

Georgopoulos, A.P., Kettner, R.E., and Schwartz, A.B. Primate motor cortex and free arm movements to visual targets in three-dimensional space. II. Coding of the direction of arm movement by a neural population. *Journal of Neuroscience*, 8, 2928. 1988.

Ghilardi, M., Ghez, C., Dhawan, V., Moeller, J., Mentis, M., Nakamura, T., Antonini, A., and Eidelberg, D. Patterns of regional brain activation associated with different forms of motor learning. *Brain Research*, 871, 127, 2000.

Glimcher, P.W., and Sparks, D.L. Movement selection in advance of action in the superior colliculus. *Nature*, 355, 542, 1992.

Goodale, M.A., Jakobson, L.S., and Keillor, J.M. Differences in the visual control of pantomimed and natural grasping movements. *Neuropsychologia*, 32, 1159, 1994.

Goodale, M.A., and Milner, A.D. Separate visual pathways for perception and action. *Trends in Neurosciences*, 15, 20, 1992.

Husain, M., Mattingley, J.B., Rorden, C., Kennard, C., and Driver, J. Distinguishing sensory and motor biases in parietal and frontal neglect. *Brain*, 123 Part 8, 1643, 2000.

Iacoboni, M. Adjusting reaches: feedback in the posterior parietal cortex. *Nature Neuroscience*, 2, 492, 1999.

Ito, M. *The Cerebellum and Neural Control.* New York, Raven, 1984.

Johnson, P.B., Ferraina, S., Bianchi, L., and Caminiti, R. Cortical networks for visual reaching: Physiological and anatomical organization of frontal and parietal arm regions. *Cerebral Cortex*, 6, 102, 1996.

Johnson, P.B., Ferraina, S., and Caminiti, R. Cortical networks for visual reaching. *Experimental Brain Research*, 97, 361, 1993.

Jones, E.G., Coulter, J.D., and Hendry, H.C. Intracortical connectivity of achitectonic fields in the somatic sensory, motor and parietal cortex of monkeys. *Journal of Comparative Neurology*, 181, 291, 1978.

Kalaska, J.F., Caminiti, R., and Georgopoulos, A.P. Cortical mechanisms related to the direction of two-dimensional arm movements: Relations in parietal area 5 and comparison with motor cortex. *Experimental Brain Research*, 51, 247, 1983.

Kalaska, J.F., Cohen, D.A.D., Hyde, M.L., and Prud'homme, M.J. A comparison of movement direction-related vs. load direction-related activity in primate motor cortex, using a two-dimensional reaching task. *Journal of Neuroscience*, 9, 2080, 1989.

Kalaska, J.F., Cohen, D.A.D., Prud'homme, M.J., and Hyde, M.L. Parietal area 5 neuronal activity encodes movement kinematics, not movement dynamics. *Experimental Brain Research*, 80, 351, 1990.

Kalaska, J.F., and Crammond, D.J. Deciding not to GO: Neuronal correlates of response selection in a GO/NOGO task in primate premotor and parietal cortex. *Cerebral Cortex*, 5, 410, 1995.

Kalaska, J.F., and Hyde, M.L. Area 4 and area 5: differences between the load direction-dependent discharge variability of cells during active postural fixation. *Experimental Brain Research*, 59, 197, 1985.

Kalaska, J.F., Scott, S.H., Cisek, P., and Sergio, L.E. Cortical control of reaching movements. *Current Opinion in Neurobiology*, 7, 849, 1997.

Kawato, M., and Gomi, H. The cerebellum and VOR/OKR learning models. *Trends in Neurosciences*, 15, 445, 1992.

Kettner, R.E., Schwartz, A.B., and Georgopoulos, A.P. Primate motor cortex and free arm movements to visual targets in three-dimensional space. III. Positional gradients and population coding of movement direction from various movement origins. *Journal of Neuroscience*, 8, 2938, 1988.

Krakauer, J.W., Ghilardi, M.F., and Ghez, C. Independent learning of internal models for kinematic and dynamic control of reaching. *Nature Neuroscience*, 2, 1026, 1999.

Lacquaniti, F., Guigon, E., Bianchi, L., Ferraina, S., and Caminiti, R. Representing spatial information for limb movement: Role of area 5 in the monkey. *Cerebral Cortex*, 5, 391, 1995.

Lacquaniti, F., Perani, D., Guigon, E., Bettinardi, V., Carrozzo, M., Grassi, F., Rossetti, Y., and Fazio, F. Visuomotor transformations for reaching to memorized targets: a PET study. *Neuroimage*, 5, 129, 1997.

Marr, D.C. *Vision.* San Francisco, W. H. Freeman, 1982.

Mazzoni, P., Bracewell, R.M., Barash, S., and Andersen, R.A. Motor intention activity in the macaque's lateral intraparietal area. I. Dissociation of motor plan from sensory memory. *Journal of Neurophysiology*, 76, 1439, 1996.

Messier, J., and Kalaska, J.F. Covariation of primate dorsal premotor cell activity with direction and amplitude during a memorized-delay reaching task. *Journal of Neurophysiology*, 84, 152, 2000.

Miall, R.C., and Wolpert, D.M. Forward models for physiological motor control. *Neural Networks*, 9, 1265, 1996.

Miller, G.A., Galanter, E., and Pribram, K.H. *Plans and the Structure of Behavior.* New York, Holt, Rinehart and Winston, 1960.

Milner, A.D., and Goodale, M.A. *The Visual Brain in Action.* Oxford University Press, 1995.

Moran, D.W., and Schwartz, A.B. Motor cortical representation of speed and direction during reaching. *Journal of Neurophysiology*, 82, 2676, 1999.

Mountcastle, V.B., Lynch, J.C., Georgopoulos, A.P., Sakata, H., and Acuna, C. Posterior parietal association cortex of the monkey: command functions for operations within extrapersonal space. *Journal of Neurophysiology*, 38, 871, 1975.

Pandya, D.N., and Kuypers, H.G.J.M. Cortico-cortical connections in the rhesus monkey. *Brain Research*, 13, 13, 1969.

Phillips, C.G. Motor apparatus of the baboon's hand. *Proceedings of the Royal Society, London*, 173, 141, 1969.

Pisella, L., Grea, H., Tilikete, C., Vighetto, A., Desmurget, M., Rode, G., Boisson, D., and Rossetti, Y. An 'automatic pilot' for the hand in human posterior parietal cortex: toward reinterpreting optic ataxia. *Nature Neuroscience*, 3, 729, 2000.

Platt, M.L., and Glimcher, P.W. Responses of intraparietal neurons to saccadic targets and visual distractors. *Journal of Neurophysiology*, 78, 1574, 1997.

Platt, M.L., and Glimcher, P.W. Neural correlates of decision variables in parietal cortex. *Nature*, 400, 233, 1999.

Powers, W.T. *Behavior: The Control of Perception*. New York, Aldine Publishing Company, 1973.

Rushworth, M.F., Nixon, P.D., and Passingham, R.E. Parietal cortex and movement. I. Movement selection and reaching. *Experimental Brain Research*, 117, 292, 1997a.

Rushworth, M.F., Nixon, P.D., and Passingham, R.E. Parietal cortex and movement. II. Spatial representation. *Experimental Brain Research*, 117, 311, 1997b.

Sakata, H., Takaoka, Y., Kawarasaki, A., and Shibutani, H. Somatosensory properties of neurons in the superior parietal cortex area 5 of the rhesus monkey. *Brain Research*, 64, 85, 1973.

Schwartz, A.B. Motor cortical activity during drawing movements: Population representation during sinusoid tracing. *Journal of Neurophysiology*, 70, 28, 1993.

Scott, S.H. Apparatus for measuring and perturbing shoulder and elbow joint positions and torques during reaching. *Journal of Neuroscience Methods*, 89, 119, 1999.

Scott, S.H., and Kalaska, J.F. Reaching movements with similar hand paths but different arm orientations. I. Activity of individual cells in motor cortex. *Journal of Neurophysiology*, 77, 826, 1997.

Sergio, L.E., and Kalaska, J.F. Changes in the temporal pattern of primary motor cortex activity in a directional isometric force vs. limb movement task. *Journal of Neurophysiology*, 80, 1577, 1998.

Shadmehr, R., and Mussa-Ivaldi, F.A. Adaptive representation of dynamics during learning of a motor task. *Journal of Neuroscience*, 14, 3208, 1994.

Snyder, L.H., Batista, A.P., and Andersen, R.A. Coding of intention in the posterior parietal cortex. *Nature*, 386, 167, 1997.

Snyder, L.H., Batista, A.P., and Andersen, R.A. Intention-related activity in the posterior parietal cortex: a review. *Vision Research*, 40, 1433, 2000.

Strick, P.L., and Kim, C.C. Input to primate motor cortex from posterior parietal cortex area 5. I. Demonstration by retrograde transport. *Brain Research*, 157, 325, 1978.

Tipper, S.P., Howard, L.A., and Houghton, G. Action-based mechanisms of attention. *Phil. Trans. R. Soc. Lond. B*, 353, 1385, 1998.

Tipper, S.P., Lortie, C., and Baylis, G.C. Selective reaching: Evidence for action-centered attention. *Journal of Experimental Psychology: Human Perception and Performance*, 18, 891, 1992.

Ungerleider, L.G., and Mishkin, M. Two cortical visual systems. In D.J. Ingle, M.A. Goodale, and R.J.W. Mansfield, Eds., *Analysis of Visual Behavior.*, 549-586. Cambridge, MA, MIT Press, 1982.

Vilis, T., and Hore, J. Central neural mechanisms contributing to cerebellar tremor produced by perturbations. *Journal of Neurophysiology*, 43, 279, 1980.

Wiener, N. *Cybernetics, or Control and Communication in the Animal and the Machine*. Paris, Hermann, 1958.

Wise, S.P., Boussaoud, D., Johnson, P.B., and Caminiti, R. Premotor and parietal cortex: corticocortical connectivity and combinatorial computations. *Annual Review of Neuroscience*, 20, 25-42, 1997.

Wise, S.P., Moody, S.L., Blomstrom, K.J., and Mitz, A.R. Changes in motor cortical activity during visuomotor adaptation. *Experimental Brain Research*, 121, 285, 1998.

Wolpert, D.M., Ghahramani, Z., and Jordan, M.I. An internal model for sensorimotor integration. *Science*, 269, 1880, 1995.

Wolpert, D.M., Goodbody, S.J., and Husain, M. Maintaining internal representations: the role of the human superior parietal lobe. *Nature Neuroscience*, 1, 529-533, 1998.

Zarzecki, P., Strick, P.L., and Asanuma, H. Input to primate motor cortex from posterior parietal cortex area 5. II. Identification by antidromic activation. *Brain Research*, 157, 331, 1978.

12 A Critique of the Pure Feedforward Model of Touch

Miguel A. L. Nicolelis, Erika Fanselow, and Craig Henriquez

CONTENTS

12.1 INTRODUCTION

For the past five decades, neurophysiological theories aimed at accounting for the exquisite tactile perceptual capabilities of mammals have been dominated by the view that the somatosensory system, the network of subcortical and cortical neurons specialized in processing somatic information, relies primarily on feedforward (FF) computations to generate a broad spectrum of sensations (e.g., fine touch, thermo sensation, pain, etc.).[29] Although there are many nuances and variations of this general model,[29,57,78,79,121] for the sake of simplicity, throughout this chapter, we will refer to this view as the feedforward model of touch. It is important to emphasize at the outset that the central assumptions made by the proponents of this classical model of touch are almost identical to those used in an equivalent feedforward model of vision. Consequently, some of the arguments presented here have also appeared in recent studies that propose interactive models of vision (for a discussion in vision see Reference 20).

The wide acceptance of the FF model of touch can be hardly overstated. One needs only to inspect the somatosensory chapters of recent medical textbooks to

0-8493-2336-3/02/$0.00+$1.50
© 2002 by CRC Press LLC

299

confirm the popularity of this view. Indeed, if one compares recent reviews on the somatosensory chapters[57,121] with those published a couple of decades ago,[29] one may conclude that either the model has proved to be very accurate or no viable and credible alternative has been proposed. In reality, both of these conclusions would be misleading. Since its first introduction in the mid-1950s,[78] considerable anatomical, physiological, and behavioral evidence has been put forward to challenge a pure feedforward view of touch. Accordingly, more than ever, the potential contribution of some of the main building blocks of an FF model of touch, concepts such as the classic receptive field, feature detectors, independent labeled lines, cortical columns, and topographic maps, have been the subject of considerable debate.[42,75,102] For instance, the discovery, almost two decades ago, that peripheral sensory deafferentations, alterations in tactile experience, or learning can lead to considerable plastic reorganization at both cortical and subcortical levels of the adult somatosensory system demonstrated that both receptive fields and somatotopic maps are adaptive and dynamic entities.[60,75]

In this chapter, our goal is to present some of the recent evidence indicating that, despite its fundamental contributions for the field of somatosensory research, a strict FF view of touch has ceased to be consistent with the available experimental data. The main message of this chapter, therefore, is that there are many reasons to argue that an alternative neurophysiological model of tactile perception is sorely needed. In our view, this new model ought to incorporate the fact that rather than being a strict feedforward network, the somatosensory system is a vastly recurrent neural network. At all processing levels of this system, neuronal ensembles receive the convergence of multiple parallel ascending projections from the periphery (Figure 12.1), as well as multiple descending feedback pathways, originated in

---

FIGURE 12.1 (opposite) (See Color Figure 12.1 in color insert.) Schematic diagram of the rat trigeminal somatosensory system. Whiskers on the rat's snout are labeled according to the row and column in which they are located. Whisker columns are labeled from 1 to 5, caudal to rostral, while whisker rows are labeled A to E, dorsal to ventral. Peripheral nerve fibers innervating single whisker follicles have their cell bodies located in the trigeminal ganglion (Vg). Here, only the projections from Vg neurons to two main subdivisions of the trigeminal brainstem complex, the principal trigeminal nucleus (PrV) and the spinal trigeminal nucleus (SpV), are illustrated. Proponents of the feedforward model of touch usually divide these projections into rapidly adapting (RA) and slowly adapting (SA) fibers, according to their physiological responses to tactile stimuli (see text). Each of these categories contains further subdivisions, which are not described here. Neurons located in these two brainstem nuclei give rise to parallel excitatory projections to the ventroposterior medial nucleus (VPM) of the thalamus. Neurons in VPM give rise to projections to layer IV of the primary somatosensory cortex (SI). A collateral of these thalamocortical projections reach the reticular nucleus (RT), whose neurons provide the main source of GABAergic inhibition to the VPM. Descending excitatory corticothalamic projections, originating in layer VI of the SI cortex, reach the VPM and the reticular nucleus of the thalamus (RTn). The assumed topographic arrangement of these projections in the VPM and the RT are illustrated in the scheme. Feedback corticofugal projections originated in layer V of the SI cortex also reach the trigeminal brainstem complex, targeting primarily the SpV subdivision.

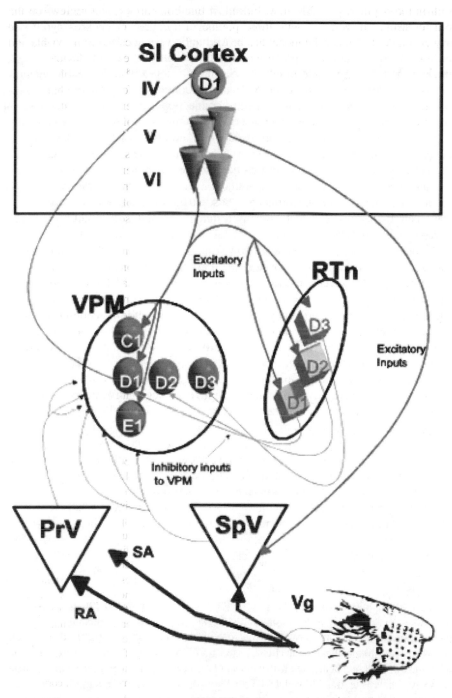

FIGURE 12.1

multiple cortical areas.[58] The main computational advantage of this architectural arrangement, we believe, is that instead of having to synthesize a new tactile view of the world every time new raw tactile information is sampled, as predicted by a pure FF model, the brain can use previously learned experiences to generate an internal model of the world.[82] This internal model could then be used to generate expectations and predictions, every time an exploratory tactile behavior is planned. Through its vast network of corticocortical connections and its massive corticofugal pathways, the central nervous system could disseminate the expectations and predictions of its touch model of the world across multiple cortical and subcortical areas. The interaction between these expectations and the raw tactile information provided by feedforward pathways may define the computations needed to continuously update this internal model, as well as to allow animals to accurately perceive the nature of any given tactile stimulus. Thus, we believe that another main mandate for a new theory of touch is to portray a much more interactive notion of tactile perception, a portrayal that exhorts the unequivocal inseparability of the somatosensory and the motor systems, which together serve as the true neuronal substrate for the emergence of the vast repertoire of tactile sensations.

A new theory of touch must also incorporate the fact that learning is the driving force of one's ability to continuously perceive old and novel tactile stimuli. Without the accumulation of past experiences through learning, the brain cannot define or refine its internal tactile model of the world throughout life. A distinction between the classical FF view of touch and the alternative model that we propose here is that, while learning can occur in the former, experience-induced learning plays an active role in perception in the latter.

Finally, in our recurrent and adaptive view of touch, there is no need for an explicit readout structure, a decoder, or a binding mechanism. The dynamic behavior and the learning rules applied by the system account for all these functions. In other words, we propose that through its distributed and dynamic interactions, the somatosensory system can use the result of previous tactile experience to interpret new tactile input conveyed by the feedforward somatosensory pathways, and create a unified perceptual experience through an interactive process. This solution is achieved by the entire system, with no single structure playing the sole role of the decoder. If one accepts this notion, and the assumption that neuronal ensembles, at different locations of the system, are capable of processing multiple features of the incoming tactile signals, it becomes clear that there is no binding problem to be solved in this model out of touch. Simply put, there is no need to bind something that has not been "broken" to start with.

As one would expect, the search for an alternative model of touch is a work in progress. We are the first to recognize that the alternative view presented here will need more time and experimental data to mature into a more definitive theory. However, we are convinced that there is considerable experimental evidence today to mount a vigorous challenge to the classic model of touch. Our objective, therefore, is to initiate the discussion that eventually will lead to the formulation of a new theory of touch. Here, our arguments in favor of a new model focus on two well-established properties of the somatosensory system, one anatomical and one physiological, that cannot be reconciled by a feedforward account of somatesthesia. In addition, we incorporate into our arguments computational principles derived from research on artificial neural networks, which suggest that recurrent somatosensory networks are more effective than feedforward models in learning and classifying new tactile experiences. These arguments have been advanced by computational neuroscientists who have argued convincingly against a pure feedforward model of perception for many years.[45,47,48,80-82] The common theme of these new models is that perception results from an interactive process, which involves the dynamic interplay of both bottom–up and top–down systems. This process requires a neuronal system endowed with a lifelong ability to learn and use past experiences to define expectations and predictions that can be tested during the process of interpreting a novel tactile encounter.

Our anatomical argument against the classic model of touch is the presence of dense and widespread cortical feedback (CFB) projections, which target all cortical and subcortical relays of the somatosensory system. In every sensory system, in all mammalian species that have been surveyed, CFB projections are denser than FF pathways that convey information from the periphery to the neocortex. Although, in this chapter, we focus our arguments primarily on the potential role of corticofugal projections that target subcortical somatosensory structures, such as the brainstem and thalamic nuclei, it is important to emphasize that a more realistic model of touch has to take into account the fact that somatosensory, as well as motor, and association cortical areas also give rise to corticocortical feedback projections.

Recent neurophysiological data obtained in the somatosensory, visual, and auditory systems clearly indicate that corticocortical, corticothalamic, and corticobulbar feedback projections can alter the sensory responses of cortical and subcortical neurons (for a review see Reference 40) that receive FF afferents from the periphery. These corticothalamic projections are invariably organized in a topographic manner, and tend to reach broader territories than their FF counterparts. Consequently, their modulatory effects can include much larger brain territories.

Research on artificial neural networks has suggested that the presence of widespread CFB projections could endow sensory systems with the capability of employing the results of previous experiences to process new incoming sensory information.[47] This would be accomplished through dynamic interactions between CFB and FF projections that reciprocally connect populations of cortical and thalamic neurons. In this scenario, descending corticofugal projections allow the implementation of more robust learning algorithms[47] than those that can be implemented in a multilayer FF neural network.[65] Yet, to date, corticofugal projections have never found a position of relevance in the classic theory of touch. This omission is particularly surprising when one realizes that most theories of pain perception, a particular type of somatic sensation, propose a central role for top–down influences, which are mediated by descending projections. These descending projections, which originate at many levels of the neuroaxis, are known to modulate the processing and transmission of nociceptive information at all intermediary relays of the spinothalamic pathway, the subdivision of the somatosensory system that is specialized in signaling the presence, location, and nature of noxious stimuli.[35] It is our contention that rather than constituting an exception, the role played by descending projections in pain perception depicts a ubiquitous computational principle of sensory information processing in the mammalian brain. We believe that this principle is likely to be a general property of all mammalian sensory systems.

The physiological argument against a pure FF model of touch is based on an extensive list of experiments that have demonstrated that the tactile responses of cortical and subcortical neurons vary significantly with the behavioral state of the animal. Thus, as animals engage in active tactile exploration, one can observe a remarkable modulation of the tactile responses of neurons distributed across all subcortical and cortical relays of the somatosensory system.[13] Behaviorally, this modulation of tactile responses is paralleled by changes in tactile discrimination thresholds, which modify the ability of subjects to respond to tactile input during an exploratory behavior.[11] This phenomenon, which has been observed in rodents, primates and human subjects, is usually referred to as sensory gating. Although the particular circuit mechanisms involved in sensory gating remain issues of debate, considerable evidence supports the hypothesis that this phenomenon is mediated centrally, and as such, it illustrates how top–down influences may play as fundamental a role in touch perception as feedforward processing. Indeed, on a closer examination, our anatomical (the presence of cortical feedback projections) and physiological (the existence of behaviorally dependent modulation of tactile responses) arguments against the pure FF model of touch are closely related. As it is described below, indirect experimental evidence has implicated CFB projections as the potential anatomical substrate for the occurrence of sensory gating at both cortical and subcortical levels of the somatosensory system.

Before describing arguments in more detail it seems appropriate to present a brief summary of the classic feedforward theory of touch. Due to space limitations, it is impossible to list and discuss all arguments in favor of this view. In fact, one should take the following description as a very broad and succinct account of a model that has been built by the contributions of many investigators. Again, we also would like to emphasize that this model has served as a powerful intellectual framework for the study of the somatosensory system.

12.2 THE FEEDFORWARD MODEL OF TOUCH

Like its counterpart model of vision, the feedforward model of touch has its origins in research conducted during a period that led to fundamental discoveries in sensory physiology.[78,79] At the time of the model's conception, sensory physiology was swept by a revolutionary experimental paradigm: the single neuron recording. By taking advantage of the ability to record extracellular activity of a single neuron, in both anesthetized and awake animals, sensory neurophysiologists began to build what continues to be the most accepted theory of how the mammalian somatosensory system operates. What follows is a brief account of this model. For a more thorough account, see the review by Dykes.[29] As expected, the main assumptions of this model clearly reflect the main neurophysiological approach used at the time the model was first conceived. Thus, in most accounts of the FF model of touch, the functional unit of computation is the single neuron, which is invariably described as a feature extractor, i.e., an element that is narrowly tuned to respond to a single feature (or a restricted set of features) of a tactile stimulus.

The central argument of the FF model of touch revolves around the well-known fact that the somatosensory system contains parallel, feedforward pathways that connect peripheral tactile receptors located throughout the body to the neocortex. These pathways carry information generated from a broad range of peripheral receptors, which are responsible for transducing different types of energies (mechanical, thermal, and chemical) that impact the body surface or are generated within the body (by proprioceptors) into trains of action potentials. According to the FF model of touch, the role of these specialized peripheral receptors is to decompose complex tactile stimuli into their primary features. This feature decomposition is carried out by the differential tuning properties of a large variety of specialized mechanical, thermal, nociceptive, polimodal, and deep receptors (also known as proprioceptors) that send their outputs to the CNS through parallel ascending somatosensory pathways. Following a series of synapses in highly segregated subregions of the spinal cord, brainstem, and thalamus, these parallel streams of information terminate in the primary somatosensory cortex.[62]

The most extreme version of the FF model purports that each group of nerve fibers carrying the output of a particular class of somatosensory receptors defines an independent, feedforward labeled-line that faithfully conveys specific tactile information all the way to the neocortex.[29,122] Thus, in addition to the well-accepted segregation between the dorsal-column/medial lemniscal system, which is specialized in conveying information from low threshold mechanoreceptors and proprioceptors, and the spinothalamic tract, which carries information from high-threshold mechanoreceptors, thermoreceptors, and nociceptors, some authors have proposed the existence of independent and parallel ascending streams that originate from different populations of rapidly or slowly adapting mechanoreceptors and terminate in the primary somatosensory cortex.[29]

Support for the view that these highly specialized parallel streams of tactile information underscore the existence of a strict labeled-line coding scheme comes from a variety of experimental observations. Implicit to this model is the assumption that local circuits within the intermediary relays of the somatosensory pathways, (the spinal cord, and several nuclei in the brainstem and thalamus) contribute little to the processing of ascending neuronal signals generated in the cutaneous periphery. According to this view, the main function of all subcortical relays of the somatosensory system is to faithfully transmit information sampled in the body's periphery to the neocortex, where all computations required for the emergence of a perceptual experience should take place. Thus, for the supporters of the FF model of touch, the observation of rapidly adapting (RA) or slowly adapting (SA) neuronal responses in the brainstem, thalamus, or even in the primary somatosensory cortex, is taken as direct evidence for the existence of segregated feedforward RA and SA pathways. Following this observation, the adaptation properties of central neurons should indicate whether they belong to either RA or SA pathways, which have their origins in the RA and SA fibers that innervate low-threshold mechanoreceptors in the animal's skin. Indeed, strict anatomical and physiological segregation schemes for RA and SA neurons have been proposed to exist in layer IV of area 3b of the primary somatosensory cortex of at least one primate species.[118]

The second important experimental observation commonly used to support the perceptual relevance of an FF model of touch is the finding that microstimulation of individual peripheral nerve fibers that innervate some classes of mechanoreceptors in humans can elicit distinct tactile perceptual experiences, which are often referred to as elementary sensations.[122] These well-localized tactile sensations are experienced by subjects once the electrical stimulation of a given peripheral fiber reaches a critical threshold level. Electrical stimulation of single Meissner and Pacini units (i.e., fibers that innervate rapidly adapting mechanoreceptors) typically elicit a sensation of vibration, whereas similar stimulation of Merkel units (innervating a slowly adapting mechanoreceptor) can produce a sensation of sustained touch or pressure. Parametrical increase in the electrical stimulus intensity is often followed by the report of additional tactile sensations by the subject, since more and more tactile fibers are recruited by a stronger stimulus.[56] Indeed, further increase in stimulus intensity can lead to paresthesias and even pain, likely because of recruitment of nociceptive fibers.

Additional experimental evidence clearly highlights the limits to which one can employ the findings described in the previous paragraph to support a strict FF view of the somatosensory system. First, no study to date has demonstrated the existence of a natural tactile stimulus capable of selectively activating either RA or SA mechanoreceptors. Instead, what is observed in reality is that both classes of mechanoreceptors tend to respond to commonly used tactile stimuli, particularly when there is relative motion between the manipulandum and the skin surface. Second, the assumption that local circuits, at each relay station of the somatosensory, are incapable of altering incoming afferent signals is clearly not supported by the experimental evidence. To disprove this rather simplistic assumption, one needs only to point out that reduction in local inhibitory feedback, such as obtained by local infusion of GABA antagonists, leads to significant physiological changes in cortical and subcortical somatosensory neurons,[66] which include changes in firing properties, such as response adaptation, and enlargement of neuronal receptive fields. In fact, local changes in the inhibitory tone or even in afferent-driven inhibition are believed to in part account for immediate receptive field reorganization that can be induced in the brainstem, thalamus, and cortex following a peripheral deafferentation.[33] Thus, one can argue that the interplay of multiple local and extrinsic afferents that converge on brainstem, thalamic, and cortical neurons, as well as the peculiar intrinsic biophysical properties of these somatosensory neurons, are likely to influence the firing adaptation properties of these cells. We conclude that the presence of RA (phasic) and SA (tonic) tactile responses across the somatosensory pathway cannot be used as the sole criterion to infer that there are segregated feedforward labeled lines originating from each of the categories of mechanoreceptors. Phasic (RA) tactile responses are much more commonly encountered in central relays of the somatosensory system (ranging from 60 to 95% of the neurons) than in peripheral somatosensory fibers (62 to 71% of the single first order fibers).[29] Moreover, RA neurons are also more often identified than SA neurons in the central nervous system.[29] Since the disparity in the frequency of RA and SA neurons along the somatosensory system is usually higher than that observed in peripheral nerves, one can argue that local circuit interactions, such as the ones provided by local inhibitory feedback, may play a role in converting some of the original SA afferent signals into RA neuronal responses.

The nature of ascending somatosensory responses may also be altered by other modulatory pathways, such as the noradrenergic, serotoninergic, and cholinergic afferents that converge at each relay station of the somatosensory system. It is also likely that cortical feedback projections (see below) could contribute for the transformation of SA into RA responses in subcortical structures simply by modulating the inhibitory tonus provided by local interneurons.

As mentioned above, the elegant results from the microneurography studies in humans have often been used as the most decisive evidence in favor of the labeled-line coding scheme of touch. However, though they may look compelling at first glance, there are several caveats that diminish their relevance as evidence in favor of a pure feedforward theory of touch. First of all, it is not surprising that stimulation of somatosensory fibers leads to some type of tactile sensation. The key question is whether the elementary sensations reported in these studies bear any resemblance to the actual percepts experienced by subjects engaged in active tactile discrimination tasks. Thus, the first concern one may have in interpreting the evidence generated in these experiments is the validity of using such an artificial stimulus (electrical microstimulation of single fibers) to categorize the perceptual capabilities of human subjects. It is safe to say that the human somatosensory system did not evolve to experience a single fiber stimulus and, as a consequence, the perceptual experiences elicited by such an uncommon stimulus should be far from the norm. The experimental evidence actually supports this prediction.[121] At a critical level of microstimulation that presumptively activates only one tactile fiber, subjects report feeling a well localized tactile sensation, which resembles the original receptive field of the stimulated fiber. However, the same subjects often report that during this single-fiber stimulation they experience an odd, almost exotic mechanical sensation, which seems very unusual to them.[121] Indeed, these subjects cannot relate this sensation to any real mechanical stimulus that they normally experience in real life.[121] Subjective accounts like these support the observation that no natural tactile stimulus known to somatosensory physiologists is capable of selectively activating just a subpopulation of cutaneous mechanoreceptors, let alone a single somatosensory fiber. This is an important issue, particularly if one realizes that only a fraction of the single fibers that are stimulated electrically can elicit very distinct perceptual experiences.[121] Indeed, stimulation of a considerable number of Ruffini units and even a subpopulation of Meissner, Merkel, and Pacini units produces no elementary tactile sensations whatsoever. Although the relevance of these negative results have been downplayed over the years, one only needs to examine them under a different framework to find support for the view that normal tactile perception emerges not by the transmission along a labeled line but rather through the integration across multiple ascending, as well as descending pathways. The fact that the elementary sensations produced by microstimulation of individual mechanoreceptor fibers are rather simple and unusual clearly distinguishes them from the type of complex, but familiar, types of tactile perceptual experiences that subjects experience when faced with real life tactile stimuli.

Another important point that is often neglected is the fact that the elementary sensations reported in the microstimulation studies result from a passive delivery of the tactile stimulus. Although this is not an issue for most proponents of the FF model of touch, those who defend a more interactive model of tactile perception would emphasize the integral role active movement contributes to the emergence of tactile percepts. Ethological evidence supports the notion that natural tactile perception emerges as the result of active exploration, which normally requires the engagement of the cutaneous periphery in manipulative behaviors that allow animals to actively scan the attributes of tangible objects. Although animals can perceive stimuli passively delivered to tactile organs, it is known that tactile discrimination occurs through the use of stereotyped behaviors, such as hand movements in primates and whisking in rodents. Since movement is known to modulate the activity of somatosensory neurons at all levels of the neuroaxis (see below), one would predict that engagement in an active tactile discrimination task would likely change the nature of the elementary sensations produced by microstimulation of single afferent fibers. In summary, even though one cannot deny the fact that electrical stimulation of individual somatosensory fibers elicits a conscious experience that can be described as an elementary tactile sensation, the role of these sensations in generating normal tactile percepts is far from clear. Certainly, viewed under the prism that we favor, the occurrence of the production of elementary tactile sensations by microstimulation of single peripheral tactile fibers sheds little light on whether labeled lines really play the dominant role in normal tactile perception that the proponents of the FF model of touch have postulated.

On their way to the neocortex, somatosensory pathways make synapses in a series of subcortical nuclei in the spinal cord, brainstem, and thalamus.[62] In each of these intermediary relays, as well as in the somatosensory cortex, one can readily identify a topographic representation of the animal's body surface. An important feature of these maps is that they are somewhat distorted, since more neuronal tissue is used to represent body regions with high densities of low-threshold mechanoreceptors (such as the hand in humans and non-human primates, and the whiskers in rodents). In each species, the distortion of these somatotopic representations also reflects the fact that body regions with high densities of mechanoreceptors invariably constitute the most important tactile organ used by animals for active tactile discrimination (e.g., hands and peri-oral regions in primates, whiskers in rodents, etc.).

Although one cannot deny the conspicuous presence of topographic maps of the body surface in every mammalian species, the precise role of topography in tactile perception is still open for debate.[102] Indeed, the formulation of the main tenants of the FF theory of touch precedes the discovery, two decades later, that these maps are highly dynamic structures that can undergo considerable plastic reorganization throughout life.[59] The plastic potential of both cortical and subcortical representations in the adult somatosensory system, which was not predicted by the FF model of touch, can no longer be ignored by any model of touch. This omission is particularly egregious if one takes into account new experimental evidence suggesting that plastic reorganization of cortical and thalamic maps, following limb amputation in humans, may account for the occurrence of a vivid tactile illusion known as phantom limb sensation.[103] In its most perverse and paradoxical form, this phantom limb sensation can be accompanied by excruciating chronic pain in a part of the body that no longer exists.[63] While information about location of a tactile stimulus could be coded and read out by stacks of topographic maps, it is important to emphasize that the lack of a precise somatotopic representation does not preclude information about stimulus location being extracted from populations of neurons located in a cortical area. For example, recent multi-electrode recordings in primates have revealed that information about stimulus location can be readily extracted, on a single-trial basis, from ensembles of neurons located in the secondary somatosensory cortex (SII) and area 2 of the parietal cortex, two regions in which one observes much less well-defined topographic maps than in the primary somatosensory (SI) cortex.[92] Interestingly, due to a degree of overlap in the timing of SI, SII, and area 2 tactile responses, stimulus location could be derived almost simultaneously in all three cortical areas. In other words, the location of a tactile stimulus can be resolved by populations of somatosensory neurons, which define highly distributed representations.[92]

We believe that in the process of defining a new theory of touch, the potential physiological role of somatotopic maps in general, as well as other modular neuronal structures, such as the barrels, barreloids, and barrelets that are observed throughout the trigeminal system of rodents, will have to be revisited. We tend to favor the notion that topographic maps primarily reflect the result of the self-organizing process that is responsible for wiring up the somatosensory system during early stages of development. Thus, although somatotopic representations may impose important constraints on the type of dynamic interactions and encoding schemes that can be implemented at certain levels of the somatosensory system, they do not necessarily preclude the existence of other representation schemes, even at the level of the thalamus and primary SI cortex. Recent experimental evidence suggests that the temporal domain of tactile neuronal responses, synchronous firing, and correlated neuronal activity could also play a role in tactile information processing[38,85] that is independent of the topographic relationships.

12.3 THE POTENTIAL ROLE OF CORTICOTHALAMIC FEEDBACK PROJECTIONS IN TACTILE INFORMATION PROCESSING

The FF model of touch completely ignores the fact that, like all other mammalian sensory systems, the somatosensory system also contains massive corticocortical and corticofugal feedback projections. For instance, in primates, feedforward somatosensory pathways terminate in four distinct somatosensory areas located in the anterior parietal cortex (areas 3a, 3b, 1, and 2). These areas are connected through feedforward corticocortical projections. Projections from the anterior parietal cortex also reach motor cortical areas in the frontal lobe, the secondary somatosensory cortex, and the somatosensory and multi-modal cortical areas of the posterior parietal cortex.[61] What is often ignored in any model of touch is that cortical neurons located in frontal motor areas and in the posterior somatosensory areas give rise to massive corticocortical feedback projections that target somatosensory fields in the anterior parietal cortex. Anterior parietal areas are also reciprocally connected through feedforward and feedback projections. In addition, corticofugal projections which originate in the infragranular layers of the primary and higher order somatosensory cortical areas project to all intermediary subcortical relays (i.e., spinal cord, brainstem, and thalamic nuclei) of the somatosensory system.[61] Indeed, once the full domain of these feedback projections is considered, the somatosensory system can only be defined as a highly recurrent network, in which multiple feedback projections are intertwined with several parallel feedforward pathways.

The importance of these corticofugal projections to any neurophysiological model of touch can be illustrated by a brief description of the anatomical organization and the physiological effects mediated by these pathways. Recently, the anatomical organization of corticothalamic projections has been investigated in great detail in rodents. As in every other mammalian species,[1,4,27,110] feedback projections from several somatosensory (e.g., SI, SII, PV, etc.) cortical areas [7,19] converge on neurons located in primary and secondary thalamic nuclei (e.g., VPM, POM, and ZI) of the trigeminal system of rodents. Studies in mice[51,53] and rats[7,26] have shown that these corticothalamic projections terminate primarily in the distal dendrites of these thalamic neurons.[100] In the case of the ventral posterior medial (VPM) nucleus, the primary thalamic relay of the trigeminal system, these corticothalamic projections are organized in a topographic manner[7,27,51,130] (see Figure 12.1). In this arrangement, corticothalamic projections originating from layer VI neurons, which are located under a particular cortical barrel[127] (e.g., barrel D1 in layer IV), terminate on thalamic neurons located across the thalamic barreloids [123] that define the representation of a whisker arc or column (e.g., barreloids A1, B1, C1, D1, and E1) in the VPM (see Figure 12.1). Corticothalamic projections from layer VI also reach the reticular nucleus (RT) of the thalamus,[98,100] the main source of GABAergic inhibition in the rat VPM.[99] These cortical-RT projections are also organized in a topographic arrangement, which seem to be orthogonal to that observed in the VPM nucleus.[51,52] Thus, axons from layer VI neurons located under a given cortical barrel (e.g., C1) target neurons located across the representation of a whisker row (e.g., C1, C2, C3, and C4) in the RT nucleus.[51,52] Neurons located in secondary somatosensory thalamic nuclei, such as the posterior medial nucleus (POM), also receive corticothalamic terminals,[51,53] albeit these are primarily derived from pyramidal neurons located in layer V of the somatosensory cortex. The morphology of corticothalamic terminals also varies according to whether they terminate in the primary (e.g., VPM) or secondary thalamic relay (e.g., POM).[52,53]

Physiological studies have shown that corticothalamic projections are primarily excitatory and likely employ glutamate as their main neurotransmitter.[119,120] The glutamate released from these corticothalamic terminals acts on NMDA, AMPA, and metabotropic receptors located in the distal dendrites of thalamic neurons.[73,105,120] Activation of NMDA, AMPA, and metabotropic receptors by *in vitro* stimulation of corticothalamic axons produces long-lasting, slow-rising EPSPs in the thalamus.[106,119] Based on some of these findings, corticothalamic-mediated activation of metabotropic receptors has been suggested to produce the modulation of neuronal firing in the VPM nucleus.[106,119] For instance, it is conceivable that the slowly rising depolarization produced by activation of corticothalamic projections could allow thalamic neurons to reach firing threshold in the presence of subthreshold synaptic input. In addition, corticothalamic afferents could also contribute to the slow activation of a low-threshold calcium conductance that underlies the production of bursts of action potentials by thalamic neurons.[110]

Despite a wealth of anatomical, pharmacological, and *in vitro* physiological information, the role played by corticothalamic projections in tactile information processing has remained elusive. For instance, penicillin-induced epileptic discharge in the cat somatosensory cortex[96] and cortical spreading depression in the rat cortex[2] were found to induce a depression of sensory evoked responses in the thalamus. In another series of experiments, carried out in both anesthetized and awake preparations, Yuan et al.[128,129] reported that lidocaine-induced inactivation of SI cortex resulted in reduced thalamic responses to electrocutaneous stimulation without any effect on the spontaneous activity, stimulus threshold, response latency, and receptive fields of the same thalamic neurons. Other studies, however, have reported a facilitatory influence of SI cortex on evoked thalamic discharges[3,4] using cortical spreading depression[125] or electrical stimulation.[5]

It is likely that part of the confusion in the literature arises because corticothalamic pathways can mediate both a monosynaptic excitatory and a dysynaptic inhibitory (via RT nucleus) postsynaptic potential in the thalamus. Thus, depending on how the cortex is stimulated or blocked, a variety of facilitatory and inhibitory response effects could be induced in the thalamus. Thus, by microstimulation of small territories of the SI cortex, Shin and Chapin[115] described a range of thalamic effects, in addition to an overall suppressive influence of thalamic sensory responses, that depended upon the topographic location of neurons in the ventral posterior nucleus of the thalamus.

In our hands, pharmacological block of SI cortical activity by focal infusion of the GABA$_A$ agonist muscimol, and consequent silencing of pools of cortical neurons that give rise to corticofugal projections to the thalamus and brainstem, produced a series of physiological effects in the rat VPM.[66] First, we observed that blocking cortical activity altered both the short- and long-latency components of the tactile responses of VPM neurons. The end result of these modifications was the demonstration that corticofugal projections contribute to the definition of the complex spatiotemporal structure[66] of the RFs of VPM neurons. These results were obtained by using traditional single-whisker stimuli. When more complex tactile stimuli were used in our experiments, we observed that the ability of VPM neurons to integrate complex tactile stimuli (e.g., multi-whisker deflections) in a non-linear way was also significantly reduced by a pharmacological block of cortical activity. Both supra- and sub linear summation of multi-whisker stimuli [41] was reduced in these experiments.[39] Overall, these findings not only support the hypothesis that corticothalamic projections may mediate both facilitatory and suppressing effects on thalamic neurons, but they also suggest that the action of these corticofugal projections may also depend on the type of tactile stimulus provided to the somatosensory system. As we will see below, there is indirect evidence that the physiological contribution of these descending pathways to tactile information processing may also depend on the behavioral state of the animal.

Further evidence for the functional relevance of corticofugal projections in the rat somatosensory system was obtained in studies carried out in our laboratory to evaluate the contribution of corticofugal projections to the ability of subcortical neurons to express unmasking of novel tactile responses following a peripheral deafferentation.[66] This reorganization process, which we dubbed immediate or acute plasticity, is known to trigger a system-wide reorganization of the somatotopic maps located at cortical, thalamic, and brainstem levels.[33] The most conspicuous effect of this immediate reorganization is the shifting of receptive fields of individual neurons away from the deafferented region due to the unmasking of neuronal tactile responses that were not present before the peripheral block. Interestingly, such unmasking tends to occur almost simultaneously in the brainstem, thalamus, and cortex.[33] In a recent series of experiments, we observed that blocking of neuronal activity in the infragranular layers of the SI cortex, a procedure that silences the projecting neurons that give rise to corticobulbar and corticothalamic feedback projections, reduces by almost 50% the number of VPM thalamic neurons that exhibit unmasking of tactile responses following a partial and reversible peripheral deafferentation.[66] Although plastic reorganization in the VPM nucleus is still observed after cortical inactivation, its spatial extent is reduced significantly. These findings have been confirmed and extended further by the recent demonstration that the immediate, but not the late phase of plastic reorganization in the ventral posterior lateral nucleus (the thalamic relay for somatosensory fibers from the rest of the body), are reduced or eliminated by removal of corticofugal projections.[97]

Hebbian-based rules of synaptic plasticity have been proposed to account for the occurrence of adult sensory plasticity in a feedforward model of touch.[76] Despite this addition, a strict feedforward view is clearly at odds with the evidence implicating cortical feedback projections in the early stages of thalamic plasticity. Thus, the available empirical evidence seems to indicate that any attempt to apply a pure FF model of the somatosensory system to account for the phenomenon of adult plasticity will be short lived.

Further support for the functional relevance of descending corticofugal projections comes from the observation that these projections have been demonstrated to affect the physiological properties of several other subcortical relays of the somatosensory system. For instance, block of neuronal activity in the SI cortex has been reported to eliminate most of the tactile responses of neurons located in the POM nucleus of the thalamus.[28] In addition, corticobulbar projections have also been shown to influence the physiological properties of neurons located in the brainstem nuclei that relay ascending somatosensory information to the thalamus.[55] For instance, removal of corticofugal projections in rats increases the responsiveness of neurons in spinal trigeminal brainstem complex to whisker stimuli.[55]

Overall, the results reviewed above make a compelling case for the need to incorporate recurrent corticofugal projections as an integral part of a comprehensive and realistic model of touch. Indeed, the recurrent nature of the somatosensory system strengthens our hypothesis that the mammalian somatosensory system relies on highly distributed neuronal interactions, which emerge from the dynamic interplay of multiple ascending and descending pathways, to represent tactile information.[85-88,90,92-94] Although the concept of distributed processing is not new, and many investigators have proposed schemes based on population coding,[25,30,31,37,49,81,109] this distributed encoding scheme has recently attracted the attention of neuroscientists because of the successful application of artificial neural networks in pattern recognition problems.[6,44,46] In a distributed coding scheme, divergent neural connections ensure that specific units of information are not held in single or small groups of neurons, but instead are widely distributed, or "encoded," by large neural ensembles located at multiple cortical and subcortical levels of the system.[49] Consequently, each neuron contributes in some way to processing of most of the information handled by the network. In line with this hypothesis, a series of studies in our lab, as well as other labs,[42,64,71,77,85,89,92,101] has begun to re-examine traditional views of information encoding by the somatosensory system. Anatomical evidence in favor of a distributed model includes the fact that ascending feedforward (FF) somatosensory pathways that carry information from the periphery to the SI cortex exhibit different degrees of divergence,[18,19,54,68-70,99,104,124] which contribute to the large multiwhisker RFs observed in the VPM and SI.[42,86] Thus, the effects of even small but incremental changes at each processing level of the pathway (e.g., from brainstem to thalamus) would tend to multiply through successive relays and could be markedly amplified by the time they reached the cortex. In addition, wide-field sensory inputs, such as high-threshold mechanical and noxious stimuli, which are transmitted through paralemniscal pathways, could also converge on cortical neurons. These effects could be further amplified by corticocortical connections within the SI and between the SI and other cortical areas.[12,32,91] In this context, the existence of massive divergent corticofugal feedback projections to all subcortical somatosensory relay nuclei provide almost unlimited opportunity for increasing the ultimate radius of influence from a single sensory event.[80] In this model, single neurons would not function as single feature detectors to serve as the functional unit of the system. Instead, neurons would work as part of ensembles that are capable of representing and processing multiple tactile attributes of a given complex stimulus simultaneously.

Massive corticofugal projections, that reach somatosensory relay structures located in the thalamus, brainstem, and the spinal cord, could offer the anatomical substrate for massive neural ensembles dedicated for tactile information processing. Such structures could be formed by somatosensory, motor, limbic, and association cortical areas, and would influence the activity of neurons located in subcortical centers, even before mechanoreceptors in the skin are activated by a tactile stimulus. According to this view, corticofugal feedback projections could incorporate subcortical nuclei into the computational processes required for the emergence of tactile percepts. Although rarely discussed in the literature, reciprocal loops between cortical and thalamic nuclei could also mediate a different type of corticocortical communication, in which thalamic networks could be employed to transform convergent signals from one or more cortical areas and then disseminate the result of this transformation to vast cortical territories. Such an interactive view of the somatosensory system would predict that top–down influences would be capable of modulating the activity of subcortical neurons during different behavioral states. But is there evidence for the existence of such top–down influences in the somatosensory system? In the next section, we describe a well-known phenomenon that may provide the key for unraveling the primordial physiological role played by corticofugal feedback on tactile perception and, hence, serve as the basis for mounting a formidable challenge to the FF model of touch.

12.4 SENSORY GATING OF NEURAL RESPONSES DURING ACTIVE TACTILE EXPLORATION

Our second fundamental argument against the feedforward model of touch is entrenched in the repeated demonstration that tactile neuronal responses are substantially affected by the motor activity that underlies the behaviors employed by different species to explore their surrounding environment.[16,17,34,83] A large body of studies, carried out in many species, indicates that, during these different exploratory behaviors, the magnitude and latencies of tactile responses, as well as the manner in which the brain responds to complex tactile stimuli, can change considerably. Thus, in rats, reductions in responses to tactile stimuli during motor activity have been observed in SI,[13,15,114] the ventral posterior lateral thalamus (VPL),[113,114] and the dorsal column nuclei (DCN).[112] Similarly, in cats, medial lemniscus sensory responses elicited by stimulation of the radial nerve are reduced in magnitude during limb movement.[24,43] Primates also show modulations in SI[17,83,84] prior to and during motor movement. Alterations in sensory responses during movement have also been observed in human evoked potential studies.[21,23,67]

These observations imply that the nervous system is capable of dynamically altering how cortical and subcortical neurons respond to a tactile stimulus, depending on the behavioral context in which such a stimulus is presented to the animal. The crux of this argument, therefore, lies in the hypothesis that the emergence of the broad spectrum of natural tactile sensations experienced by mammals results from a much more intimate association between the somatosensory and motor systems than postulated before by previous neurophysiololgical theories of touch.

But what is the significance of endowing the somatosensory system with the capability of altering the type of tactile information that can reach the cortex during motor activity? First, the alterations in response magnitude may allow tactile and proprioceptive information pertinent to execution or completion of the movement to be selectively enhanced. Thus, during the execution of a planned motor act, reciprocal interactions between the motor and the somatosensory cortices would ensure that certain types of input are gated in while others are gated out. This may be necessary in order to allow the movement to occur as planned, without interference from extraneous sensory feedback. In support of this idea, Chapin and Woodward[14] showed that tactile responses, across the cortical and subcortical relays of the somatosensory system, can be inhibited or enhanced at different epochs of the step cycle in locomoting rats. According to these authors, irrelevant tactile responses, such as those caused by the movement itself, would be selectively gated out at certain times during the movement, while sensory information that would describe certain movement epochs (e.g., footfall) would be enhanced.

There is strong evidence in the literature supporting the hypothesis that this selective modulation of tactile responses is mediated by descending efferent activity from the motor cortex or other central motor nuclei. The most relevant experimental finding supporting the occurrence of centrally mediated gating of tactile information is the observation that sensory responses across the somatosensory pathway can be reduced as much as 100 ms prior to the initiation of a given movement. These findings, which have been obtained in both monkeys[17,84] and cats,[24,43] strongly suggest that reductions in tactile responses in the cortical and subcortical relays of the somatosensory system do not result from alterations in tactile or proprioceptive feedback generated by the movement itself. Instead, they may be related to the central motor command that is generated hundreds of milliseconds prior to the movement onset. Further support for this view comes from the observation that microstimulation of the motor cortex can reduce tactile neuronal responses throughout the somatosensory system. For example, Shin and Chapin showed that stimulation of the forepaw region in the MI cortex, prior to electrical stimulation to the forepaw, led to a 43% suppression of tactile responses in the thalamus[114] and 8% reduction in the dorsal column nuclei.[112]

Another set of experiments has demonstrated that the amount of tactile response modulation varies across the different intermediary relays of the ascending somatosensory pathways. For instance, in rats, SI and VPL tactile responses to forepaw stimulation have been shown to decrease 71% and 31%, respectively, when the stimuli are delivered during the animal's locomotion.[13,114] These findings illustrate the general observation that the magnitude and frequency of somatosensory gating effect tends to increase as one ascends through the somatosensory system. As such, this observation further supports the hypothesis that modulatory signals derive from central neural networks responsible for generating the motor command.

Despite robust evidence favoring the existence of a central mechanism for modulating tactile neuronal responses, in some cases one can also demonstrate that feedforward mechanisms contribute to the alteration of cortical and subcortical tactile responses during the execution of exploratory behavior. For example, Schmidt et al.[108] have shown that when anesthesia was applied to one or more sensory nerves of the hand, inhibition of tactile stimulation that normally occurred when subjects moved the finger being stimulated was reduced (i.e., there was less gating of the response) by up to 70%. This led to the conclusion that a significant portion of the gating effect, but not all, was caused by afferent sensory stimulation generated in the periphery by the movement itself. Support for the existence of peripheral mechanisms of gating has also been provided by Chapman et al.,[17] who reported that tactile responses in SI to stimulation of the medial lemniscus or the thalamus were not reduced prior to movement, but only during the execution of the movement. These authors reported that somatosensory gating occurring at the level of the dorsal column nuclei (DCN) was caused by central modulation, since it occurred prior to the movement. However, in their hands, any additional gating at higher levels of the somatosensory system was caused by motor-induced peripheral afferent activity. Another series of experiments partially supported this view by showing that while passive movements do not cause tactile gating in the DCN, they can induce a certain degree of gating in the thalamus (VPLc) and SI.[17] Thus, even though there is substantial evidence for centrally mediated modulation of tactile responses, one cannot discard the possibility that proprioceptive and tactile afferent signals, generated during the execution of movements, contribute to the gating of tactile responses observed at higher levels of the somatosensory system.

It is important to emphasize that noradrenergic, serotoninergic, and cholinergic projections, which originate in different locations of the brainstem and diencephalon, and target all intermediary relays of the somatosensory system, could also contribute to the central modulation of tactile neuronal responses during different behavioral states,[72,126] as they do in other sensory systems.[74] Over the last three decades, one of the most elegant examples of multi-disciplinary research in neuroscience has indicated that some of these modulatory systems play a fundamental role in the control of the ascending flow of nociceptive information from the periphery that is used for the perception of pain.[35] The demonstration that physiological or pharmacological activation of these descending modulatory projections can block the ascending flow of nociceptive information through the spinothalamic system and produce maintained analgesia has revolutionized our understanding of pain perception. Since pain belongs to the spectrum of tactile sensation that all mammals experience, these observations offer more experimental support for our contention that top–down influences cannot be ignored by any theory aimed at describing the neurophysiological basis of tactile perception.

In line with this hypothesis, recent studies in the trigeminal system of awake, freely moving rats have corroborated and extended our conviction that top–down influences, such as those mediating the phenomenon of "somatosensory gating," play a crucial role in the emergence of tactile perception. In these experiments, simultaneous, multi-site chronic recordings were employed to monitor the activity of large populations of single cortical, thalamic, and brainstem somatosensory neurons, while rats moved freely in a behavioral box. Initially, these experiments allowed us to investigate how the expression of different behaviors (e.g., awake immobility, active whisking, and moving without whisker movements) could influence the physiological properties of populations of cortical and subcortical neurons in freely behaving animals.[34] Subsequently, the same experimental paradigm was used to measure how similar tactile stimuli are processed under different behavioral conditions across the rat somatosensory system.

The first important result, which argued against a simple FF view of the somatosensory system, was the observation that complex and dynamic corticothalamic interactions tend to precede any active tactile discrimination in freely behaving rats. We observed that as awake rats assume an immobile posture (i.e., standing on the forepaws without producing any whisker or other major body movements), most of the neurons in the SI cortex and VPM thalamus start producing rhythmic bursts of action potentials, which are translated into 7–12Hz rhythmic oscillations.[34,85] In the vast majority of the analyzed events, these 7–12Hz rhythmic oscillations initiate in the whisker area of the rat SI cortex (the barrel fields) (Figure 12.2). After a few tens of milliseconds, these oscillations appear in the VPM nucleus and later on they can be observed in the spinal nucleus (but not in the principal) of the trigeminal brainstem complex (Figures 12.2 and 12.3). Importantly, these oscillations were never detected in the rat trigeminal ganglion, suggesting that they are generated centrally. Further analysis revealed that these 7–12Hz thalamocortical oscillations

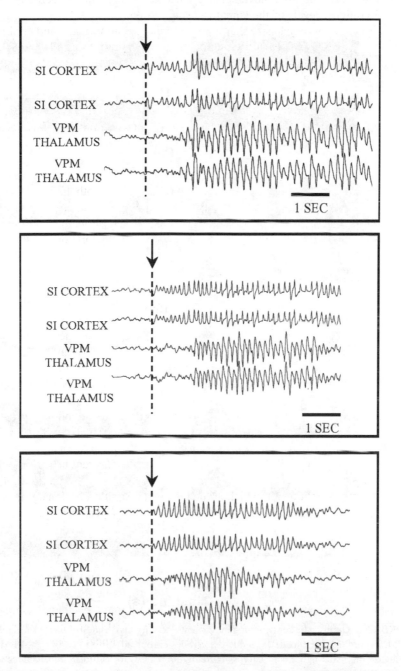

FIGURE 12.2 Simultaneous field-potential recordings in VPM and SI reveal that 7–12 Hz oscillations tend to appear first in the SI cortex and only later are evident in the somatosensory thalamus.

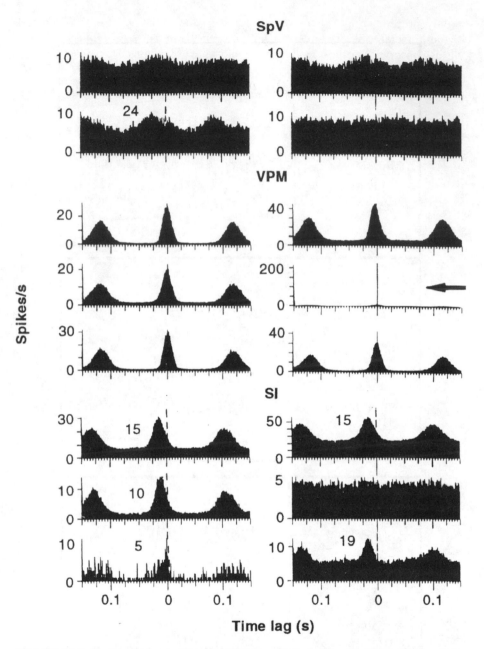

FIGURE 12.3 During whisker twitching movements, activity in SI and SpV phase leads activity in VPM. Recordings were made from microelectrodes chronically implanted in SpV, VPM, and SI in awake rats. This figure depicts cross-correlograms (CCs) for neurons in these three areas, during 7–12Hz whisker twitching movements, which were accompanied by μ-oscillations. The CCs are centered on the VPM neuron depicted by the arrow, and the numbers above the CCs show the number of milliseconds by which a given SpV or SI neuron phase led the VPM reference neuron. It can be seen that activity in SI and SpV phase led activity in VPM.

usually precede, by hundreds of milliseconds, the initiation of small amplitude rhythmic facial whisker twitching (WT) movements in the same frequency range.[85] We also observed that the initiation of WT movements modulated these oscillations.[85] Thus, soon after the onset of WT movements, rats invariably started to produce slower (4–6 Hz) rhythmic whisker protractions, which had much larger amplitudes than the WT movements. Previous behavioral studies have indicated that rats use these large rhythmic whisker movements to discriminate the tactile attributes of objects.[8] In fact, whisking is present in most rodents and is considered the most important exploratory behavior of rats. In our experiments, we also documented that as soon as the animal started to produce these slower and larger whisker movements, the 7–12 Hz thalamocortical oscillations disappeared.[85]

Altogether, these observations were very reminiscent of a similar phenomenon originally described by Gastaut in human scalp EEG recordings carried out in the 1950s.[36] Since its original discovery, both EEG and magnetoencephalographic (MEG) recordings have been used to demonstrate the occurrence of widespread 10Hz oscillations, originating in the hand representation of the primary somatosensory cortex of the vast majority of healthy human subjects and non-human primates.[95] In the EEG literature, these oscillations were named motor (μ) rhythm, since its main characteristic is to appear during awake immobility and disappear as soon as the subject starts any hand movement, the most important tactile exploratory behavior of primates. The existence of these similarities led us to postulate that the 7–12HZ oscillations that precede, and are modulated by, whisker movements in rats are equivalent to the μ rhythm of primates.

The functional role of μ oscillations in primates and rodents is still not clear. Although MEG recordings have clearly indicated that this preparatory rhythm is present in most normal human subjects, no clear consensus has been reached regarding the potential functional role played by these oscillations. Because rats use rhythmic whisker movements as their main tactile exploratory behavior, the presence of the μ rhythm in this species led us to propose that these thalamocortical oscillations could prepare the somatosensory system for the imminent onset of a cycle of tactile exploration. According to this hypothesis, during the occurrence of 7–12 Hz oscillations, tactile information would continue to flow from the VPM to the somatosensory cortex. However, because during these oscillations, a significant percentage of VPM neurons are producing bursts, it is conceivable that these neurons would have difficulty in faithfully transmitting complex spatiotemporal patterns of tactile information, which are likely to be generated when rats use their whiskers to actively explore an object. Instead, we proposed[85] that these 7–12Hz oscillations could be used to enhance or even maximize the ability of the somatosensory system to detect the presence of tactile stimuli, either during awake immobility or during the production whisker twitching movements. In other words, these 7–12Hz oscillations could represent an expectation signal, a template which would be produced by the rat somatosensory system in anticipation of whisking. In our view, this expectation signal could be generated as part of central motor program and be disseminated to most of the somatosensory system through corticofugal projections.

In our model of the rat somatosensory system, once the presence of a tactile stimulus is detected by an immobile rat (or during WT movements), the animal initiates the production of larger amplitude and slower (4–6Hz) rhythmic whisker protractions, which are used for a more detailed tactile exploration of objects. As the animal's behavior changes, so does the physiological setting of the thalamocortical loop. During the execution of these rhythmic large amplitude whisker movements, thalamocortical oscillations vanish and VPM neurons switch to a tonic firing mode. In this physiological state, VPM neurons have higher spontaneous firing rates and can faithfully represent and transmit complex spatiotemporal patterns of tactile inputs to the SI cortex. It is important to emphasize, however, that during this active tactile exploration, corticothalamic projections can still mediate important interactive computations at the level of the thalamus. Thus, in our interactive view of the somatosensory system, the thalamus is not considered as a simple passive relay. Instead, through its reciprocal interactions with different cortical areas, the somatosensory thalamus, which includes the tactile portion of the reticular nucleus, could participate in a variety of computations, such as non-linear summation of tactile stimuli,[41,111] signal segmentation through ressonant interactions, template matching, and error generation. Similar to the recurrent models of Grossberg[47] and Mumford,[82] the somatosensory thalamus could function as the site where incoming afferent tactile information is compared with cortically stored templates that resume the previous tactile experience of the animal. As feedforward and feedback projections may be required for the definition of the complex spatiotemporal structure of receptive fields in the rat VPM,[66] ensembles of these neurons could participate in template-matching operations and other computations on afferent tactile signals.

In order to test some of these assumptions, another series of experiments was carried out in our laboratory. In these experiments, multi-site chronic recordings were carried out while a nerve cuff electrode was used to provide consistent stimulation to the infraorbital (IO) nerve, the nerve that carries tactile information from the vibrissae to the central nervous system, as rats switched between a series of behavioral states.[34] In the first series of experiments, individual electrical stimuli, which produced neuronal responses that mimic those obtained by mechanical stimulation of multiple facial whiskers, were delivered to the IO nerve while rats were immobile, or when they produced the two different types of whisker movements described above. The cortical (SI) and thalamic (VPM) sensory responses elicited by the electrical stimuli were then compared. As predicted by previous studies, the magnitudes of the neural responses in SI and VPM neurons were substantially reduced during the production of rhythmic whisker movements. Interestingly, this reduction was not observed when the same animals engaged in behaviors that did not include the whisker movements (e.g., movement of the head or body). These findings corroborate work in cats[24] and in humans,[107,108] in which decreases in sensory responsiveness were most robust when the sensory stimulus was applied to the part of the body engaged in a tactile exploration, as compared to adjacent digits or contralateral limbs. Thus, as previously suggested by many authors, the phenomenon of somatosensory gating appears to be fairly topographically specific, since it occurs only during motor activity used for active tactile exploration, rather than following any action involved in increasing the general arousal level of the animal.

In addition to these findings, previous studies have shown that the presence of one stimulus can alter the ability of cortical and subcortical neurons to respond to

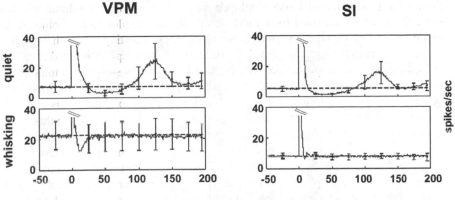

FIGURE 12.4 Activity levels in VPM and SI following peripheral stimulation differ depending on the behavioral state of an animal. Individual electrical pulses were presented to the infraorbital nerve in awake, freely moving rats and responses to this stimulation were recorded from chronically implanted microwires in VPM and SI. When the animal was in a state of quiet immobility, the initial excitatory response was followed by a period of suppressed firing, during which activity went below pre-stimulus baseline levels (dotted line). This period of suppressed firing was followed by a late excitatory component at approximately 125 post-stimulus. In contrast, during exploratory whisking behavior, the period of suppressed firing was substantially shorter in VPM and non-existent in SI, and there was no late excitatory component in either area. Error bars represent ±SEM. The initial excitatory peaks have been clipped in order to show the other components of the traces more clearly.

a subsequent stimulus for a period of time.[116,117] Evidence from our experiments and other studies[9,10] suggests that the ability of one tactile response to modulate the magnitude of a subsequent one is substantially decreased during motor activity (Figure 12.5). For example, when two tactile stimuli were presented with an inter-stimulus interval of 25–75 ms in the absence of any whisker movements (i.e., awake immobility), the response to the second stimulus was significantly reduced (Figure 12.5). However, during periods in which the same rats produced whisker movements, the response of VPM and SI neurons to the second stimulus was not statistically different in magnitude from the first at any inter-stimulus interval tested (Figure 12.5). Further examination of these results indicated that these effects paralleled a change in the amount of post-excitatory inhibition that follows the first IO stimulus in different behavior states (immobility vs. whisker movements). Thus, in the absence of any movement of the whiskers, we observed the occurrence of a long period of reduced firing, following the presentation of the first tactile stimulus.[50,116,117] However, this period is substantially shorter (in VPM) or non-existent (in SI) during the presence of exploratory whisker movements (Figure 12.4), suggesting that motor activity-related changes in post-stimulus inhibition could account for the differential responses to paired stimuli we observed in different behavioral conditions.

Overall, the results of these experiments suggest that during different behavioral states, different types of thalamocortical transmission may occur (in this case, awake immobility vs. whisking) and that these different modes of transmission may serve

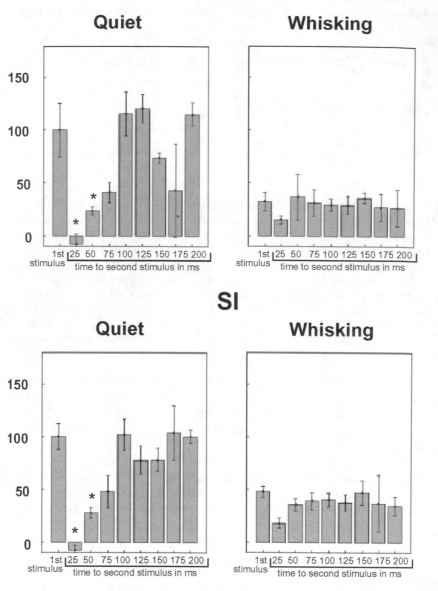

FIGURE 12.5

different perceptual purposes. Thus, differences in cortical and subcortical tactile response characteristics, from periods of whisker immobility to periods of whisker movements, suggest that the somatosensory system can shift from a state of high sensitivity for detecting individual punctate stimuli (i.e., during immobility and thalamic bursting), to a state in which the system can process with high fidelity the complex incoming tactile afferent information that are generated by the active tactile exploratory behavior employed by the animal to probe its surrounding environment.

Although there is no definitive proof that corticofugal projections play a fundamental role in either the recruitment of VPM neurons into 7–12 Hz oscillations during awake immobility, or in the switching of VPM neurons from bursting to tonic firing mode, several indirect observations can be used to build a strong case in favor of this hypothesis. First, in the vast majority of our recordings, 7–12 Hz oscillations clearly initiate in the SI cortex and only later appear in the VPM thalamus. Likewise, in all other species in which similar oscillations have been reported, they were found to originate at the cortical level. The hypothesis that corticofugal projections provide the anatomical substrate for recruiting the thalamus into a massive wave of synchronous activity is also supported by the observation that removal of corticothalamic projections significantly reduces or completely abolishes the synchronization of neuronal firing across the thalamus.[22]

The potential contribution of corticothalamic projections to the switching in firing mode of thalamic neurons is also supported by several indirect observations. Since corticothalamic axons terminate in the distal dendrites of VPM neurons, and can exert their direct excitatory effects through metabotropic receptors, they could provide the type of slow depolarization synaptic events that are required for activating the low-threshold calcium conductance that endows thalamic neurons with the ability to fire in bursts. De-inactivation of this calcium conductance, which requires hyperpolarization of VPM neurons, could also be achieved by corticothalamic projections acting through the reticular nucleus, which provides GABAergic innervation to thalamic relay neurons.

Though many more experiments are required to fully demonstrate the computations carried out by the interplay of corticofugal and ascending somatosensory pathways, the central assumption of our argument remains valid. The type of dynamic thalamocortical interactions described above cannot be explained by a simple feedforward description of the somatosensory system. As seen above, changes in behavioral state significantly alter the responses to tactile stimuli across cortical and subcortical levels of the somatosensory system. These studies have demonstrated that neuronal response properties can be altered on the order of seconds, as animals

---

FIGURE 12.5 (opposite) Responses to peripheral stimulation differ depending on the behavioral state of an animal. Recordings were made from multiple chronically implanted microwires in awake, freely moving rats. Stimulation was provided to a nerve cuff electrode implanted around the infraorbital nerve. Stimuli were presented in pairs with interstimulus intervals (isi) ranging from 25–200ms. This figure demonstrates two effects we observed by looking at the responses to these stimuli during two different behaviors, quiet immobility and exploratory whisking. First, during whisking, responses to the first stimulus in the pairs were smaller than those during the quiet state. This indicates that during the whisking state there is an overall gating of responses to ascending stimuli. The second effect was that when animals were in a state of quiet immobility, responses to the second stimulus in a pair were suppressed if the isi was 25–75 ms. In contrast, during active, exploratory whisking behavior, the responses were not significantly suppressed for any isi, compared to the response to the first stimulus in the pair. Error bars represent ±SEM.

switch from one behavioral state to the next. The possibility of altering the manner in which somatosensory neurons respond to the same tactile stimuli under different circumstances confers a high degree of adaptability to the animals, since it may allow them to filter information in different ways, as required by the situation with which they are involved. This rapid, behavior-dependent adaptation may also provide the somatosensory system with more flexibility for detection of a wider range of stimuli, or allow preferential detection of certain types of stimulation under different circumstances.

12.5   THE COMPUTATIONAL ARGUMENT IN FAVOR OF A RECURRENT VIEW OF TOUCH

Perhaps the most important computational advantage gained by having a recurrent model of touch is the demonstration that learning can be much more effective and general in these networks than in classical feedforward network architectures. An interesting discussion of this issue has been presented by Grossberg in a recent review article.[47] By comparing two types of artificial networks (feedforward self-organizing and recurrent networks), Grossberg notes that classical feedforward self-organizing feature maps that make use of lateral inhibition are capable of stable learning. Such stability, however, is not observed when the environmental inputs become numerous and involve a large number of categories to be classified. Indeed, Grossberg has demonstrated that for arbitrary environments with multiple inputs, feedforward networks tend to forget previously learned patterns. In contrast, the inclusion of feedback connections acts to provide the necessary reinforcement, such that learning remains robust even as the environment is modified. Recurrent network models, such as those based on Grossberg's Adaptive Resonance Theory (ART, Figure 12.6),[47] accomplish this by reinforcing selected input patterns through top–down pathways. Here, the neurons in the top layer create a firing pattern consistent with an expectation of the sensory input. In the absence of bottom–up input, such output can act to sensitize or prime the input layer by either bringing cells closer to threshold or by suppressing cells whose activity is not expected. In the presence of bottom–up input, the output from the top layers serves to test or match the output from the bottom–up neurons through feedback and amplification of the desired features of the input. Through feedback, learning is enhanced by minimizing the possibility that spurious inputs, inconsistent with expectations, are left unfiltered and therefore capable of destabilizing previously learned patterns and behaviors. In addition, adaptive algorithms can be implemented, as in ART, to allow the modification of stored expectations or the creation of new ones when significant mismatches are encountered during the comparison of the ascending raw input and the existing descending expectation templates. Thus, the occurrence of mismatches can trigger the creation of a new template or stimulus class. Indeed, the ability to generalize as well as create new classifications is a hallmark of models derived from ART and pattern theory. These two types of models differ, however, in the way they represent and store the templates that guide the definition of the descending expectation signal (for a review see Reference 82).

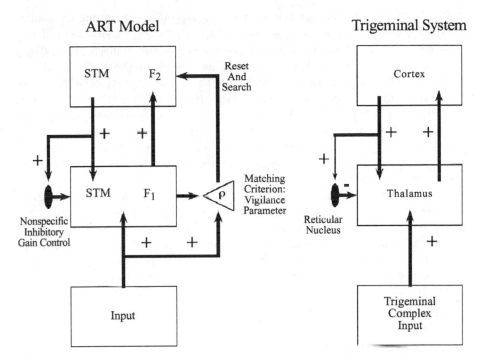

ART Model   Trigeminal System

FIGURE 12.6 (Left Panel) A prototypical ART model circuit adapted from Grossberg.[47] A short-term memory (STM) activation pattern evoked across feature detectors in level F_1 by ascending input is transmitted to level F_2 where a compressed STM of the ascending patterns is created. Learned expectations are matched through top–down pathways against bottom–up input evoked patterns in F_1; the difference in this matching, relative to the "vigilance parameter," ρ, is used to initiate a search for a more appropriate or novel recognition code. (Right Panel) Grossberg's ART model provides a theoretic framework that fits well our view of the functioning of the trigeminal, somatosensory system. The appropriateness of this comparison extends beyond the superficial similarities illustrated in the figure between the ART model's input layer, non-specific inhibitory gain control, F_2 and F_1 layers and the trigeminal system's input, reticular nucleus, thalamus, and cortex. In addition to these structural similarities, it is the dynamic interplay between ascending spatiotemporal input patterns and descending expectations, captured in both systems to generate adaptive memories that is complementary and appealing.

Another important computational limitation of a FF model of touch is that if one accepts that perception of a complex tactile stimulus is achieved only through hierarchical feedforward computations carried out by parallel labeled lines, one would have to postulate the existence of a readout mechanism that could bind the different primary attributes of the stimulus (shape, texture, temperature, texture, softness, velocity, direction, etc.) into a single percept. Although the binding problem has been the topic of intense debate in the visual system, this issue is perhaps more critical in a feedforward model of touch. That is because in addition to having to fuse a variety of primary stimulus attributes (e.g., location, intensity, orientation, direction, shape, etc.) that are similar to those integrated by the visual system, the somatosensory system also needs to take into account whether such an stimulus is harmful or not. By contributing to the definition of the sensation of pain, the spinothalamic system and its relay structures, which are specialized in conveying information about noxious stimuli, as well as temperature, to the brain, provide yet another level of complexity for the definition of a unified tactile percept. As any of us can attest, the presence of pain can considerably alter a tactile experience. To this date, no solution has been proposed and despite the arguments provided by those who believe that convergence of feedforward projections may suffice to bind individual features into a single percept, there is no experimental evidence that this is the case. Indeed, all evidence available so far goes against the existence of a single readout structure in the visual or in any other sensory system.

In our recurrent model of touch, individual neurons encode information about multiple features of a stimulus, instead of being responsible for representing the presence of a select feature. In addition, highly distributed and recurrent neuronal interactions underlie the identification of a perceptual category. Under these assumptions, the binding problem ceases to be a major issue. In fact, one could argue that in a true distributed and recurrent model of touch, one in which information about a sensory stimulus is not fragmented into separate labeled lines and that neurons do not work as strict feature detectors, there is no binding problem.

12.6 CONCLUSIONS

Although the classic feedforward model of touch has provided a fundamental blueprint for the development of somatosensory research in the last five decades, a variety of experimental findings and theoretical arguments demonstrate that this model no longer offers a useful account on how tactile perception may emerge in the mammalian brain. Instead, anatomical, physiological, and computational arguments favor the hypothesis that tactile perception emerge through interactive and recurrent interactions between the multiple cortical and subcortical levels that define the mammalian somatosensory system. Central to this recurrent model of touch is the experimental demonstration that the massive corticofugal projections, that originate in the neocortex and reach most of the subcortical structures that form the somatosensory system, may play as relevant a role in tactile information processing as the parallel feedforward pathways of this system. In addition, we propose that a recurrent model of touch has to acknowledge unequivocal inseparability of the somatosensory and motor systems, which together serve as the true neuronal substrate of tactile perception. Finally, a recurrent model of touch offers a more biologically relevant account of the computations required for learning to occur throughout life.

ACKNOWLEDGMENTS

This chapter describes research supported by grants from DARPA-ONR (N00014-98-1-0676), NSF IBN-99-80043, and NIH DE-11121-01 to M.A.L.N.

REFERENCES

1. Adams, N.C., Lozsadi, D.A., and Guillery, R.W., Complexities in the thalamocortical and corticothalamic pathways, *European Journal of Neuroscience*, 9, 204, 1997.
2. Albe-Fessard, D., Condes-Lara, M., Kesar, S., and Sanderson, P., Tonic cortical controls acting on spontaneous and evoked thalamic activity. In G. Macchi, A. Rustioni, and R. Spreafico, Eds., *Somatosensory Integration in the Thalamus: A Reevaluation Based on New Methodological Approaches,* Elsevier, Amsterdam, 1983.
3. Andersen, P., Junge, K., and Sveen, O., Cortico-thalamic facilitation of somatosensory impulses, *Nature*, 214, 1011, 1967.
4. Andersen, P., Junge, K., and Sveen, O., Cortifugal facilitation of thalamic transmission. *Brain, Behavior, Evolution*, 6, 170, 1972.
5. Anderson, P., Eccles, J.C., and Sears, T.A., Cortically evoked depolarization of primary afferent fibers in the spinal cord, *Journal of Neurophysiology*, 27, 63, 1964.
6. Bishop, C.M., *Neural Networks for Pattern Recognition.* Clarendon Press, Oxford, 1995.
7. Bourassa, J., Pinault, D., and Deschenes, M., Corticothalamic projections from the cortical barrel field to the somatosensory thalamus in rats: a single-fibre study using biocytin as an anterograde tracer, *European Journal of Neuroscience* 7, 19, 1995.
8. Carvell, G.E. and Simons, D.J., Biometric analyses of vibrissal tactile discrimination in the rat. *Journal of Neuroscience*, 10, 2638, 1990.
9. Castro-Alamancos, M.A. and Connors, B.W., Short-term synaptic enhancement and long-term potentiation in neocortex. *Proceedings of the National Academy of Sciences U.S.A.*, 93, 1335, 1996.
10. Castro-Alamancos, M.A. and Connors, B.W., Spatiotemporal properties of short-term plasticity in sensorimotor thalamocortical pathway of the rat. *Journal of Neuroscience*, 16, 2767, 1996.
11. Chapin, J.K., Modulation of cutaneous sensory transmission during movement: possible mechanisms and biological significance. In S.P. Wise, Ed., *Higher Brain Function: Recent Explorations of the Brain's Emergent Properties*, John Wiley & Sons, New York, 181, 1987.
12. Chapin, J.K., Sadeq, M., and Guise, J.L.U., Corticocortical connections within the primary somatosensory cortex of the rat, *Journal of Comparative Neurology*, 263, 326, 1987.
13. Chapin, J.K. and Woodward, D.J., Modulation of sensory responsiveness of single somatosensory cortical cells during movement and arousal behaviors, *Experimental Neurology*, 72, 164, 1981.
14. Chapin, J.K. and Woodward, D.J., Somatic sensory transmission to the cortex during movement: gating of single cell responses to touch, *Experimental Neurology*, 78, 654, 1982.
15. Chapin, J.K. and Woodward, D.J., Somatic sensory transmission to the cortex during movement: phasic modulation over the locomotor step cycle, *Experimental Neurology*, 78, 670, 1982.
16. Chapman, C.E., Bushnell, M.C., Miron, D., Duncan, G.H., and Lund, J.P., Sensory perception during movement in man, *Experimental Brain Research*, 68, 516, 1987.
17. Chapman, C.E., Jiang, W., and Lamarre, Y., Modulation of lemniscal input during conditioned arm movements in the monkey, *Experimental Brain Research*, 72, 316, 1988.

18. Chiaia, N.L., Rhoades, R.W., Bennett-Clarke, C.A., Fish, S.E., and Killackey, H.P., Thalamic processing of vibrissal information in the rat. I. Afferent input to the medial ventral posterior and posterior nuclei, *Journal of Comparative Neurology*, 314, 201, 1991.

19. Chmielowska, J., Carvell, G.E., and Simons, D.J., Spatial organization of thalamo-cortical and corticothalamic projection systems in the rat SmI barrel cortex, *Journal of Comparative Neurology*, 285, 325, 1989.

20. Churchland, P.S., Ramachandran, V.S., and Sejnowski, T.J., A critique of pure vision. In C. Koch and J. Davis, Eds., *Large-Scale Neuronal Theories of the Brain*, MIT Press, Cambridge, MA, 1, 1993.

21. Cohen, L.G. and Starr, A., Localization, timing, and specificity of gating of soma-tosensory evoked potentials during active movement in man, *Brain*, 110, 451, 1987.

22. Contreras, D., Destexhe, A., Sejnowski, T.J., and Steriade, M., Control of spatiotem-poral coherence of a thalamic oscillation by corticothalamic feedback, *Science*, 274, 771, 1996.

23. Coquery, J.M., Changes in somaesthetic evoked potentials during movement, *Brain Research*, 31, 375, 1971.

24. Coulter, J.D., Sensory transmission through lemniscal pathway during voluntary movement in the cat, *Journal of Neurophysiology*, 37, 831, 1974.

25. Deadwyler, S.A. and Hampson, R.E., The significance of neural ensemble codes during behavior and cognition, *Annual Review of Neuroscience*, 20, 217, 1997.

26. Deschenes, M., Bourassa, J., and Pinault, D., Corticothalamic projections from layer V cells in rat are collaterals of long-range corticofugal axons, *Brain Research*, 664, 215, 1994.

27. Deschenes, M., Veinante, P., and Zhang, Z.W., The organization of corticothalamic projections: reciprocity vs. parity, *Brain Research Reviews*, 28, 286, 1998.

28. Diamond, M.E., Armstrong-James, M., Budway, M.J., and Ebner, F.F., Somatic sensory responses in the rostral sector of the posterior group, POm and in the ventral posterior medial nucleus, VPM of the rat thalamus: dependence on the barrel field cortex, *Journal of Comparative Neurology*, 319, 66, 1992.

29. Dykes, R.W., Parallel processing of somatosensory information: A theory, *Brain Research Reviews*, 6, 47, 1983.

30. Erickson, R.P., Stimulus coding in topographic and non-topographic afferent modal-ities: On the significance of the activity of individual sensory neurons, *Psychological Review*, 75, 447, 1968.

31. Erickson, R.P., A neural metric, *Neuroscience & Biobehavioral Reviews*, 10, 377, 1986.

32. Fabri, M. and Burton, H., Ipsilateral cortical connections of primary somatic sensory cortex in rats, *Journal of Comparative Neurology*, 311, 405, 1991.

33. Faggin, B.M., Nguyen, K.T., and Nicolelis, M.A., Immediate and simultaneous sensory reorganization at cortical and subcortical levels of the somatosensory system, *Proceedings of the National Academy of Sciences U.S.A.*, 94, 9428, 1997.

34. Fanselow, E. and Nicolelis, M., Behavioral modulation of tactile responses in the rat soamtosensory system, *Journal of Neuroscience*, 19, 7603, 1999.

35. Fields, H.L. and Heinricher, M.M., Anatomy and physiology of a nociceptive mod-ulatory system, *Philosophical Transactions of the Royal Society of London*, B308, 361, 1985.

36. Gastaut, H., Etude electrocorticographique de la reativite des rhytmes rolandiques., *Re. Neurol., Paris*, 87, 176, 1952.

37. Georgopoulos, A.P., Swartz, A.B., and Ketter, R.E., Neuronal population coding of movement direction, *Science*, 233, 1416, 1986.
38. Ghazanfar, A., Stambaugh, C., and Nicolelis, M., Putative strategies for encoding tactile information by somatosensory thalamocortical ensembles, *Journal of Neuroscience*, in press, 2000.
39. Ghazanfar, A.A., Krupa, D.J., and Nicolelis, M.A.L., Tactile processing by thalamic neural ensembles: the role of cortical feedback, *Society for Neuroscience Abstracts*, 1797, 1997.
40. Ghazanfar, A.A. and Nicolelis, M., The space time continuum in mammalian sensory pathways. In R. Miller, Ed., *Time and the Brain*, Hardwood Academic Publishers, Sydney, 1999.
41. Ghazanfar, A.A. and Nicolelis, M.A.L., Nonlinear Processing of Tactile Information in the Thalamocortical Loop, *Journal of Neurophysiology*, 78, 506, 1997.
42. Ghazanfar, A.A. and Nicolelis, M.A.L., Spatiotemporal properties of layer V neurons of the rat primary somatosensory cortex, *Cerebral Cortex*, 9, 348, 1999.
43. Ghez, C. and Lenzi, G.L., Modulation of sensory transmission in cat lemniscal system during voluntary movement, *Pflugers Archiv — European Journal of Physiology*, 323, 273, 1971.
44. Grossberg, S., Adaptive pattern classification and universal recording: II. Feedback, expectation, olfaction, illusions, *Biological Cybernetics*, 23, 187, 1976.
45. Grossberg, S., How does a brain build a cognitive code?, *Psychological Review*, 87, 1, 1980.
46. Grossberg, S., Nonlinear neural networks: principles, mechanisms, and architectures, *Neural Networks*, 1, 17, 1988.
47. Grossberg, S., The link between brain, learning, attention, and consciousness, *Consciousness and Cognition*, 8, 1, 1999.
48. Grossberg, S. and Somers, D., Synchronized oscillations during cooperative feature linking in a cortical model of visual perception, *Neural Networks*, 4, 453, 1991.
49. Hebb, D.O., *The Organization of Behavior: A Neuropsychological Theory*, John Wiley and Sons, New York, 1, 1949.
50. Hellweg, F.C., Schultz, W., and Creutzfeldt, O.D., Extracellular and intracellular recordings from cat's cortical whisker projection area: thalamocortical response transformation, *Journal of Neurophysiology*, 40, 463, 1977.
51. Hoogland, P.V., Welker, E., and Van der Loos, H., Organization of the projections from barrel cortex to thalamus in mice studied with Phaseolus vulgaris-leucoagglutinin and HRP, *Experimental Brain Research*, 68, 73, 1987.
52. Hoogland, P.V., Welker, E., Van der Loos, H., and Wouterlood, F.G., The organization and structure of the thalamic afferents from the barrel cortex in the mouse; a PHA-L study. In M. Bentivoglio and R. Spreafico, Eds., *Cellular Thalamic Mechanisms*, Elsevier Science Publishers BV, Amsterdam, 151, 1988.
53. Hoogland, P.V., Wouterlood, F.G., Welker, E., and Van der Loos, H., Ultrastructure of giant and small thalamic terminals of cortical origin: A study of the projections from the barrel cortex in mice using *Phaseolus vulgaris* leuco-agglutinin, PHA-L, *Experimental Brain Research*, 87, 159, 1991.
54. Jacquin, M.F., Chiaia, N.L., Haring, J.H., and Rhoades, R.W., Intersubnucleus connections within the rat trigeminal brainstem complex, *Somatosensory and Motor Research*, 7, 399, 1990.
55. Jacquin, M.F., Wiegand, M.R., and Renehan, W.E., Structure-function relationships in rat brain stem subnucleus interpolaris. VIII. Cortical inputs, *Journal of Neurophysiology*, 64, 3, 1990.

56. Johansson, R.S., Hager, C., and Riso, R., Somatosensory control of precision grip during unpredictable pulling loads: II. Changes in load force rate, *Experimental Brain Research*, 89, 192, 1992.

57. Johnson, K.O., Hsiao, S.S., and Twombly, I.A., Neural mechanisms of tactile form recognition. In M.S. Gazzaniga, Ed., *The Cognitive Neurosciences*, MIT Press, Cambridge, MA, 253, 1995.

58. Kaas, J.H., Somatosensory system, In G. Paxinos, Ed., *The Human Nervous System*, Academic Press, San Diego, 813, 1990.

59. Kaas, J.H., Plasticity of sensory and motor maps in adult mammals, *Annual Review of Neuroscience*, 14, 137, 1991.

60. Kaas, J.H., Plasticity of sensory and motor maps in adult mammals, *Annual Review of Neuroscience*, 14, 137, 1991.

61. Kaas, J.H., Merzenich, M.M., and Killackey, H.P., The organization of the somatosensory cortex following peripheral nerve damage in adult and developing mammals, *Annual Review of Neuroscience*, 6, 325, 1983.

62. Kaas, J.H. and Pons, T.P., The somatosensory system of primates. In H.D. Steklis and J. Erwin, Eds., *Comparative Primate Biology*, Vol. 4, Alan R. Liss, New York, 421, 1988.

63. Katz, J. and Melzack, R., Pain "memories" in phantom limbs: review and clinical observations, *Pain*, 43, 319,1990.

64. Kleinfeld, D. and Delaney, K.R., Distributed representation of vibrissa movement in the upper layers of somatosensory cortex revealed with voltage-sensitive dyes [published erratum appears in *Journal of Comparative Neurology*, 1997, Feb 24; 378 (4:594)], *Journal of Comparative Neurology*, 375, 89, 1996.

65. Kohonen, T., *Self-Organizing Maps*, Springer, New York, 1997.

66. Krupa, D.J., Ghazanfar, A.A., and Nicolelis, M.A.L., Immediate thalamic sensory plasticity depends on corticothalamic feedback. *Proceedings of the National Academy of Sciences U.S.A.,* 96, 8200, 1999.

67. Lee, R.G. and White, D.G., Modification of the human somatosensory evoked response during voluntary movement, *Electroencephalography and Clinical Neurophysiology*, 36, 53, 1974.

68. Lin, C.S., Nicolelis, M.A., Schneider, J.S., and Chapin, J.K., A major direct GABAergic pathway from zona incerta to neocortex [see comments], *Science*, 248, 1553, 1990.

69. Lu, S.-H. and Lin, R.C.S., Thalamic afferents of the rat barrel cortex: a light- and electron-microscopic study using Phaseolus vulgaris Leucoagglutinin as an anterograde tracer, *Somatosensory and Motor Research*, 10, 1, 1993.

70. Lu, S.M. and Lin, C.S., Cortical projection patterns of the medial division of the nucleus posterior thalami in the rat, *Society of Neuroscience Abstracts,* 12, 1434, 1986.

71. Masino, S.A. and Frostig, R.D., Quantitative long-term imaging of the functional representation of a whisker in rat barrel cortex, *Proceedings of the National Academy of Sciences United States of America*, 93, 4942, 1996.

72. McCormick, D.A. and Pape, H.-C., Noradrenergic and serotonergic modualtion of a hyperpolarization-activated cation current in thalamic relay neurones, *Journal of Physiology*, 431, 319, 1990.

73. McCormick, D.A. and von Krosigk, M., Corticothalamic activation modulates thalamic firing through glutamate "metabotropic" receptors, *Proceedings of the National Academy of Sciences U.S.A.,* 89, 2774, 1992.

74. McLean, J. and Waterhouse, B.D., Noradrenergic modulation of cat area 17 neuronal responses to moving visual stimuli, *Brain Research*, 667, 83, 1994.

75. Merzenich, M.M., Kaas, J.H., Wall, J.T., Nelson, R.J., Sur, M., and Felleman, D.J., Topographic reorganization of somatosensory cortical areas 3b and 1 in adult monkeys following restricted deafferenation, *Neuroscience*, 8, 33, 1983.

76. Merzenich, M.M. and Sameshima, K., Cortical plasticity and memory, *Current Opinion in Neurobiology*, 3, 187, 1993.

77. Moore, C.I. and Nelson, S.B., Spatio-temporal subthreshold receptive fields in the vibrissa representation of rat primary somatosensory cortex, *Journal of Neurophysiology*, 80, 2882, 1998.

78 Mountcastle, V., Modality and topographic properties of single neurons of cats' somatic sensory cortex, *Journal of Neurophysiology*, 20, 408, 1957.

79. Mountcastle, V., Neural mechanisms in somesthesia, In V. Mountcastle, Ed., *Medical Physiology*, Vol. I, C.V. Mosby, St. Louis, 307, 1974.

80. Mumford, D., On the computational architecture of the neocortex. I. The role of the thalamo-cortical loop, *Biological Cybernetics*, 65, 135, 1991.

81. Mumford, D., On the computational architecture of the neocortex. II. The role of corticocortical loops, *Biological Cybernetics*, 66, 241, 1992.

82. Mumford, D., Neuronal Architectures for pattern-theoretic problems. In A. D. Koch, C. Koch, and J. Davis, Eds., *Large-Scale Neuronal Theories of the Brain*, MIT Press, Cambridge, MA, 125, 1994.

83. Nelson, R.J., Responsiveness of monkey primary somatosensory cortical neurons to peripheral stimulation depends on 'motor-set,' *Brain Research*, 304, 143, 1984.

84. Nelson, R.J., Activity of monkey primary somatosensory cortical neurons changes prior to active movement, *Brain Research*, 406, 402, 1987.

85. Nicolelis, M.A., Baccala, L.A., Lin, R.C., and Chapin, J.K., Sensorimotor encoding by synchronous neural ensemble activity at multiple levels of the somatosensory system, *Science*, 268, 1353, 1995.

86. Nicolelis, M.A. and Chapin, J.K., Spatiotemporal structure of somatosensory responses of many neuron ensembles in the rat ventral posterior medial nucleus of the thalamus, *Journal of Neuroscience*, 14, 3511, 1994.

87. Nicolelis, M.A., Fanselow, E.E., and Ghazanfar, A.A., Hebb's dream: the resurgence of cell assemblies, *Neuron*, 19, 219, 1997.

88. Nicolelis, M.A., Katz, D., and Krupa, D.J., Potential circuit mechanisms underlying concurrent thalamic and cortical plasticity, *Reviews of Neuroscience*, 9, 213, 1998.

89. Nicolelis, M.A., Lin, R.C., Woodward, D.J., and Chapin, J.K., Dynamic and distributed properties of many neuron ensembles in the ventral posterior medial thalamus of awake rats, *Proceedings of the National Academy of Sciences U.S.A.*, 90, 2212, 1993.

90. Nicolelis, M.A.L., Beyond maps: a dynamic view of the somatosensory system, *Brazilian Journal of Medical and Biological Research*, 29, 401, 1996.

91. Nicolelis, M.A.L., Chapin, J.K., and Lin, R.C.S., Ontogeny of Corticocortical Projections of the Rat Somatosensory Cortex, *Somatosensory and Motor Research*, 8, 193, 1991.

92. Nicolelis, M.A.L., Ghazanfar, A.A., Stambaugh, C.R., Oliveira, L.M.O., Laubach, M., Chapin, J.K., Nelson, R.J., and Kaas, J.H., Simultaneous encoding of tactile information by three primate cortical areas, *Nature Neuroscience*, 1, 621, 1998.

93. Nicolelis, M.A.L., Lin, R.C.S., Woodward, D.J., and Chapin, J.K., Induction of immediate spatiotemporal changes in thalamic networks by peripheral block of ascending cutaneous information, *Nature*, 361, 533, 1993.

94. Nicolelis, M.A.L., Oliveira, L.M.O., Lin, R.C.S., and Chapin, J.K., Active tactile exploration influences the functional maturation of the somatosensory system, *Journal of Neurophysiology*, 75, 2192, 1996.

95. Niedermeyer, E., The normal EEG of the waking adult, In E. Niedermeyed and F. Lopez da Silva, Eds., *Electroencephalography. Basic Principles, Clinical Applications, and Related Fields*, Williams and Wilkins, Baltimore, 131, 1993.

96. Ogden, T.E., Cortical control of thalamic somatosensory relay nuclei, *Electroencephalography and Clinical Neurophysiology*, 12, 621, 1960.

97. Parker, J. and Dostrovsky, J., Cortical involvement in the induction, but not expression, of thalamic plasticity, *Journal of Neuroscience*, 19, 8623, 1999.

98. Pinault, D., Bourassa, J., and Deschenes, M., The axonal arborization of single thalamic reticular neurons in the somatosensory thalamus of the rat, *European Journal of Neuroscience*, 7, 31, 1995.

99. Pinault, D. and Deschenes, M., Projection and innervation patterns of individual thalamic reticular axons in the thalamus of the adult rat: a three-dimensional, graphic, and morphometric analysis, *Journal of Comparative Neurology*, 391, 180, 1998.

100. Pinault, D., Smith, Y., and Deschenes, M., Dendrodendritic and axoaxonic synapses in the thalamic reticular nucleus of the adult rat, *Journal of Neuroscience*, 17, 3215, 1997.

101. Polley, D.B., Chen-Bee, C.H., and Frostig, R.D., Varying the degree of single-whisker stimulation differentially affects phases of intrinsic signals in rat barrel cortex, *Journal of Neurophysiology*, 81, 692, 1999.

102. Purves, D., Riddle, D.R., and LaMantia, A.-S., Iterated patterns of brain circuitry, or how the cortex gets its spots, *Trends in Neuroscience*, 15, 362, 1992.

103. Ramachandran, V.S., Behavioral and magnetoencephalographic correlates of plasticity in the adult human brain, *Proceedings of the National Academy of Sciences U.S.A.*, 90, 10413, 1993.

104. Rhoades, R.W., Belford, G.R., and Killackey, H.P., Receptive-field properties of rat ventral posterior medial neurons before and after selective kainic acid lesions of the trigeminal brain stem complex, *Journal of Neurophysiology*, 57, 1577, 1987.

105. Salt, T.E. and Eaton, S.A., Functions of ionotropic and metabotropic glutamate receptors in sensory transmission in the mammalian thalamus, *Progress in Neurobiology*, 48, 55, 1996.

106. Salt, T.E. and Turner, J.P., Modulation of sensory inhibition in the ventrobasal thalamus via activation of group II metabotropic glutamate receptors by 2R, 4R-aminopyrrolidine-2, 4-dicarboxylate, *Experimental Brain Research*, 121, 181, 1998.

107. Schmidt, R.F., Schady, W.J., and Torebjork, H.E., Gating of tactile input from the hand. I. Effects of finger movement, *Experimental Brain Research*, 79, 97, 1990.

108. Schmidt, R.F., Torebjork, H.E., and Schady, W.J., Gating of tactile input from the hand. II. Effects of remote movements and anaesthesia, *Experimental Brain Research*, 79, 103, 1990.

109. Sejnowski, T.J., Koch, C., and Churchland, P.S., Computational Neuroscience, *Science*, 241, 1299, 1988.

110. Sherman, S.M. and Guillery, R.W., Functional organization of thalamocortical relays, *Journal of Neurophysiology*, 76, 1367, 1996.

111. Shigemi, S., Ichikawa, T., Akasaki, T., and Sato, H., Temporal characteristics of response integration evoked by multiple whisker stimulations in the barrel cortex of rats, *Journal of Neuroscience*, 19, 10164, 1999.

112. Shin, H.C. and Chapin, J.K., Mapping the effects of motor cortex stimulation on single neurons in the dorsal column nuclei in the rat: direct responses and afferent modulation, *Brain Research Bulletin*, 22, 245, 1989.

113. Shin, H.C. and Chapin, J.K., Modulation of afferent transmission to single neurons in the ventroposterior thalamus during movement in rats, *Neuroscience Letters*, 108, 116, 1990.

114. Shin, H.C. and Chapin, J.K., Movement induced modulation of afferent transmission to single neurons in the ventroposterior thalamus and somatosensory cortex in rat, *Experimental Brain Research*, 81, 515, 1990.

115. Shin, H.-C. and Chapin, J.K., Mapping the effects of SI cortex stimulation on somatosensory relay neurons in the rat thalamus: direct responses and afferent modulation, *Somatosensory and Motor Research*, 7, 421, 1990.

116. Simons, D.J., Temporal and spatial integration in the rat SI vibrissa cortex, *Journal of Neurophysiology*, 54, 615, 1985.

117. Simons, D.J. and Carvell, G.E., Thalamocortical response transformation in the rat vibrissa/barrel system, *Journal of Neurophysiology*, 61, 311, 1989.

118. Sur, M., Wall, J. and Kaas, J., Modular distribution of neurons with slowly adapting and rapidly addapting responses in area 3b of somatosensory cortex in monkeys, *Journal of Neurophysiology*, 51, 724, 1984.

119. Turner, J.P. and Salt, T.E., Characterization of sensory and corticothalamic excitatory inputs to rat thalamocortical neurones *in vitro*, *Journal of Physiology, London*, 510, 829, 1998.

120. Turner, J.P. and Salt, T.E., Group III metabotropic glutamate receptors control corticothalamic synaptic transmission in the rat thalamus *in vitro* [In Process Citation], *Journal of Physiology, London*, 519 Pt 2, 481, 1999.

121. Vallbo, A., Single-afferent neurons and somatic sensation in humans, In M. Gazzaniga, Ed., *The Cognitive Neurosciences*, MIT Press, Cambridge, MA, 237, 1995.

122. Vallbo, A.B. and Wessberg, J., Organization of motor output in slow finger movements in man, *Journal of Physiology*, 169, 673, 1993.

123. Van der Loos, H., Barreloids in mouse somatosensory thalamus, *Neuroscience Letters*, 2, 1, 1976.

124. Veinante, P. and Deschenes, M., Single- and multi-whisker channels in the ascending projections from the principal trigeminal nucleus in the rat, *Journal of Neuroscience*, 19, 5085, 1999.

125. Waller, H.J. and Feldman, S.M., Somatosensory thalamic neurons: effects of cortical depression, *Science*, 157, 1074, 1967.

126. Waterhouse, B.D., Border, B., Wahl, L., and Mihailoff, G.A., Topographic organization of rat locus coeruleus and dorsal raphe nuclei: distribution of cells projecting to visual system structures, *Journal of Comparative Neurology*, 336, 345, 1994.

127. Woolsey, T.A. and Van der Loos, H., The structural organization of layer IV in the somatosensory region, SI of mouse cerebral cortex: the description of a cortical field composed of discrete cytoarchitectonic units, *Brain Research*, 17, 205, 1970.

128. Yuan, B., Morrow, T.J., and Casey, K.L., Responsiveness of ventrobasal thalamic neurons after suppression of SI cortex in the anesthetized rat, *Journal of Neuroscience*, 5, 2971, 1985.

129. Yuan, B., Morrow, T.J., and Casey, K.L., Cortifugal influences of S1 cortex on ventrobasal thalamic neurons in the awake rat, *Journal of Neuroscience*, 6, 3611, 1986.

130. Zhang, Z.-W. and Deschenes, M., Projections to layer VI of the postermedial barrel field in the rat: a reappraisal of the role of corticothalamic pathways, *Cerebral Cortex*, 8, 428, 1998.

13 The Changeful Mind: Plasticity in the Somatosensory System

Sherre L. Florence

CONTENTS

13.1 INTRODUCTION

Over the past decade, the literature on plasticity has grown and there now are diverse and manifold descriptions of somatosensory plasticity. In some studies, plastic changes are invoked by modifications in behavior, through training or alterations in the environment. However, in the vast majority of cases, evidence for plasticity comes from deafferentation studies, where peripheral sensory afferents are either silenced or eliminated. Both experimental paradigms produce shifts in the balance of activity from the peripheral sensory inputs, and a common assay is to measure changes in sensory maps, or the map substrates. At first pass, the breadth and diversity of the descriptions may seem overwhelming, but after consideration of the data, what emerges is a sense that some basic features of reorganization are common across experimental paradigms, levels of the system, and species.

The most reliable feature of somatosensory plasticity is that reorganization nearly always involves shifts of representation among near neighbors in the somatotopic map. This is by no means a new idea,[1,2] but as additional experimental data emerge, it is becoming apparent that neighbor relationships are not as simple as they seem. Specifically, near neighbors in the sensory representation do not all have equal access to the deprived territory. This is particularly apparent from deafferentation studies where patterns of sensory deprivation are graded from mild to severe. The data suggest that there is a hierarchy within the central maps that dictate which inputs are preferred. Moreover, there seems to be alternative possibilities depending on the pattern of denervation.

Another basic feature of plasticity in somatosensory systems is that the full potential for reorganization emerges over an extended time course. Rapidly induced changes are apparent immediately after sensory deprivation, but characteristically involve only limited extents of the sensory representation. Somewhat more extensive modifications appear gradually over the subsequent weeks after the sensory manipulation. Much later, further remodeling may occur, and these long-term changes can span large magnitudes of the sensory representation.

The mechanisms that subserve the remodeling constitute a third common characteristic of plasticity in the somatosensory system. Different mechanisms typically are associated with the various time courses of reorganization: rapid changes mediated by synaptic unmasking, more slowly developing changes attributed to NMDA receptor mechanisms, and late-emerging changes produced by new growth. Additionally, since the plasticity mechanisms act at multiple levels of the pathway, there is potential for modulation of somatosensory information at each of the stations.

In the present chapter, the evidence for these common features of plasticity in the somatosensory system will be reviewed. Literature on the effects of peripheral denervation will be emphasized, since in the majority of studies, plasticity in somatosensory system is induced by removal or silencing of sensory inputs. However, presumably the same principles apply to plasticity induced by behavioral training or after sensory enrichment. Also, since much of the plasticity research has been concentrated on cortex, the discussion will be devoted predominantly to effects of deafferentation at the cortical level. Finally, work on primates will be emphasized. It should be stressed that the patterns of reorganization that emerge in any given situation may reflect the history of the individual or features of emphasis unique to certain species. Thus, outcomes across individuals and species may differ, despite similarities in the basic processes that underlie the plasticity.

13.2 TOPOGRAPHIC HIERARCHY

A hallmark of somatosensory cortical plasticity is that topography dictates patterns of reorganization, with near neighbors in the sensory map prominently involved in the remodeling after peripheral deprivation. The most common type of reorganization is a reactivation of the region deprived of its dominant inputs by other intact inputs that are physically adjacent to the deprived representation. However, this is not, by any means, the rule. The source of the reactivating inputs can vary with different types of injury. Moreover, there appears to be a relatively rigid pecking order that governs which inputs are favored in the competition for the deprived space. The process by which inputs reactivate sensory representations after peripheral denervation is rather like the "if–then" clause used in computer programming. If any of the dominant inputs to a region are spared by the deafferentation procedure, they have preferential access to the deprived zone. If all dominant inputs are completely eliminated, then alternative inputs dominate the reactivation. Even among the alternatives, there seems to be a hierarchy so that some alternates are preferred over others. The evidence for such a pecking order is discussed below.

13.2.1 DOMINANT INPUTS

For the most part, experimental protocols for sensory denervation completely eliminate the dominant inputs to a sensory representation. For example, nerve transection produces a robust and verifiable deprivation to a readily delineated region of the sensory epithelium, particularly in the core region of the deprived zone away from borders where the peripheral innervation from other sensory nerves may partially overlap the territory of the transected nerve.[3] Amputation is an even better example of an easily demarcated sensory denervation.[4] However, in other cases, it can be difficult to completely eliminate all sensory inputs to the targeted portion of the sensory map. This is particularly troublesome in studies that involve transection of the spinal tracts. The optimal strategy dictates that the spinal gray is left intact as much as possible, and thus sensory fibers that course near the gray matter often escape transection. However, these unwanted outcomes from studies of the effects of spinal cord injury have proven to be informative for questions regarding the hierarchy of deprivation-induced takeovers.

The data suggest compellingly that if even a few dominant sensory inputs are spared, their influence is amplified and the skin surface that they innervate takes over the full extent of the deprived territory. Jain et al.[5] carefully documented cortical reactivation mediated by remaining inputs. In monkeys that had cervical dorsal column transection, tracers were injected into the skin near the peripheral nerve endings of forelimb sensory nerves to label the central sensory terminations. In normal monkeys, this technique yields patterns of labeling that reflect the precise topography of inputs from the skin of the forelimb.[6-8] In monkeys with spinal cord transection, the tracer should be unable to cross the transection site in the dorsal column, and no terminal labeling should be apparent in the dorsal column nuclei, *if the transection is complete.* Another outcome would be expected if some of the sensory afferents that ascend in the dorsal columns are spared. Axons labeled by the subcutaneous injections would be observed crossing the lesion site in the dorsal spinal cord, and some labeling in the spared afferent terminals would be expected in the dorsal column nuclei. Both indices of spared sensory afferents were described by Jain et al.[5] in monkeys that had incomplete dorsal column transection.

Using electrophysiological recording techniques, Jain and colleagues[5] also mapped the region of primary somatosensory cortex (area 3b) affected by the dorsal column transection. The forelimb representation of primary somatosensory cortex of normal monkeys has been well characterized.[9-11] As shown schematically in Figure 13.1, digit 1 is represented lateral-most and the other digits are represented

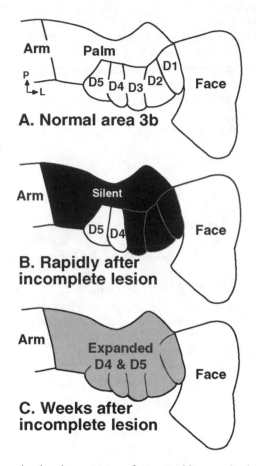

FIGURE 13.1 Schematic showing patterns of topographic organization of somatosensory cortical area 3b of owl monkeys after dorsal column transection. **A.** Normal somatotopy of the hand representation, with the orderly sequential map of the digits situated between the face representation (lateral) and the more proximal aspects of the forelimb (medial). The small islands of representation of the dorsal skin of the hand, that typically are situated medial and lateral to the digits, are not shown. **B.** Complete dorsal column section at cervical levels deprives the hand representation of all activating inputs. However, if the lesion is incomplete so that some sensory afferents are spared, neurons in area 3b that receive activation from spared inputs continue to respond robustly to sensory stimulation. In this example, a few inputs from digits 4 and 5 were spared. Over the course of weeks, the influence of the spared inputs expands. In a monkey studied 5 weeks after partial dorsal column transection (**C**), neurons throughout the full extent of the hand representation responded to spared inputs predominantly from digits 4 and 5. Adapted from Reference 5. The changes in cortical map organization have been overlaid on a summary drawing of the structural isomorph of the hand (Jain et al.[24]) so that the extent of change can be evaluated in the context of the normal hand representation. P, posterior; L, lateral.

sequentially more medially. Each of the glabrous digit representations occupies a large block of cortical tissue that has a precise internal organization. For example, the proximal skin of each digit is represented caudally, and a distal-ward progression across the skin of the digit is represented by an anterior progression through area 3b. The palm of the hand is represented caudal to the digits. Since these features of organization are highly consistent in normal monkeys, changes after sensory perturbations can be readily interpreted. In the study by Jain et al.[5] if dorsal column afferents had been spared by the spinal manipulation, as evidenced by labeled axons traversing the lesion, as well as terminal labeling in the brain stem, the portions of the map in hand cortex that corresponded to the preserved inputs contained neurons that still responded briskly to sensory stimulation after the lesion. Typically, the spared inputs stemmed from the medial hand (digits 4 and 5 and adjacent palm), so only the medial portion of the hand representation contained responsive neurons initially after the injury (Figure 13.1B). However, by about 5 weeks after onset of the deprivation, the medial hand representation had expanded and occupied the full extent of deprived cortex (Figure 13.1C). Thus, the functional efficacy of the spared inputs was enormously magnified. The explanation for this type of reactivation is not fully understood; however, it may involve the unmasking or potentiation of previously silent connections. Moreover, although the expansion is observed in cortex, it need not arise there. It could occur at any level in the relay.

Indeed, Darian-Smith and Brown[12] recently have shown data to suggest that potentiation of spared peripheral afferents may contribute to cortical reorganization. The dorsal rootlets containing sensory afferents from the digits in monkeys were mapped, and those containing input from digits 1 and 2 were lesioned. Immediately afterward, the complete deactivation of the sensory representations of digits 1 and 2 in somatosensory cortex was verified. Yet, 2–7 months later, electrophysiological recordings in somatosensory cortex revealed nearly normal somatotopic representations of the presumably deafferented digits. Responses to stimulation of digits 1 and 2 also had emerged in the dorsal rootlets immediately adjacent to those that had been lesioned, even though no response to those digits was originally detected in the preserved rootlets. No doubt, these peripheral afferents were present prior to the injury, but went undetected. Perhaps, they were weak or originally responsive to non-tactile modalities. However, during the ensuing months after elimination of the dominant inputs from digits 1 and 2, activation of the residual afferents by tactile stimulation was potentiated and they developed sufficient capacity to dominate the deprived sensory representation. Importantly, changes at peripheral levels of the somatosensory pathway that may involve relatively small numbers of afferents can render changes that appear to be robust at higher-order stations.

Even if all dominant inputs are deactivated, but return by virtue of nerve regeneration after injury, they eventually reclaim their original territory in the sensory maps. This may seem like a logical outcome, given that most of the original connections in the denervated territories likely remain after the injury. However, during the period that intervenes from the deafferenting injury to reinnervation by the regenerating nerve, the alternative inputs become functionally pre-eminent, and would seem to have a competitive advantage over inputs that were deprived by the injury. Nonetheless, the alternative inputs are unsuccessful at maintaining the newly acquired territory when the original inputs are reestablished. This sequence has been best characterized in monkeys that had transection[13-15] or crush[16] of the median nerve which innervates the palmar surface of the thumbward half of the hand. During the interval after the injury but before the nerve reinnervates the skin of the hand, the deprived territory in somatosensory cortex is not silent. Instead, studies of reorganization after median nerve cut demonstrate that deprived neurons are taken over by inputs from the dorsal (hairy) surface of the thumbward half of the hand.[1,2,17-21] The period of regeneration can take several months so the newly expressed, dorsal hand inputs have ample time to establish strong connections with their targets. Even still, if the median nerve is allowed to regenerate, neurons in the median nerve territory become non-responsive to dorsal hand stimulation and instead reacquire receptive fields on the median nerve hand.[13-16] Thus, there appears to be an inherent preference for the original sensory inputs, even if they undergo protracted periods of functional deprivation.

13.2.2 LATENT INPUTS

If all dominant inputs are removed, other previously silent, or "latent," inputs come to activate some of the deprived neurons in somatosensory cortex. In general, the source of the latent inputs is a near neighbor in the sensory map. For example, in macaque monkeys, if a portion of a digit is amputated, denervated neurons in area 3b take on new receptive fields on the portion of the digit left intact (Figure 13.2B, see also Reference 22). If the entire digit is amputated, most of the deprived neurons in area 3b are reactivated by inputs from the adjacent digits.[4,23,24] If all the digits are lost, the neurons in the deprived zone are reactivated by inputs from the palm (Figure 13.2C, see also Reference 25). If the entire hand is amputated, the deprived

---

FIGURE 13.2 (opposite) Schematic showing patterns of topographic organization in area 3b of macaque monkeys after amputation of part or all of the forelimb. **A.** The topographic organization of the hand representation in area 3b typical of normal macaque monkeys.[10,11] Large numerals 1–5 refer to the representations of digits 1–5, and thin lines indicate borders between representations. **B.** The representation of the hand in a monkey that lost the middle and distal phalanges of digit 2. The top panel shows the approximate extent of the denervated region in area 3b (black fill). The bottom panel is a schematic rendering of the pattern of reorganization that emerged. Arrows indicate that adjacent representations expanded into the deprived zone. Adapted from Manger et al.[22] **C.** After amputation of all the digits on one hand for treatment of injury, the region of area 3b where the digits are normally represented is taken over by a large representation of the palm of the hand. Conventions are the same as described for B. Adapted from Florence et al.[25] **D.** After amputation at the level of the wrist, the region where the hand would be represented in normal monkeys is taken over by representations of the wrist. Conventions are the same as described for B. Adapted from Florence and Kaas.[26] **E.** After amputation at the mid-humeral level of the arm, the deprived region is taken over by an expanded representation of the stump of the arm medially and the face laterally. Conventions are the same as described for B. Adapted from Reference 25. C, caudal; D, dorsum of the hand (dors.); F, face; FA, forearm; M, medial; P, palm; p2, proximal phalange of digit 2.

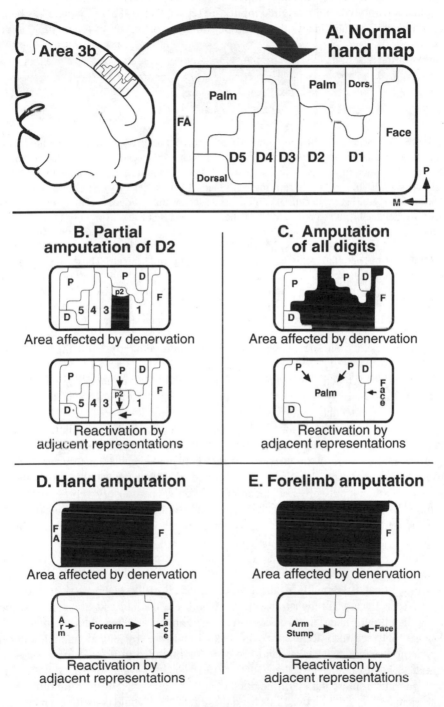

FIGURE 13.2

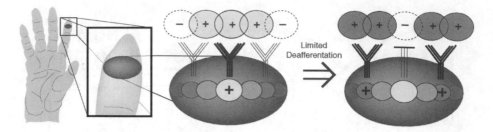

FIGURE 13.3 Schematic illustrating the potential for and limitations on receptive field reorganization based on divergent afferent inputs. To the far left is a cartoon of the glabrous hand showing the region on the distal phalange of D2 that is magnified in the panels to the right. Sensory stimuli applied to this skin region (illustrated by series of small circles within the shaded skin region) activate central neurons in a manner that produces partially overlapping receptive fields (shown as larger circles to the top). Because of the divergence of the inputs, stimulation to the skin site indicated by a plus (+) produces excitatory potentials (indicated by a + sign in the receptive fields) over a larger extent than would be predicted if there was a simple one-to-one relationship between skin and the central targets. Thus even if sensory inputs are deafferented, indicated by large X in the panel to the right, central neurons with alternate sources of excitatory inputs will not be functionally silenced. However, if the denervation eliminates all sensory inputs, as for the central neuron in the panel on the right, the neuron will be deactivated permanently or until new inputs are formed.

cortical representation is reactivated by inputs predominantly from the wrist and forearm (Figure 13.2D, see also Reference 26). In this series of examples, the pecking order for reactivation in area 3b is dictated by the extent of the denervation, so that as progressively more and more inputs are removed, the next nearest neighbor in the map provides the major source of new activation.

A common explanation for the dominance of near neighbors as alternate influences on deprived neurons in area 3b is that the thalamocortical afferent terminals are divergent so there is topographic overlap between adjacent sensory representations (Figure 13.3, see also References 27 and 28). Under normal circumstances, the overlapping inputs are not apparent, and only one sensory representation is dominant. However, if the dominant source of activation is deprived, the latent inputs gain strength and become the new governing influence, presumably through disinhibition of the latent inputs (see below).

Some of the topographic hierarchies that emerge in cortex after peripheral denervation cannot be accounted for simply by topographic overlap among the thalamocortical afferents. In squirrel monkeys, if all of the glabrous surface of the hand is deactivated, through transection of both the median and ulnar nerves, the dorsal hand representation takes over the large zone of affected neurons in area 3b (Figure 13.4; see also Reference 17). The dominance by dorsal hand inputs is surprising because the dorsal hand has a minor presence in normal hand maps in primates. For example, less than 10% of the map in normal squirrel monkeys is devoted to the representation of the dorsal hand.[29] Thus, a takeover of the full extent of the hand representation by inputs from the dorsal hand would constitute a manifold enlargement in the dorsal hand representation. If such an expansion was subserved

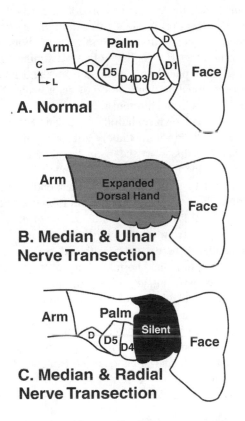

FIGURE 13.4 Patterns of topographic organization in somatosensory cortical area 3b after transection of nerves to the hand in owl or squirrel monkeys. **A.** The normal somatotopy of the hand representation. The representations of the hairy, dorsal skin of the digits and hand typically lie lateral and medial to the digit representations as shown. **B.** After transection of both the median and ulnar nerves to denervate the glabrous representations of the digits and palm, the full extent of the deprived zone is reactivated by remaining inputs from the hairy skin. Adapted from Reference 17. **C.** After transection of the median and radial nerves to eliminate both the glabrous and dorsal inputs from the lateral hand, the deprived cortical zone is only partially reactivated and a large non-responsive region (black) persists (from Reference 57). The changes in cortical map organization have been overlaid on a summary drawing of the structural isomorph of the hand (see Reference 24) so that the extent of change can be evaluated in the context of the normal hand representation. M, medial; R, rostral.

by divergent thalamocortical afferents, it would require that afferents relaying information about the dorsal hand project to all neurons in the hand representation. Since there are considerably fewer neurons in the dorsal hand representation of VP compared to those that relay information from the glabrous hand, each afferent carrying dorsal hand information would need to span large distances in area 3b to cover the full extent of the hand representation. However, no evidence of such expansive afferents have been found at the single axon level,[30,31] except for one report of a zone of terminal labeling in somatosensory cortex of a macaque monkey that spanned

8–9 mm mediolaterally across the face representation after a bulk injection in VPM.[28] Since bulk injections can label projections from multiple neurons, the widespread label does not support the contention that individual arbors have such massive divergence. Moreover, no other studies have obtained similar results.[30-32]

The alternative and more plausible explanation for the dramatic enlargement of the dorsal hand representation after denervation of glabrous hand is that reactivation takes place at earlier stations in the afferent relay. Xu and Wall[33,34] have found that neurons in the glabrous hand representation of the cuneate nucleus acquire new receptive fields on the dorsal skin immediately after median nerve transection. The rapid reactivation likely results from the near-neighbor arrangement of afferent terminals from the skin of the hand in the cuneate nucleus; inputs from the dorsal hand are directly adjacent to those from the corresponding part of the glabrous hand.[7,8] Afferents from glabrous D1 terminate adjacent to, and perhaps partially overlap afferents from, dorsal D1 in the cuneate nucleus of monkeys. Thus, because of the convergence of inputs from the two surfaces of the hand, cuneate neurons probably have the potential to be driven by either dorsal or glabrous inputs. Presumably, the glabrous inputs dominate normally; however, when the glabrous inputs are placed at a competitive disadvantage, such as occurs after median nerve transection, the afferents from the dorsal hand become potent enough to activate the target neurons. This shift in functional efficacy of the dorsal hand inputs would yield a new pattern of representation in the cuneate nucleus, consisting of a dorsal hand takeover of the entire hand representation. The relay neurons would transfer the new pattern of representation to VP, and after local adjustments that may occur in VP, the dorsally dominant hand map would be relayed to area 3b.

13.2.3 OTHER INPUTS

Other types of reorganization are not readily explained by existing connections, at any level of the pathway. The most dramatic example is that the face representation takes over hand cortex when all inputs from the forelimb are denervated. Such large-scale expansion was first shown by Pons and colleagues.[35] Somatosensory cortex was mapped in macaque monkeys that had long-standing loss of all sensory inputs from the upper extremity as a result of dorsal rhizotomy at the cervical level. Throughout the large extent of cortex where the forelimb representation normally is situated, neurons had acquired new receptive fields. The vast majority of receptive fields were on the chin of the face. Thus, the reorganized map consisted of a greatly expanded representation of the chin. More evidence of the capacity of face inputs to reactivate hand cortex came from studies in monkeys that suffered injuries to the forelimb that ultimately resulted in surgical amputation of some portion of the limb.[25,26] The level of the amputations varied considerably from animal to animal, so that in some monkeys the amputation occurred distally, below the elbow, and in others, nearly the entire arm had to be removed. If only the hand was amputated, much of the deprived cortex was taken over by wrist and forearm inputs (Figure 13.2D); however, there may have been some expansion of the face at the lateral border of the deprived zone. At the other extreme, if most of the arm was amputated, much of the denervated zone was reoccupied by an expanded representation of the face (Figure 13.2E). A similar takeover of forelimb cortex by the face also has been reported in monkeys that had dorsal column transection.[5]

What is the mechanism for face expansion? Such large-scale takeover by face inputs cannot be accounted for simply by divergent thalamocortical afferents. In the macaque monkeys that had dorsal rhizotomy, the face representation expanded to take over a mediolateral extent in cortex of 11 millimeters or more.[35] The largest thalamocortical afferent arborization observed in area 3b of a macaque monkey spanned a total extent of about 2.5 mm, and most others were much smaller.[31] Thus, the extents of thalamocortical afferents are too limited to accommodate changes that occupy many millimeters of cortical territory. Similarly, there is no explanation for the massive takeover of hand cortex by face inputs, based on existing subcortical connections. The sensory inputs from the face terminate in the trigeminal nucleus, which is lateral to the cuneate nucleus, and there is no overlap among face and hand inputs normally.[36] Since there is no evidence of a normally existing, structural framework that could allow for the widespread activation of hand cortex by face stimulation, the alternative explanation had to be considered, that new connections formed somewhere in the pathway so that face inputs had access to neurons normally activated by the hand.

To explain the takeover of forelimb neurons by the face representation, the potential for sprouting of primary sensory afferents from the face into the deprived hand representation of the cuneate nucleus was considered.[36] As will be discussed in more detail in a subsequent section, sprouting has been described at multiple levels of the pathway in monkeys after peripheral sensory manipulations. In monkeys that had either forelimb amputation or dorsal column section, bilateral injections into the face to reveal the primary sensory afferent projection patterns produced label contralateral to the limb denervation that was restricted to the trigeminal nucleus. This is the normal projection target for face inputs. However, the labeled face projection extended into the cuneate nucleus on the side ipsilateral to the denervation, and labeled areas that normally only receive inputs from the hand.[6-8] The expanded projection to the cuneate nucleus was sparse, but might have had enough functional potency to activate the deprived neurons in the cuneate nucleus. If the new connections were effective, the representation of the face would expand to take over the deprived zone in the cuneate nucleus. Although no studies yet have evaluated the electrophysiological effects of limb denervation at the level of the dorsal column nuclei in monkeys, the data from recent studies of VP of monkeys with long-standing forelimb denervation support that claim that the redistributed peripheral afferents have physiological impact. In VP of monkeys that had arm amputation[37] and in monkeys with dorsal rhizotomy,[38] the face representation extended beyond the cell-free septa that normally separates the face and hand subnuclei into the deprived hand representation. Presumably, the expanded projection from the trigeminal nucleus to the cuneate nucleus generated a functional map of the face that was relayed to the hand subnucleus of VP. In turn, the new map in VP would be relayed to somatosensory cortex, and the consequence would be widespread reactivation of hand cortex by face inputs.

Another question that is considerably more difficult to address is why inputs from the face are favored in the deprived hand representation of the primary somatosensory relay over other inputs? The gracile nucleus, which receives primary sensory afferents from the lower trunk and hindlimb is just medial to the cuneate nucleus, and just as closely adjacent to the cuneate nucleus as the trigeminal nucleus. In fact, the gracile and cuneate are separated by a less distinctive border than the cuneate and trigeminal nuclei. Thus, sprouting of gracile neurons into the deprived cuneate nucleus would seem to be equally as likely as sprouting of face neurons. Because there is no functional evidence of hindlimb expansion, the possibility of new growth from the gracile nucleus into the cuneate nucleus has not been explored in monkeys. If such sprouting occurs, the inputs presumably have no physiological impact. Indeed, in rats that had forelimb amputation early in life, hindlimb inputs sprout into the cuneate nucleus, but their impact is largely suppressed by GABA-mediated inhibition in S1.[39,40] At present, the explanation for the apparent preference for face inputs as new sources of activation after forelimb deafferentation in the relay to area 3b remains a mystery.

There is less extensive information about hierarchies among sensory representations at other levels of the pathway in monkeys, but the limited data suggest that the propensity for takeover by the face representation after forelimb deafferentation is not shared by some higher order cortical areas. Pons and colleagues[41] mapped the second somatosensory cortical area, SII, in macaque monkeys that had a lesion throughout the forelimb representation of cortical SI. Six to eight weeks after the lesion, neurons in the deprived forelimb representation responded to stimulation of the foot and lower leg, not the face. In SII of macaques, the foot and leg representation is adjacent to the forelimb on the anterolateral side, so it is not entirely surprising that the foot inputs reactivated the deprived zone. However, other adjacent representations did not expand similarly. For example, the face and head representations are medially adjacent to the forelimb in SII, yet there appeared to be only minimal enlargement of the head representation. This suggests that hindlimb inputs may be the preferred alternate in hand cortex of SII.

13.3 STAGES OF PLASTICITY

In the preceding section, the patterns of reorganization that were described represented endpoints of a progressive process that occurs over a time course of months and, in some cases, years. The endpoints are achieved by a succession of changes that evolve after deafferentation. For example, some new features of organization usually are apparent immediately, and additional changes emerge during the subsequent weeks and months. The importance of the temporal distinctions for the different stages of plasticity is that they would not be expected if all plasticity was explained by one plasticity mechanism, or by modulation of changes that appear rapidly. Instead, the distinctive temporal signature of each stage indicates that each is subserved by a different cellular process, or mechanism. As a cautionary note, however, at any given point in the plasticity sequence, the observed pattern of reorganization no doubt represents an integration of changes produced by multiple plasticity mechanisms. Thus, the plasticity process may be more akin to a progressive evolution of changes that emerge over an extended time-continuum rather than discernible epochs during which changes occur separated by periods of little plasticity.

13.3.1 CHANGES THAT APPEAR RAPIDLY

The emergence of new cortical receptive fields after the suppression or removal of the dominant excitatory inputs can be observed instantaneously after a peripheral manipulation. The most elegant demonstrations of this immediate reactivation require that neuronal recordings be maintained at the same site during an inactivation experiment to rule out interpretation errors that could result from small differences in electrode placement before and after the denervation. In the first study of this kind, Calford and Tweedale[42,43] implanted electrodes in primary somatosensory cortex of flying foxes. In initial experiments to test the reproducibility of the implanted electrodes in normal animals, little change in receptive field location was observed over the course of 3–4 weeks. However, when a digit on the forelimb was amputated, new receptive fields on nearby, intact parts of the forelimb became apparent immediately after the procedure at some recording sites. Critically, these new fields appeared only at sites where the initial receptive field was restricted to the amputated digit. At sites where the dominant activating inputs remained intact, no changes were observed. The rapid emergence of new receptive fields in primary somatosensory cortex of monkeys was described subsequently by Calford and Tweedale using this same strategy.[23] They also showed that if the dominant sensory inputs are deactivated transiently, by subcutaneous lidocaine injections, the original receptive field re-emerged when the anesthetic effects of the lidocaine dwindled and displaced the field that appeared during deactivation.

Another approach to detect rapidly emerging changes in receptive fields after sensory deactivation involves high-resolution mapping. First, in the intact animal, the cortical representation of the sensory surface is mapped. Then, the dominant excitatory inputs to a delimited portion of the mapped zone are deactivated, usually by nerve transection or digit amputation. Promptly thereafter, a second map is made of the sensory representation, usually with electrodes placed in approximately the same locations as before the deactivation. Because of possible errors in electrode placement, it cannot be certain that the same neurons are studied before and after the denervation. Thus, the procedure is not as compelling as the same-site recordings described above. However, the advantage of the mapping procedure is that it can provide insight about the topographic distributions of the acute alterations. The results from the mapping studies indicate that neurons that immediately acquire new receptive fields after sensory deactivations characteristically are positioned at the edge of the deactivated zone, near adjacent sensory representations where activating inputs remain intact (Figure 13.3; see also References 2, 34, and 44). Neurons that are distant from functionally intact inputs typically are not reactivated within a short time frame (Figure 13.3).

Subcortical relays in the lemniscal pathway also have potential for dynamic adjustments in receptive field size and location. The strategy of maintaining recordings at the same site to evaluate effects of deactivation on receptive field organization has been applied in non-primates at multiple levels of the somatosensory pathway, including the ventroposterior nucleus[45-48] and the dorsal column nuclei of the brain stem.[48-53] At all levels, neurons deprived of their dominant sensory activation can acquire new receptive fields on nearby intact sensory surfaces just as described for cortex, although it is not certain what contribution each of the subcortical relays make to the immediately observed changes found in cortex (see below for more discussion).

There are a few important exceptions to the well-established notion that latent inputs are expressed rapidly after sensory denervation. The most recent comes from a study of the effects of peripheral cold blockade on cuneate neurons in cats by Zhang and Rowe.[54] No evidence of new receptive fields was obtained during the period of cold-induced deafferentation. At present there is no explanation for the results, which seem to be in conflict with nearly all other reports. However, limited changes may have been overlooked. For example, if the blockade eliminated both dominant and latent inputs to the majority of the neurons being studied, then only those at the perimetry of the cold blockade, where silent inputs were spared, would acquire new receptive fields immediately. Alternatively, perhaps the time course of the blockade was too short. Tests for new receptive fields began 5 minutes after the cold blockade was established and were continued for no longer than 30 minutes; however, some studies have reported that acute changes only begin to emerge 15–30 minutes after deactivation.[46,53] Northgrave and Rasmusson[55] also found no new receptive fields in the cuneate nucleus of raccoons when stimulation was applied to non-deprived skin adjacent to the denervated zone, called "off-focus" stimulation. The emphasis in this paper was on inhibitory interactions in the cuneate nucleus, and it appears that only high-threshold stimuli (squeezes) were applied to off-focus sites. Perhaps the activation produced by the high-threshold stimuli suppressed any weak excitatory influences that may have resided in off-focus tactile inputs.

13.3.2 CHANGES THAT DEVELOP OVER AN INTERMEDIATE TIME COURSE

In the large majority of studies on the functional effects of sensory denervation in the somatosensory system, the emphasis is on changes that appear a few days after deactivation or within the subsequent weeks and months. The deprivation must be maintained throughout the period of observation, via some manipulation that separates the peripheral inputs from the ascending relays or by making a lesion in a portion of the sensory representation at an early station in the somatosensory pathway. These types of experiments have been performed in a wide range of species using diverse deafferentation strategies. The pattern of reorganization that appears and the time required for the changes to emerge can vary considerably, depending on the manipulation used to invoke the plasticity and the size of the deprived representation centrally. However, two general statements hold true for nearly all the results. First, the extent of reactivation increases with time after sensory deprivation. Thus, there is more extensive reactivation apparent a few weeks after the deprivation than is observed immediately, and still more reactivation is present months later (Figure 13.5). Second, there are spatial limits to the extent of reactivation that can be mediated by intermediate mechanisms. These conclusions recur in

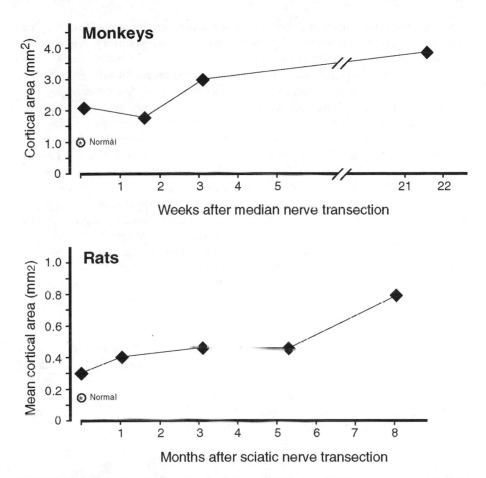

FIGURE 13.5 Plots showing the changes with time in the size of intact sensory representation that dominates deprived cortex after median nerve transection in a monkey (top) and after sciatic nerve transection in rats (bottom). In the case of the monkey, the size of the dorsal skin representation for the lateral half of the hand was measured from Figure 7 of Reference 2. Evidence for the change in size of the saphenous nerve representation after sciatic nerve cut in rats is from Figure 2 of Reference 67. In both species, the representations double in size from the control values immediately after deafferentation. The representations slowly increase further in size over the course of the following weeks and months, until all or nearly all the denervated neurons are reactivated.

virtually all cases where somatosensory deactivation has been followed over extended time frames.

In primates, much of the information on time course of reactivation comes from studies of the effects of median nerve transection, to denervate the lateral half of the palmar hand. The emergence of new receptive fields on the dorsal surface of the hand may require many weeks before the full extent of potential change is expressed (Figure 13.5). Initially, neurons only along the borders of the deactivated zone are responsive to cutaneous stimulation of adjacent intact parts of the hand.[2,18,34,44]

Gradually over time, the zone of reactivation extends further and further into the core of the denervated representation and ultimately encompasses most or all of the once-deprived representation.[1,2,20,21,56]

Another feature of reorganization that evolves over time after the nerve transection is the topographic precision within the reactivated zone. As originally reported by Merzenich's group,[2] the new receptive fields that occupy deprived cortex are large initially, but gradually decrease in size. Churchill and colleagues[21] have proposed that this receptive field refinement is attributable to a "consolidation" process whereby the most useful synaptic inputs are extracted through a use-dependent process of selection from all available excitatory inputs. In their study, monkeys that were sacrificed 2–4.5 months after median nerve transection were compared with others sacrificed more than 11 months after the same manipulation. The deprived zone was reactivated at the earliest time points, but topography was crude. At the later time point, topographic consolidation had occurred and receptive fields were refined and progressed in a more orderly manner. There were few details about the time course of the consolidation, since many months had ensued between the two time points studied; however, work from the Dykes group in cats that had forelimb denervation suggests that topographic refinements occur gradually over time (see below).

Evidence that there can be spatial limits to the cortical reorganization, even after weeks and months, comes from deafferentation strategies where both the palmar and dorsal surfaces of the hand are denervated so that both the dominant and latent inputs to the central representation are silenced. One such manipulation involves transection of combinations of the nerves to the hand. The median and radial nerves relay sensory information from both the glabrous and hairy surfaces of the lateral half of the hand, and the ulnar and radial nerves relay information from the medial half of the hand. After transection of either combination of nerves, a large region of the affected cortex remained silent for up to 11 months, the longest time point examined (see Reference 57; see also Figure 13.4). There may have been some enlargement of the remaining hand representation along the border of the deprived zone, but not enough to provide new sources of activation to the full extent of the deprived region in the cortical map. Even in early postnatal monkeys, after transection of combined nerves to the hand followed by long recovery, small regions of area 3b remain unresponsive to tactile stimulation.[58] The persistence of silent cortex over intermediate time courses suggested that the spatial extent of the deprived cortex was greater than could be reactivated by the available plasticity mechanisms.

Further evidence that the potential for reactivation is spatially limited within a few weeks or months after sensory deprivation comes from studies where one or more digits on the hand are amputated. Digit amputations produce patterns of deprivation centrally much like transection of nerves to the dorsal and palmar surfaces of the hand, in that both the dominant and latent inputs are removed. If only one digit is amputated in monkeys, the extent of the cortical representation that would be affected is only about a millimeter in width. When the electrophysiological recording experiments were performed weeks or months later, the deprived zone was completely reactivated by inputs from the amputated stump or from the adjacent digits.[4,22,24] In contrast, if two or more adjacent digits are amputated, the denervated zone would involve a cortical extent of 2 millimeters or more, some of which remain silent for months following the injury.[4] With much longer survival times (e.g., years), no evidence of the deactivated cortical neurons remain,[24] even if all the digits are amputated.[25] These reactivations likely reflect different mechanisms than those that subserve plasticity over an intermediate time frame, and will be discussed in the subsequent section.

The most extensive denervations of the forelimb representation have been produced by cervical dorsal rhizotomy[35] and by dorsal column transection,[5] which eliminates all or most of the sensory inputs from the forelimb. The only comprehensive evaluation of the time course of reactivation in cortex after large-scale deactivation was done by Jain et al.[5] and the outcome reinforces the notion that there are spatial limits to the extent of cortex that can be reactivated over an intermediate time frame. If all sensory inputs from the forearm are removed, neurons in area 3b remain unresponsive to tactile stimulation for many months.[5,59] Presumably, there are limited expansions of the intact sensory representations along the border of the deprived zone, yet the large central portion of the affected region is rendered unresponsive. Very slowly developing changes eventually bring about extensive reactivation of deprived cortex, but as for above, these data will be discussed in the subsequent section.

Only a small number of studies have looked carefully at the temporal sequence of reactivation in non-primates, but the data that have emerged from these few are consistent with the scenario proposed for primates. Kelehan and Doetsch[60] examined the effects of digit amputation on the somatosensory cortex of raccoons at 1 hour, 1 week, 2 weeks, 4 weeks, and 36 weeks after the injury. In raccoons, the digit representations in somatosensory cortex are greatly magnified, much as in primates, so that the progressive reactivation of the extensive zone of cortex deprived by the amputation was readily apparent. However, because of the vast extent of the deprived zone produced by the amputation, some non-responsive neurons persisted even at the longest survival time. The explanation for this is the same as for monkeys. The latent inputs from adjacent sensory representations do not span the full extent of a digit representation, so that the central core zone has no alternative source of activation after digit amputation. This conclusion is consistent with the findings of Rasmusson and colleagues.[61,62] Dykes and colleagues[63] studied somatosensory cortex in cats that had transection of multiple nerves in the forelimb to remove all sensory innervation to the forepaw (see also Reference 64). Reactivation appeared initially near the border between the deprived forelimb and the adjacent trunk representation and over time progressed further and further into the denervated cortex. Receptive fields at the leading edge of the reactivated cortex were large and poorly defined, and other abnormal physiological properties were apparent. However, over time, some of the most abnormal features were suppressed, much like the consolidation described by Churchill et al.[21] Even up to a year after deafferentation of the forelimb, some neurons in somatosensory cortex of cats remained non-responsive to tactile stimulation.

The time course for reactivation of somatosensory cortex has also been studied in rats after elimination of inputs from the sciatic nerve, which innervates a large portion of the hindlimb.[65-67] An important contribution of these studies is that they demonstrated the rates at which the changes occur. After sciatic nerve deafferentation, neurons in the deprived zone of cortex near the border of the intact saphenous nerve territory acquire new receptive fields on saphenous nerve skin, so that the saphaneous nerve representation expands. Quantitative evaluations of the size of the saphenous representation with time after sciatic denervation indicate that the representation expands to more than double its normal size within days of the injury (Figure 13.5; see also Reference 67). Subsequently, the saphenous nerve field again doubles in size; however, the expansion is much slower than that observed within the first few days after injury, spanning up to 8 months (Figure 13.5).

There are a few exceptions to the conclusion that new patterns of sensory representation emerge with time after peripheral denervation. In an early influential study by McMahon and Wall,[68] no evidence of receptive field reorganization was found in the dorsal column/trigeminal complex after transection of nerves to the hindfoot in adult rats. Originally, it was presumed that the outcome demonstrated the implastic nature of the dorsal column nuclei under conditions where the sensory neurons remain. However, given the more recent evidence for plasticity under even less disruptive conditions (i.e., temporary anesthetic block[49]), it is likely that small changes in receptive field organization were overlooked. This same explanation might account for the absence of new receptive fields in the trigeminal brain stem after infraorbital nerve section.[69] Another apparent exception comes from rats in which the dorsal columns were cut at thoracic levels.[70] Much of the deafferented hindlimb portion of S1 remained unresponsive to tactile stimulation, even after months of recovery.[70] Similar results were obtained in adult cats when the dorsal columns and all other ascending afferents from the hindlimb were sectioned by spinal cord transection at lower thoracic levels.[71] The emphasis in these studies was the possibility of a functional takeover of the hindlimb representation by forelimb inputs, and although a forelimb expansion was not observed there was no compelling evidence for a complete absence of reorganization. Indeed, as described by McKinley and Smith,[71] some reactivation of the deprived hindlimb neurons by trunk inputs was detected. Presumably, this occurred along the border of the deprived zone via mechanisms common to most reports of plasticity.

13.3.3 Changes that Require a Long Time Course

Initially, few studies looked at the long-term (many months to years) consequences of sensory denervation. In cases where the deafferentation spared latent inputs, the process of reorganization was complete in a matter of months.[2,17,20] In cases of more extensive denervations, whatever changes had appeared within two or so months after the initial deafferentation were thought to be all that the mature brain allowed.[4,57] No systematic inquiries challenged this presumption until Pons et al.[35] found evidence of cortical reactivation over distances of more than 11 mm in monkeys that had lived for 12 years or more after dorsal rhizotomy of spinal segments C2–T4. The procedure had eliminated all sensory input from the arm; nonetheless, neurons throughout the arm representation in somatosensory cortex had become responsive to sensory stimulation of other intact sensory surfaces. The vast majority of neurons responded to stimulation of the face.

The specific pattern of change was reminiscent of clinical reports of phantom sensations in humans with amputation of the hand or some portion of the upper limb. Characteristically, these individuals have the sensation that the missing limb is still present,[72] and some patients feel touch on the digits of their missing arm, when touched on the side of the face ipsilateral to the amputation.[73,74] This suggested that, much as described by Pons et al.[35] in monkeys with limb deafferentation, the deprived hand representation in cortex of human amputees might be activated by stimulating the face. This supposition has now been confirmed using non-invasive magnetic source imaging methods. After loss of the forelimb in humans, the region of cortex activated by the face expands to occupy much of the forelimb region.[75-81]

Reactivation of large extents of forelimb cortex also was detected in monkeys that had hand or forearm amputation.[25,26] Additional evidence for large-scale reactivations in somatosensory cortex have been reported after dorsal column section in monkeys.[5] In some cases, the deprived cortical representation becomes reactivated by an expanded representation of the face, but in some cases where arm inputs are left intact, such as after hand amputation, most of reorganization involved inputs from the remaining stump of the arm.[26] Nonetheless, the major point of the expansions that have been observed after long time courses is not the source of the new activation, but that the spatial extent of the reorganization is much more massive that previously thought possible.

The common explanation for the large-scale reactivation is that it reflects processes that evolve slowly over time. In owl monkeys that had dorsal column section, Jain and colleagues[5] found that full reactivation of hand cortex appeared by 8 months after the deactivation. The macaque monkeys that were studied after hand amputation[25,26] and those that had dorsal rhizotomy[35] had survived for years after the denervation. Thus, there was no information about the time required for the full extent of the reactivations to emerge. However, complete reorganization may take longer in higher-order primates than those reported for owl monkeys by Jain et al.[5] The cortical representation of the hand in macaque monkeys is several orders of magnitude larger than in owl monkeys; thus, the progression of reactivation may be considerably more lengthy.

There has been recent argument that the reactivation may not be a protracted process. The emergence of phantom limb sensations can be quite rapid. The regions of skin that evoked phantom limb sensations when stimulated, called "trigger zones," were apparent within 24 hours after amputation in one individual and expanded over the course of the next 8 weeks.[82] Such rapid emergence of phantom sensations would not be expected if the neurological mechanism for the sensations was a slowly developing process. The trigger zones that were apparent most rapidly after the injury involved the skin of the stump, the upper arm in the case of the one individual studied.[82] Since the upper arm is represented immediately adjacent to neurons that relay information from the forearm in the cuneate nucleus, and separated from hand neurons in the cuneate neurons by less than 200 microns (e.g., Reference 7), the earliest phantoms may result from new patterns of activation that involve only limited reactivations. Of course, the relationship between the perception of phantom limbs and the structural/functional changes observed in primary somatosensory pathway is unknown, and, thus, these arguments are purely speculative, at present.

In non-primates, most of the studies of the effects of limb deafferentation are directed toward understanding developmental mechanisms of plasticity.[40,64,71,83-85] Fewer have examined the organization of somatosensory cortex after adult deafferentation and the data are not as clear-cut as in primates. In rodents, adult forelimb amputation renders much of S1 non-responsive for up to 16 weeks (the longest time point included).[85] Yet, sprouting of hindlimb inputs into denervated cuneate nucleus has been reported within 2 weeks after forelimb denervation through cervical dorsal rhizotomy in adult rats.[86] Perhaps the new inputs to cuneate require longer time courses to activate target neurons and initiate new patterns of activation in the ascending pathway, or perhaps GABA mechanisms suppress the new inputs, much as reported by Lane et al.[84] in rats that had early neonatal forelimb amputation. Up to a year after forelimb denervation in cats, neurons at only about half the recording sites had acquired new receptive fields,[63] but longer time courses were not examined. Finally, in one raccoon that had a forelimb amputation, reactivation of the full extent of the massive forelimb cortex was reported.[87] The injury had occurred at some unknown time prior to capture, but appeared to have been a long-standing injury. Thus, the mechanism of reactivation in this animal may have been akin to that described above for primates.

13.4 MECHANISMS OF PLASTICITY

13.4.1 ADJUSTMENTS IN SYNAPTIC EFFICACY

As originally suggested by Wall[88] the rapid emergence of new receptive fields probably reflects the unmasking of normally suppressed excitatory inputs. Presumably, the latent inputs were suppressed by the activity of the dominant inputs, but become functionally viable when the suppression was released. The premise to this notion is that the full complement of sensory inputs that converge on individual neurons is not expressed under normal conditions (Figure 13.6); there is more divergence and convergence of thalamocortical inputs than is expressed normally in the functional maps of somatosensory cortex of primates.[27,28] The dominant inputs presumably are those that were reinforced through correlated patterns of sensory-driven activity.[89] However, the process is highly dynamic and the dominance of those inputs reflects on-line adjustments in the relative strengths of diverse excitatory and inhibitory inputs (for review, see Reference 90). The latent inputs that are not expressed are often referred to as a subthreshold "fringe." The rapid expression of the inputs in this fringe region is attributed to a reduction of the excitatory drive of inhibitory cortical neurons, called "afferent-driven inhibition."[91] When antagonists of the inhibitory neurotransmitter, GABA, are administered intracortically to eliminate inhibitory influences, somatosensory receptive fields become markedly larger than when inhibition was intact. The rapid acquisition of new receptive fields after peripheral sensory denervation seems to occur simultaneously at multiple levels of the pathway[34,48] and it is assumed that the changes at each level are mediated by local activity-dependent modulation of GABAergic inhibitory processes.[48]

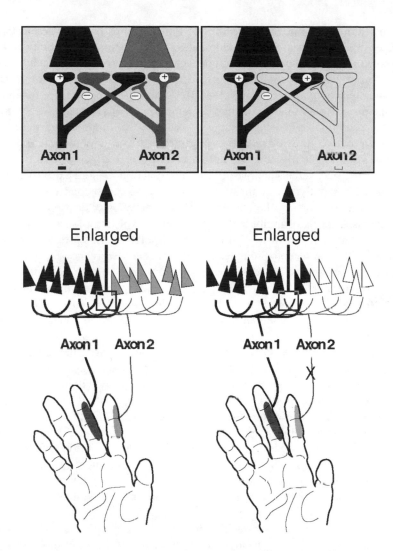

FIGURE 13.6 Schematic illustration showing the role of afferent-driven inhibition in shaping receptive field organization in the somatosensory system. The cartoon to the left shows the normal condition where receptive fields are dictated by the balance between competing intact sensory afferent inputs. Afferents from different sensory sources can terminate on the same neurons in a somatosensory relay station; however, the presence of both inputs may not be expressed. As shown in the enlarged drawing on the left (top), the dominant input to the neuron on the left (black fill) comes from axon 1. The input from axon 2 to that same neuron is inhibited by an afferent collateral from axon 1. Conversely, axon 2 dominates the neuron on the right (gray fill) and the influence of axon 1 is suppressed. However, if axon 2 is silenced (panel on right), the inhibition onto the collateral from axon 1 is removed (right,top), and the neuron that originally responded to stimulation of axon 2 acquires a new receptive field that reflects the input from axon 1 (black fill).

The second phase in cortical plasticity that develops over the course of days and weeks subsequent to sensory deprivation leads to somewhat more extensive map reorganization, as well as what appears to be the consolidation of the topographic order within reorganized cortex.[21] The substrate for this more slowly developing plasticity probably involves Hebbian-type mechanisms for synaptic strengthening of previously existing, latent inputs.[92] The inputs presumably are part of the subthreshold fringe, but may comprise a wider fringe than can be expressed rapidly following reduction of afferent-driven inhibition. Although it is often not stated explicitly, the assumption is that because the scale of the reorganization that emerges days and weeks after denervation is more extensive that that which appears rapidly, the connections that subserve the second wave of reorganization were too weak to be expressed immediately after deafferentation, even with decreased levels of inhibition. Over the course of time, these latent connections are potentiated and slowly gain functional potency. This process presumably involves the NMDA receptor complex (for review, see Reference 93), since NMDA receptor blockade during the recovery period eliminated denervation-induced plasticity in somatosensory cortex.[94] There are a host of other modulators of synaptic strength that also may contribute to the final outcome after sensory denervation[95] but consideration of all potential mechanisms for modulation is outside the scope of the present chapter.

Although most of the changes that have been described in the rapid and intermediate stages of reorganization after sensory denervation have been ascribed to existing connections, there may be local changes in anatomy within the intermediate time frame. Even with limited structural modifications, existing connections appear to be sufficient for reactivations that span a linear extent of only about 1.5 mm in cortex. This "distance limit" was first proposed by Merzenich et al.[4] and is now well established. Thus, in monkeys, if the denervation produces a zone of deactivation that spans more than 3 millimeters of cortical space (i.e., more than 1.5 mm from alternate intact inputs medial and lateral to the deprived zone), neurons in the central core of the deprived representation presumably are outside the reach of alternative cortical inputs and will remain deactivated, at least for months after the injury.[4,5,57] There are exceptions, naturally, such as the widespread reactivation of the glabrous hand representation by hairy hand inputs in monkeys (see earlier discussion); but, for the most part, topographic overlap among existing connections in the primate somatosensory system appears to be limited.

Rausell and colleagues[28] argue this conclusion and suggest that a small percentage of projections from VP to area 3b in normal monkeys span large domains of cortex and could account for the widespread plastic changes observed after peripheral injury. However, if such widespread afferents exist, presumably they would be expressed within a few weeks or months after sensory denervations that remove dominant inputs. Yet, as discussed earlier, data from deafferentation studies show that extensive denervations render portions of somatosensory cortex silent for many months (see above). Thus, there is little experimental support for the tenet that plasticity mechanisms that involve only existing connections can account for large-scale reactivations.

13.4.2 Sprouting

There had been a long-held view that axon growth, often called sprouting, only occurred in the developing brain. In the mature brain, sprouting was necessarily limited to preserve the stability of brain function, or so it was thought. Over the last decade, however, evidence has accumulated that sprouting can occur in the adult somatosensory system and recent reviews have documented the findings in detail (for reviews, see References 96 and 97).

One process that can initiate new growth is injury. For example, damage to peripheral sensory nerves apparently triggers the upregulation of genes that normally are associated with developmental growth (i.e., GAP-43) in the injured dorsal root ganglion cells. The growth factors are transferred to the injured peripheral axons, to facilitate repair; however, growth factors also are transported to the central terminals of the injured neurons and result in the growth of new connections in the brain.[98,99]

This is not the only means by which sprouting is induced. Neurons apparently can sprout new processes without being prompted to grow by injury. In monkeys that had limb amputation, subcutaneous injections of anatomical tracers were made into the arm proximal to the amputation to label their input patterns in the cuneate nucleus.[26] Matched injections were made into the opposite, intact arm for comparison. Inputs from the arm normally terminate dorsal to those from the hand; however, in the amputee monkeys the labeled sensory afferents from the arm occupied the hand and digit portions of the nucleus, presumably by growing into the deprived region.[26] Importantly, this change in pattern of connections involved uninjured afferent inputs. The sprouting of trigeminal inputs from the face in the cuneate nucleus after hand amputation or dorsal column section,[36] described in an earlier section, is another example of sprouting of connections that are not directly injured by deafferentation. Sprouting of new connections from uninjured neurons also occurs at the cortical level in the somatosensory system. In monkeys that had limb amputation, horizontal connections in the reorganized cortical zone are more extensive than normal.[25] Neurons in the deprived cortical zones extend outside their normal territory and project to nearby non-deprived cortex. Conversely, neurons in non-deprived cortical zones sprout into the deprived region of cortex.

What might be the impetus for the formation of new connections by intact, non-injured neurons after peripheral deactivation? The widespread loss of sensory activation could have far-reaching effects on neurons in the somatosensory pathway. Vacant synaptic sites on the post-synaptic neurons may release molecular signals to signal synaptic availability, such as signal could stimulate nearby axons to grow, particularly if other local factors favored axon growth. The production of growth inhibitors that normally suppress neurite growth (see Reference 100 for review), may be down-regulated or blocked after injury, so that the growth of new connections is no longer prevented. This is all largely speculation since the molecular requirements for new growth in the central nervous system are still being unraveled.[101] However, considerable research effort is being directed toward this issue, and eventually we may be in a position to foster growth where clinical outcomes can be improved.

The functional impact of new activating inputs to denervated regions of the brain may be substantial, and this is why evidence for sprouting has been heralded with considerable interest. New inputs to deprived neurons in the cuneate nucleus would provide an alternate source of activation for deprived cuneate neurons. In turn, the new pattern of activation in the brainstem would be relayed to the ventroposterior nucleus in the thalamus, and subsequently to somatosensory cortex where new cortical growth likely contributed further to the reorganization. Moreover, at each level of the relay there would be potential for integration of the new patterns of organization and consolidation with other sensory inputs. For a number of years, there was no empirical evidence to support this assumption; however, as discussed below, recent data suggest that plasticity mechanisms at all levels of the somatosensory pathway contribute to the final outcome.

13.4.3 INTEGRATION ACROSS MULTIPLE LEVELS OF THE PATHWAY

It has only been in the last few years that investigators are systematically evaluating the impact of plastic changes across multiple levels of the somatosensory pathway. One impetus for such evaluations in the primate literature was to determine whether the new patterns of connections that form in the cuneate nucleus after limb denervation could provide activation for the new sensory maps that evolve in cortex. If the new inputs to the deprived cuneate neurons are functionally potent, their impact should be evident at the next level in the processing sequence, in VPL. In turn, if the information in VPL is relayed faithfully to cortex, the functional properties of neurons at both levels should be comparable.

Experiments to explore these relationships were performed in monkeys that had long-standing amputation of the forelimb.[37] In the region of VPL, where the hand would normally activate neurons, new receptive fields were apparent, either on the stump of the arm, the face, or both. Presumably, the excitatory relay from the cuneate nucleus was the source of the new activation for the deprived thalamic neurons, and in turn, the explanation for the altered efferent patterns from the cuneate nucleus was that new connections from the stump and face had formed and acquired physiological impact.

New patterns of organization in the somatosensory thalamus after forelimb injury have been observed in other primate studies. In monkeys that had cervical dorsal rhizotomy, despite considerable shrinkage and cell loss in the portion of VPL related to the forelimb,[102] the deprived zone had become reactivated by inputs from the face and trunk.[38] Also in monkeys that had transection of multiple nerves to the hand, neurons throughout the deprived portion of VPL were reactivated by alternative inputs.[103] In humans being treated for pain after limb amputation, recordings in the thalamus revealed neurons in the deprived portion of ventroposterior nucleus that were activated by stimulation of the limb stump.[104,105] In fact, a variety of deafferentating injuries in humans produce reorganization in the somatosensory thalamus.[106-108] Taken together, the results support the argument that sensory reactivations after peripheral sensory deprivation are initiated subcortically.

A question that follows from the reports of subcortical reorganization is the extent to which changes at lower levels are reflected in cortex. To make inferences about the contribution of the subcortical inputs to the patterns of functional reorganization in cortex, the ideal approach is to compare the functional properties of neurons at multiple levels in the same animal after a sensory manipulation. In the few studies where such restricted comparisons across levels of the system were made in monkeys, the general representational features of the reorganized zones subcortically and in cortex were similar.[17,34,37,103] This indicates that the changes observed in cortex reflect at least some of the aspects of reorganization that were initiated subcortically. Data from subcortical and cortical comparisons in non-primates have been interpreted similarly.[48,49,109,110]

Not all aspects of the response properties observed subcortically were exhibited in cortex, however. For example, in monkeys that had limb amputation, there were more split receptive fields involving both the face and the stump and larger stump-only fields in VPL compared to area 3b.[37] Similarly, in monkeys that had transection of the median and ulnar nerves, receptive fields in the reorganized portion of the cuneate nucleus were significantly larger than those in the corresponding portion of area 3b.[34] Even in normal monkeys, these differences exist.[29] These findings suggest that cortical mechanisms of plasticity filter the information relayed from thalamus, and perhaps repress the information that is functionally irrelevant. A similar interpretation explains differences in the response properties in the brain stem and in cortex of rats that had forelimb amputation early in life. Lane and colleagues[39] found that only 5% of the neurons in reorganized cortex had the unusual characteristic of responsive zones on both the forelimb stump and the hindlimb, whereas 41% of the receptive fields in the cuneate nucleus demonstrated this unusual response property. They went on to show that the decrease in the number of abnormal receptive fields is mediated by intracortical inhibition.[40] When GABA antagonists were administered to cortex, while recording in the forelimb cortical representation of the rats that had early forelimb amputation, many more cortical receptive fields that included the hindlimb appeared. Taken together, the data indicate that subcortical mechanisms of plasticity can provide for new sources of sensory activation after peripheral denervation; however, the final pattern of representation that emerges in somatosensory cortex results from the subcortical contributions as well as intracortical processes of synaptic selection.

13.5 CONCLUSIONS

In the somatosensory system, some principles of reorganization are common to nearly all descriptions of plasticity after peripheral sensory deprivation. The recurrence of these same principles, despite different manipulations used to induce reorganization and differences in outcomes, suggests that the system adapts to peripheral modulations using a only few simple strategies.

Three common principles are discussed in the present chapter based on evidence from studies in the primate somatosensory system. The principles are summarized below.

1. The source of new sensory inputs that reactivate neurons in a denervated cortical map can vary with different types of injury. If dominant inputs are spared, they have preferential access to the deprived zone. If all dominant inputs are eliminated, then near neighbors in the sensory representation provide a new source of activating inputs. However, neighboring representations do not all have equal potency in the reactivation. Dorsal hand inputs dominate deprived hand cortex, even if only a portion of the glabrous inputs are removed. Face inputs reactivate forelimb cortex, at least at early cortical levels, even though the representation of the trunk and hindlimb have a similar near-neighbor relationship with the forelimb representation as the face.

2. The evolution of the somatosensory plasticity process encompasses months and, in some cases, years. Some changes emerge immediately after sensory deafferentation, but these are limited in magnitude. Additional, more extensive changes appear during the subsequent weeks and months. The most dramatic reactivations do not appear until many months, or perhaps years, after a peripheral denervation.

3. Multiple mechanisms account for the full spectrum of changes produced by peripheral sensory deactivation. The disinhibition of existing, functionally viable inputs probably accounts for immediate reactivations after sensory denervation. More slowly developing changes are attributed to the potentiation of weak connections, predominantly through NMDA-receptor mediated mechanisms. After very long-standing sensory deprivation, the growth of new connections may occur and initiate reactivations that can involve dramatically large extents. Finally, mechanisms of plasticity can affect multiple levels of the somatosensory pathway, so that the final outcomes likely reflect a compendium of network-wide adjustments as well as local-circuit modifications.

REFERENCES

1. Merzenich, M. M. et al., Topographic reorganization of somatosensory cortical areas 3b and 1 in adult monkeys following restricted deafferentation, *Neuroscience*, 8, 33, 1983.
2. Merzenich, M. M. et al., Progression of change following median nerve section in the cortical representation of the hand in areas 3b and 1 in adult owl and squirrel monkeys, *Neuroscience*, 10, 639, 1983.
3. Wall, J. T., Nepomuceno, V., and Rasey, S. K., Nerve innervation of the hand and associated nerve dominance aggregates in the somatosensory cortex of a primate (squirrel monkey), *Journal of Comparative Neurology*, 337, 191, 1993.
4. Merzenich, M. M. et al., Somatosensory cortical map changes following digit amputation in adult monkeys, *Journal of Comparative Neurology*, 224, 591, 1984.
5. Jain, N., Catania, K. C., and Kaas, J. H., Deactivation and reactivation of somatosensory cortex after dorsal spinal cord injury, *Nature*, 1997.
6. Florence, S. L., Wall, J. T., and Kaas, J. H., The somatotopic pattern of afferent projections from the digits to the spinal cord and cuneate nucleus in macaque monkeys, *Brain Research*, 452, 388, 1988.

7. Florence, S. L., Wall, J. T., and Kaas, J. H., Somatotopic organization of inputs from the hand to the spinal grey and cuneate nucleus of monkeys with observations on the cuneate nucleus of humans, *Journal of Comparative Neurology*, 286, 48, 1989.
8. Florence, S. S., Wall, J. T., and Kaas, J. H., Central projections from the skin of the hand in squirrel monkeys, *Journal of Comparative Neurology*, 311, 563, 1991.
9. Merzenich, M. M. et al., Double representation of the body surface within cytoarchitectonic areas 3b and 1 in "SI" in the owl monkey (Aotus trivirgatus), *Journal of Comparative Neurology*, 181, 41, 1978.
10. Nelson, R. J. et al., The representations of the body surface in postcentral somatosensory cortex in (Macaca fascicularis), *Journal of Comparative Neurology*, 192, 611, 1980.
11. Pons, T. P. et al., Consistent features of the representation of the hand in area 3b of macaque monkeys, *Somatosensory Research*, 4, 309 1987.
12. Darian-Smith, C. and Brown, S., Functional changes at periphery and cortex following dorsal root lesions in adult monkeys, *Nature Neuroscience*, 3, 476, 2000.
13. Paul, R. L., Goodman, H., and Merzenich, M. M., Alterations in mechanoreceptor input to Brodmann's areas 1 and 3 of the postcentral hand area of *Macaca mulatta* after nerve section and regeneration, *Brain Research*, 39, 1, 1972.
14. Wall, J. T. et al., Functional reorganization in somatosensory cortical areas 3b and 1 of adult monkeys after median nerve repair: Possible relationships to sensory recovery in humans, *Journal of Neuroscience*, 6, 218, 1986.
15. Florence, S. L. et al., Sensory afferent projections and area 3b somatotopy following median nerve cut and repair in macaque monkeys, *Cerebral Cortex*, 4, 391, 1994.
16. Wall, J. T., Felleman, D. J., and Kaas, J. H., Recovery of normal topography in the somatosensory cortex of monkeys after nerve crush and regeneration, *Science*, 221, 771, 1983.
17. Garraghty, P. E. and Kaas, J. H., Large-scale functional reorganization in adult monkey cortex after peripheral nerve injury, *Proceedings of the National Academy of Sciences U.S.A.*, 88, 6976, 1991.
18. Kolarik, R. C., Rasey, S. K., and Wall, J. T., The consistency, extent, and locations of early-onset changes in cortical nerve dominance aggregates following injury of nerves to primate hands, *Journal of Neuroscience*, 14, 4269, 1994.
19. Schroeder, C. E. et al., Electrophysiological evidence for overlapping dominant and latent inputs to somatosensory cortex in squirrel monkeys, *Journal of Neurophysiology*, 74, 722, 1995.
20. Schroeder, C. E., Seto, S., and Garraghty, P. E., Emergence of radial nerve dominance in median nerve cortex after median nerve transection in an adult squirrel monkey, *The American Physiological Society*, 77, 522, 1997.
21. Churchill, J. D. et al., Somatotopic consolidation: a third phase of reorganization after peripheral nerve injury in adult squirrel monkeys, *Experimental Brain Research*, 118, 189, 1998.
22. Manger, P. R., Woods, T. M., and Jones, E. G., Plasticity of the somatosensory cortical map in macaque monkeys after chronic partial amputation of a digit, *Proceedings of the Royal Society of London B.*, 263, 933, 1996.
23. Calford, M. B. and Tweedale, R., Immediate expansion of receptive fields of neurons in area 3b of macaque monkeys after digit denervation, *Somatosensory and Motor Research*, 8, 249, 1991.
24. Jain, N., Catania, K. C., and Kaas, J. H., A histologically visible representation of the fingers and palm in primate area 3b and its immutability following long-term deafferentations, *Cerebral Cortex*, 8, 227, 1998.

25. Florence, S. L., Taub, H. B., and Kaas, J. H., Large-scale sprouting of cortical connections after peripheral injury in adult macaque monkeys, *Science*, 282, 1117, 1998.

26. Florence, S. L. and Kaas, J. H., Large-scale reorganization at multiple levels of the somatosensory pathway follows therapeutic amputation of the hand in monkeys, *Journal of Neuroscience*, 15, 8083, 1995.

27. Rausell, E. and Jones, E. G., Extent of intracortical arborization of thalamocortical axons as a determinant of representational plasticity in monkey somatic sensory cortex, *Journal of Neuroscience*, 15, 4270, 1995.

28. Rausell, E. et al., Extensive divergence and convergence in the thalamocortical projection to monkey somatosensory cortex, *Journal of Neuroscience*, 18, 4216, 1998.

29. Xu, J. and Wall, J. T., Functional organization of tactile inputs from the hand in the cuneate nucleus and its relationship to organization in the somatosensory cortex, *Journal of Comparative Neurology*, 411, 369, 1999.

30. Garraghty, P. E. et al., The arbors of axons terminating in the middle cortical layers of somatosensory area 3b in owl monkeys, *Somatosensory and Motor Research*, 6, 401, 1989.

31. Garraghty, P. E. and Sur, M., Morphology of single intracellularly stained axons terminating in area 3b of macaque monkeys, *Journal of Comparative Neurology*, 294, 583, 1990.

32. Darian-Smith, C., Darian-Smith, I., and Cheema, S. S., Thalamic projections to sensorimotor cortex in the macaque monkey, *Journal of Comparative Neurology*, 299, 17, 1990.

33. Xu, J. and Wall, J. T., Rapid changes in brainstem maps of adult primates after peripheral injury, *Brain Research*, 774, 211, 1997.

34. Xu, J. and Wall, J. T., Evidence for brainstem and supra-brainstem contributions to rapid cortical plasticity in adult monkeys, *Journal of Neuroscience*, 19, 7578, 1999.

35. Pons, T. P. et al., Massive cortical reorganization after sensory deafferentation in adult macaques, *Science*, 252, 1857, 1991.

36. Jain, N. et al., Growth of new brainstem connections in adult monkeys with massive sensory loss, *Proceedings of the National Academy of Sciences*, 97, 5546, 2000.

37. Florence, S. L., Hackett, T. A., and Strata, F., Cortical and subcortical contributions to neural plasticity after limb amputation, *Journal of Neurophysiology*, 83, 3154, 2000.

38. Jones, E. G. and Pons, T. P., Thalamic and brainstem contributions to large-scale plasticity of primate somatosensory cortex, *Science*, 282, 1121, 1998.

39. Lane, R. D. et al., Lesion-induced reorganization in the brainstem is not completely expressed in somatosensory cortex, *Proceedings of the National Academy of Sciences U.S.A.*, 92, 4264, 1995.

40. Lane, R. D., Killackey, H. P., and Rhoades, R. W., Blockade of GABAergic inhibition reveals reordered cortical somatotopic maps in rats that sustained neonatal forelimb removal, *Journal of Neurophysiology*, 77, 2723, 1997.

41. Pons, T. P., Garraghty, P. E., and Mishkin, M., Lesion-induced plasticity in the second somatosensory cortex of adult macaques, *Proceedings of the National Academy of Sciences U.S.A.*, 85, 5279, 1988.

42. Calford, M. B. and Tweedale, R., Immediate and chronic changes in responses of somatosensory cortex in adult flying-fox after digit amputation, *Nature*, 332, 446, 1988.

43. Calford, M. B. and Tweedale, R., Acute changes in cutaneous receptive fields in primary somatosensory cortex after digit denervation in adult flying fox, *Journal of Neurophysiology*, 65, 178, 1991.

44. Silva, A. C. et al., Initial cortical reactions to injury of the median and radial nerves to the hands of adult primates, *Journal of Comparative Neurology*, 366, 700, 1996.
45. Nicolelis, M. A. et al., Induction of immediate spatiotemporal changes in thalamic networks by peripheral block of ascending cutaneous information, *Nature*, 361, 533, 1993.
46. Shin, H.-C. et al., Responses from new receptive fields of VPL neurones following deafferentation, *NeuroReport*, 7, 33, 1995.
47. Alloway, K. D. and Aaron, G. B., Adaptive changes in the somatotopic properties of individual thalamic neurons immediately following microlesions in connected regions of the nucleus cuneatus, *Synapse*, 22, 1, 1996.
48. Faggin, B. M., Nguyen, K. T., and Nicolelis, M. A., Immediate and simultaneous sensory reorganization at cortical and subcortical levels of the somatosensory system, *Proceedings of National Academy of Sciences U.S.A.*, 94, 9428, 1997.
49. Panetsos, F., Nunez, A., and Avendano, C., Local anaesthesia induces immediate receptive field changes in nucleus gracilis and cortex, *NeuroReport*, 7, 150, 1995.
50. Panetsos, F., Nunez, A., and Avendano, C., Electrophysiological effects of temporary deafferentation on two characterized cell types in the nucleus gracilis of the rat, *European Journal of Neuroscience*, 9, 563, 1997.
51. Pettit, M. J. and Schwark, H. D., Receptive field reorganization in dorsal column nuclei during temporary denervation, *Science*, 292, 2054, 1993.
52. Pettit, M. J. and Schwark, H. D., Capsaicin-induced rapid receptive field reorganization in cuneate nucleus, *Journal of Neurophysiology*, 75, 1117, 1996.
53. Klein, B. G., White, C. F., and Duffin, J. R., Rapid shifts in receptive fields of cells in trigeminal subnucleus interpolaris following infraorbital nerve transection in adult rats, *Brain Research*, 779, 136, 1998.
54. Zhang, S. P. and Rowe, M. J., Quantitative analysis of cuneate neurone responsiveness in the cat in association with reversible, partial deafferentation, *Journal of Physiology*, 505, 769, 1997.
55. Northgrave, S. A. and Rasmusson, D. D., The immediate effects of peripheral deafferentation on neurons of cuneate nucleus in raccoons, *Somatosensory and Motor Research*, 13, 103, 1996.
56. Wall, J. T., Huerta, M. F., and Kaas, J. H., Changes in the cortical map of the hand following postnatal median nerve injury in monkeys: I. Modification of somatotopic aggregates, *Journal of Neuroscience*, 12, 1992.
57. Garraghty, P. E. et al., Pattern of peripheral deafferentation predicts reorganizational limits in adult primate somatosensory cortex, *Somatosensory and Motor Research*, 11, 109, 1994.
58. Wall, J. T., Huerta, M. F., and Kaas, J. H., Changes in the cortical map of the hand following postnatal ulnar and radial nerve injury in monkeys: organization and modification of nerve dominance aggregates, *Journal of Neuroscience*, 12, 3456, 1992.
59. Bioulac, B. and Lamarre, Y., Activity of postcentral cortical neurons of the monkey during conditioned movements of a deafferented limb, *Brain Research*, 31, 427, 1979.
60. Kelehan, A. M. and Doetsch, G. S., Time-dependent changes in the functional organization of somatosensory cerebral cortex following digit amputation in adult raccoons, *Somatosensory Research*, 2, 49, 1984.
61. Rasmusson, D. D., Reorganization of raccoon somatosensory cortex following removal of the fifth digit, *Journal of Comparative Neurology*, 205, 313, 1982.
62. Rasmusson, D. D., Webster, H. H., and Dykes, R. W., Neuronal response properties within subregions of raccoon somatosensory cortex 1 week after digit amputation, *Somatosensory and Motor Research*, 9, 279, 1992.

63. Dykes, R. W., Avendano, C., and Leclerc, S. S., Evolution of cortical responsiveness subsequent to multiple forelimb nerve transections: An electrophysiological study in adult cat somatosensory cortex, *Journal of Comparative Neurology*, 354, 333, 1995.

64. Kalaska, J. and Pomeranz, B., Chronic paw denervation causes an age-dependent appearance of novel responses from forearm in "paw cortex" of kittens and adult cats, *Journal of Neurophysiology*, 42, 618, 1979.

65. Wall, J. T. and Cusick, C. G., Cutaneous responsiveness in primary somatosensory (SI) hindpaw cortex before and after partial hindpaw deafferentation in adult rats, *Journal of Neuroscience*, 4, 1499, 1984.

66. Wall, J. T. et al., Cortical organization after treatment of a peripheral nerve with ricin: an evaluation of the relationship between sensory neuron death and cortical adjustments after nerve injury, *Journal of Comparative Neurology*, 277, 578, 1988.

67. Cusick, C. G. et al., Temporal progression of cortical reorganization following nerve injury, *Brain Research*, 537, 355, 1990.

68. McMahon, S. B. and Wall, P. D., Plasticity in the nucleus gracilis of the rat, *Experimental Neurology*, 80, 195, 1983.

69. Waite, P. M. E., Rearrangement of neuronal responses in the trigeminal system of the rat following peripheral nerve section, *Journal of Physiology*, 352, 425, 1984.

70. Jain, N., Florence, S. L., and Kaas, J. H., Limits on plasticity in somatosensory cortex of adult rats: Hindlimb cortex is not reactivated after dorsal column section, *Journal of Neurophysiology*, 73, 1537, 1995.

71. McKinley, P. A. and Smith, J. L., Age-dependent differences in reorganization of primary somatosensory cortex following low thoracic (T_{12}) spinal cord transection in cats, *Journal of Neuroscience*, 10, 1429, 1990.

72. Melzack, R., Phantom limbs and the concept of a neuromatrix, *Trends in Neuroscience*, 13, 88, 1990.

73. Ramachandran, V. S., Rogers-Ramachandran, D., and Stewart, M., Perceptual correlates of massive cortical reorganization, *Science*, 258, 1159, 1992.

74. Ramachandran, V. S., Behavioral and magnetoencephalographic correlates of plasticity in the adult human brain, *Proceedings of National Academy of Science U.S.A.*, 90, 10413, 1993.

75. Halligan, P. W. et al., Thumb in cheek? Sensory reorganization and perceptual plasticity after limb amputation, *Neuroreport*, 4, 233, 1993.

76. Halligan, P. W., Marshall, J. C., and Wade, D. T., Sensory disorganization and perceptual plasticity after limb amputation: a follow-up study, *NeuroReport*, 5, 1341, 1994.

77. Elbert, T. et al., Extensive reorganization of the somatosensory cortex in adult humans after nervous system injury, *NeuroReport*, 5, 2593, 1994.

78. Yang, T. T. et al., Sensory maps in the human brain, *Nature*, 368, 592, 1994.

79. Flor, H. et al., Phantom-limb pain as a perceptual correlate of cortical reorganization following arm amputation, *Nature*, 375, 482, 1995.

80. Flor, H. et al., Cortical reorganization and phantom phenomena in congenital and traumatic upper-extremety amputees, *Experimental Brain Research*, 119, 205, 1998.

81. Knecht, S. et al., Reorganizational and perceptional changes after amputation, *Brain*, 119, 1213, 1996.

82. Doetsch, G. S., Progressive changes in cutaneous trigger zones for sensation referred to a phantom hand: a case report and review with implications for cortical reorganization, *Somatosensory and Motor Research*, 14, 6, 1997.

83. Killackey, H. P. and Dawson, D. R., Expansion of the central hindpaw representation following fetal forelimb removal in the rat, *European Journal of Neuroscience*, 1, 210, 1989.

84. Lane, R. D. et al., Source of inappropriate receptive fields in cortical somatotopic maps from rats that sustained neonatal forelimb removal, *Journal of Neurophysiology*, 81, 625, 1999.

85. Pearson, P. P., Li, C. X., and Waters, R. S., Effects of large-scale limb deafferentation on the morphological and physiological organization of the forepaw barrel subfield (FBS) in somatosensory cortex (S1) in adult and neonatal rats, *Experimental Brain Research*, 128, 315, 1999.

86. Sengelaub, D. R. et al., Denervation-induced sprouting of intact peripheral afferents into the cuneate nucleus of adult rats, *Brain Research*, 769, 256, 1997.

87. Rasmusson, D. D., Turnbull, B. G., and Leech, C. K., Unexpected reorganization of somatosensory cortex in a raccoon with extensive forelimb loss, *Neuroscience Letters*, 55, 167, 1985.

88. Wall, P. D., The presence of ineffective synapses and the circumstances which unmask them, *Philosophical Transactions of the Royal Society, London B.*, 278, 361, 1977.

89. Stent, G. S., A physiological mechanism for Hebb's postulate of learning, *Proceedings of the National Academy of Science*, 70, 997, 1973.

90. Buonomano, D. V. and Merzenich, M. M., Cortical plasticity: from synapses to maps, *Annual Review of Neuroscience*, 21, 149, 1998.

91. Alloway, K. and Burton, H., Differential effects of GABA and bicuculline on rapidly and slowly-adapting neurons in primary somatosensory cortex of primates, *Experimental Brain Research*, 85, 598, 1991.

92. Hebb, D. O. *The Organization of Behavior: A Neuropsychological Theory*, Wiley, New York, 1949.

93. Bear, M. F., Cooper, L. N., and Ebner, F. F., A physiological basis for a theory of synapse modification, *Science*, 237, 1987.

94. Garraghty, P. E. and Muja, N., NMDA receptors and plasticity in adult primate somatosensory cortex, *Journal of Comparative Neurology*, 367, 319, 1996.

95. Kaas, J. H. and Florence, S. L. in *The Mutable Brain* (ed. Kaas, J. H.), Gorden & Breach Science Publishers, London, 2000.

96. Mendell, L. M. and Lewin, G. R., Removing constraints on neural sprouting, *Current Biology*, 2, 259, 1992.

97. Florence, S. L. and Kaas, J. H. in *Somatosensory Processing: From Single Neuron to Brain Imaging* (eds. Rowe, M. J., and Iwamura, Y.), Harwood Academic Publishers, Amsterdam, 167, 2001.

98. Woolf, C. J. et al., The growth-associated protein GAP-43 appears in dorsal root ganglion cells and in the dorsal horn of the rat spinal cord following peripheral nerve injury, *Neuroscience*, 34, 465, 1990.

99. Jain, N., Florence, S.L., and Kaas, J. H., GAP–43 expression in the medulla of macaque monkeys: changes during postnatal development and the effects of early median nerve repair, *Developmental Brain Research*, 90, 24,1995.

100. Schwab, M. E., Kapfhammer, J. P., and Bandtlow, C. E., Inhibitors of neurite growth, *Annual Reviews of Neuroscience*, 16, 565, 1993.

101. Bregman, B. S. et al., Recovery from spinal cord injury mediated by antibodies to neurite growth inhibitors, *Nature*, 378, 498, 1995.

102. Woods, T. M. et al., Progressive transneuronal changes in the brainstem and thalamus after long-term dorsal rhizotomies in adult macaque monkeys, *Journal of Neuroscience*, 20, 3884, 1999.

103. Garraghty, P. E. and Kaas, J. H., Functional reorganization in adult monkey thalamus after peripheral nerve injury, *NeuroReport*, 2, 747, 1991.

104. Davis, K. D. et al., Phantom sensations generated by thalamic microstimulation, *Nature*, 391, 385, 1998.
105. Lenz, F. A. et al., Neuronal activity in the region of the thalamic principal sensory nucleus (ventralis caudalis) in patients with pain following amputations, *Neuroscience*, 86, 1065, 1998.
106. Kiss, Z. H. T., Dostrovsky, J. O., and Tasker, R. R., Plasticity in human somatosensory thalamus as a result of deafferentation, *Stereotactic and Functional Neurosurgery*, 62, 153, 1994.
107. Lenz, F. A. et al., Characteristics of somatotopic organization and spontaneous neuronal activity in the region of the thalamic principal sensory nucleus in patients with spinal cord transection, *Journal of Neurophysiology*, 72, 1570, 1994.
108. Lenz, F. A. and Byl, N. N., Reorganization in the cutaneous core of the human thalamic principal somatic sensory nucleus (Ventral caudal) in patients with dystonia, *Journal of Neurophysiology*, 82, 3204, 1999.
109. Nicolelis, M. A., Katz, D., and Krupa, D. J., Potential circuit mechanisms underlying concurrent thalamic and cortical plasticity, *Reviews in Neuroscience*, 9, 213, 1998.
110. Kis, Z. et al., Comparative study of the neuronal plasticity along the neuraxis of the vibrissal sensory system of adult rat following unilateral infraorbital nerve damage and subsequent regeneration, *Experimental Brain Research*, 126, 259, 1999.

14 Functional Implications of Plasticity and Reorganizations in the Somatosensory and Motor Systems of Developing and Adult Primates

Jon H. Kaas

CONTENTS

14.1   INTRODUCTION

The developing and mature nervous system must solve two major problems. The brain must be able to function properly, or at least well, and it must be able to change. We will see that these two problems are aspects of the same task. I am reminded of a 1931 painting by Salvador Dali called "The Persistence of Memory." In the painting, the faces of clocks had become rather fluid, and they hung loosely and folded over tree branches. The plastic clocks reminded me that we do retain our memories over long periods of time, but the memories change and gradually become distorted. While we clearly do have brain mechanisms for storing information, or for maintaining functions, there is a struggle for stability that is only partially achieved. Better stability may be possible, but perhaps not in systems that are also designed to change and to acquire new memories, and new functions. Our brains must be capable of processing information, retrieving information, and controlling behavior in a predictable and reliable manner, and this would seem to depend on morphologically and functionally stable machinery, the circuits of the brain. And yet, the circuits cannot be completely stable if we are going to acquire new skills and abilities, and adjust to sensory loss and other impairments. We also realize, as memories fade and distort, that brain circuits change even when you do not try to change them or want them to change. It may be true that once you learn to ride a bike, you can always ride a bike; however, it is not true that skills and abilities can be maintained at a high level without practice. The professional athlete or musician fully realizes how much effort is needed just to maintain abilities. These talented individuals practice to fine-tune the neural circuits used in their professions. Experience both changes and maintains.

Practical procedures for using experience and training to achieve and maintain abilities, and recover from impairments need not depend on any knowledge of the brain, its circuits, and how they are altered. Effective procedures have been and are being empirically derived. Nevertheless, altering or maintaining brain circuits is what training, practice, and therapy is about, and in principle procedures could be improved if we understand the brain, and how it is modified. We need to learn how to modify brain circuits most effectively when we want change, and to maintain brain circuits when we do not.

In this chapter, we consider evidence that many brain circuits, perhaps all, can be modified by experience, even in adult humans and animals, although less powerfully in adults than in some stages of development. We also review some of the mechanisms of rewiring brain circuits, so that ways of using these mechanisms can be considered. This review concentrates on the somatosensory and the motor systems, where research has lead to much progress over the last 20 years. Related reviews include Buonomano and Merzenich (1998), Chino (1997), Dykes (1997), Ebner et al. (1997), Kaas (1996), Kaas et al. (1997), Nicolelis (1997), and Nudo et al. (1997). The focus is on brain plasticity that may be important in recoveries from sensory loss, brain injury, and errors in development, as well as in gaining new sensorimotor skills and perceptual abilities. Important neural aspects of learning and memory are reviewed elsewhere (Cohen and Eichenbaum, 1993; Eichenbaum, 1997; Squire and Zola, 1997; Tulving and Morkowitsch, 1997; Salmon and Butters, 1995).

14.2 THE PROBLEM OF DETECTING BRAIN PLASTICITY

Important modifications in brain circuits can be quite difficult to detect. Small changes in synaptic strength, when distributed across many synapses and neurons, can be difficult to measure and quantify. Thus, plasticity is often demonstrated under rather unnatural conditions with the assumption that similar but less obvious changes occur under more natural conditions. One form of synaptic plasticity, long-term potentiation or LTP, for example, is typically studied in living slices of brain tissue rather than in the intact brain, so that electrodes can be properly placed and variables better manipulated and controlled. Electrical stimulation of axons replaces activity induced by natural stimuli in these experiments. The tissue itself, the hippocampus, is picked for study, not only because of its critical role in the early stages of learning, but because it has a highly laminar type of tissue organization that permits such electrical stimulation. Such studies have been more difficult in the neocortex, where neurons also demonstrate LTP.

In a similar manner, we have chosen the large, orderly representation of the hand in primary somatosensory cortex (S1) of monkeys for many of our studies of cortical plasticity. The hand is represented in S1 or area 3b (using the numerical scheme of Brodmann, 1909), so that separate territories exist for each digit and pad of the palm. The digit territories proceed from digits 1–5 in a latcromedial sequence in area 3b, and each digit is represented from tip to base in a rostrocaudal sequence. The pads of the palm are arranged in order in cortex caudal to the digits. In properly prepared sections of the cortex through this representation cut parallel to the surface of the brain, the orderly arrangement of the representations of the digits and pads can even be seen (Jain et al., 1998). The neural fibers associated with each digit in cortex form an elongated oval, separated from each other by a narrow fiber-poor septum. This morphological map is laid down during early brain development, and it does not change in adulthood.

Given this orderly map of the hand (and other body parts) in the cortex, we can ask the following question: "Does it change as a result of sensory experience or deprivation?" The organization of the normal map can be determined in great detail by recording receptive fields for neurons at many places in the map (hundreds of places) with penetrating microelectrodes, and using this information to reconstruct the map. All of the neurons recorded in the territory of digit 3, for example, will have receptive fields centered on digit 3, and even have most or all of each receptive field on digit 3. After any manipulation that might alter the organization of the map, microlectrode recordings can be used to characterize the potentially altered map, and the results can be compared with previously obtained maps from normal or inexperienced monkeys, a map from the same monkey before the manipulation, and even the unchanged morphological map from the manipulated monkey. Prolonged, intense stimulation of digit 3, for example, might enlarge the territory of digit 3 at the expense of other digits (see Jenkins et al., 1990), so that neurons over a larger than normal extent of cortex would have receptive fields centered on digit 3, and this might be detected by comparing the sizes of territories for digit 3 in normal and stimulated monkeys. Other changes such as reductions or increases in receptive field sizes might also be considered, but most studies have been concerned with detecting changes in territory. Given the ability that we now have to non-invasively image evoked activity patterns in the human brain, large changes in cortical territories can even be demonstrated in humans. For example, there is evidence from functional brain imaging studies that the cortical representations of the digits of the hand used in playing string instruments are larger in skilled musicians than in non-players (Elbert et al., 1995).

Of course, there are other representations in the brain besides S1 that can be studied. Somatosensory afferents enter the spinal cord and lower brainstem to terminate on an elongated sheet of neurons in the dorsal horn of the spinal cord and its extension into the brainstem. These neurons form an elongated map of tactile and other inputs. Branches of afferents also ascend in the dorsal fiber columns of the spinal cord or travel in the trigeminal nerve of the face to terminate in the dorsal-column-trigeminal nucleus complex in the brainstem, where a second map of skin receptors occurs. Neurons in the complex then project to the opposite side of the brainstem where they ascend to the ventroposterior nucleus on the opposite side of the upper brainstem forming a third representation of skin receptors. These subcortical representations of the body were studied long ago for alterations due to nerve damage, with plastic changes and somatotopic reorganizations often being reported (see Snow and Wilson, 1991 for review). However, the results were not always very convincing, since the small brainstem structures were difficult to map accurately, and reported changes were often of proportions that were close to the error of measurement. Because more obvious results can be obtained in the larger, more accessible cortical maps, recent investigators have concentrated on cortex. In addition, cortex reflects changes relayed from brainstem structures, and cortex may be more plastic than brainstem structures.

Another possibility in studies of plasticity is to evaluate other cortical maps for changes. Area 3b is the homolog in monkeys and humans of S1 in rats and cats (Kaas, 1983), but there are other cortical representations. Just rostral to the area 3b representation of tactile receptors, area 3a represents muscle-spindle receptors. In addition, strip-like areas 1 and 2, just caudal to area 3b, represent tactile receptors (area 1) or a mixture of tactile and muscle-spindle receptors (area 2). These fields project to other representations, including the second somatosensory area, S2, and the parietal ventral somatosensory area, PV (see Kaas, 1993 for review). Some of these representations, such as areas 3a and 1, and S2, have been included in studies of plasticity, but they have neurons with larger, and often less easily defined receptive fields (e.g., 3a, 2). These representations may also be smaller in size (e.g., S2, PV). Thus, convincing data on plasticity and reorganization are more difficult to obtain, and so the number of studies on these areas has been limited. Investigators have also considered auditory, visual, and motor systems, with the most easily explored primary representation more commonly considered. Some of these studies, especially those on motor cortex, are reviewed here.

14.3 REORGANIZATION AFTER DEAFFERENTATIONS

Partial deafferentations commonly occur in humans as a result of minor cuts and accidents. Almost everyone has sustained an injury that cut a peripheral nerve, often leaving part of a skin surface numb or insensitive to light touch. More profound deafferentations follow amputations of a digit or limb, and spinal cord injuries that damage ascending afferents in the spinal cord. In the auditory system, individuals lose high-frequency hearing as they age due to a loss of sensory hair cells of the inner ear. This loss can occur earlier in life with exposure to loud, enduring sounds. Parts of the retina also degenerate as certain clinical conditions interfere with retinal blood circulation. Thus, partial losses of afferents are common, and we should know what happens to a sensory system when some of the inputs are lost.

Since the partial loss of sensory afferents can produce dramatic alterations in the organizations of sensory representations (see below), such partial deafferentations usefully demonstrate the potential for brain plasticity, especially in adults where the capacity for such plasticity has been doubted. In addition, reliable methods for producing extensive and easily measured brain plasticity provide an opportunity for discovering how plasticity is mediated. Some or many of the mechanisms of plasticity that are evoked after deafferentations may play a role in other types of brain reorganizations.

Since afferents from the skin, eye, or ear project to sensory nuclei or structures in the brainstem in an orderly almost point-to-point fashion, and this pattern of input is relayed onward to other structures in the system, partial deafferentations deprive parts of representations in every part of the system of their normal sources of activation. If the deafferentation is limited, or the receptive fields for neurons are large, as in higher-order cortical areas, the differentiation may remove only part of the normal sources of activation, and the deprived neurons might only have smaller, reduced receptive fields. Such a change does not reflect any plasticity, only the activity of the preserved inputs. Instead, plasticity is demonstrated by the appearance of inputs that were not apparent before. In most experiments, this means that the neurons acquire receptive fields in new locations on the receptor sheet, although the emergence of new receptive field properties could be considered as evidence for plasticity as well.

With deafferentations, cortical neurons commonly acquire new receptive fields. Early evidence for this came from our studies of the reorganization of area 3b of somatosensory cortex (S1) of monkeys after section of the median nerve to the thumb half of the glaborous hand (Merzenich et al., 1983). The median nerve at the level of the wrist is a sensory nerve with no motor component. Thus, this procedure deprives about half of the hand representation in cortex of activating inputs without eliminating motor control. When neurons in deprived somatosensory cortex are studied with microelectrodes some weeks after the nerve section, the neurons respond to remaining inputs on the dorsal hairy surface of the hand, which is innervated by the intact radial nerve. The change in receptive field locations from the front to the back of the hand is dramatic and obvious. Clearly, neurons do acquire new receptive fields, and the system is plastic, even in the mature brain. These early findings have been confirmed repeatedly in subsequent studies (e.g., Wall et al., 1993; Garraghty et al., 1994), so the results are not in doubt. In addition, similar reorganizations have been observed in area 3b of monkeys (Merzenich et al., 1984; Jain et al., 1998) and S1 of other mammals (see Zarzecki et al., 1993) after the loss of a digit. If digit 3 is lost, for example, cortical neurons that formerly had receptive fields on digit 3 come to have receptive fields on digits 2 or 4, or the adjoining pads of the palm.

Such reorganizations of sensory representations after a limited loss of sensory inputs also occur in the auditory and visual systems. After a partial hearing loss that is restricted to a limited frequency range, cortical neurons that formerly responded to tones in the damaged part of the cochlea came to respond to new tones that were lower or higher than those in the missing range (Rajan et al., 1993; Schwaber et al., 1993; Robertson and Irvine, 1989). After a restricted retinal lesion in cats or monkeys, deprived portions of primary visual cortex came to be activated by surrounding intact portions of the retina (Kaas et al., 1990; Gilbert and Wiesel, 1992; Darian-Smith and Gilbert, 1995; Chino et al., 1992; Heinen and Skavenski, 1991; Schmid et al., 1996).

Extremely extensive reorganizations have been found in somatosensory cortex of monkeys after major but long-standing deafferentations. The loss of a forelimb, the afferents from a complete arm, or the afferents relayed from the arm and lower body by damage in the spinal cord all extensively deprive much of the body surface representation in area 3b of monkeys. While the lateral-most portion of area 3b that represents the face and mouth retains its normal source of activation, the more medial deprived parts of area 3b slowly become responsive to the face, and any preserved area inputs from the stump (Florence and Kaas, 1995; Pons et al., 1991; Jain et al., 1997). Comparable changes appear to occur in humans with amputations or spinal cord injury (see Flor et al., 1998). Because the extensive reorganizations take months to emerge (Jain et al., 1997), it seems likely that they depend on the growth of new connections (see below).

Finally, deafferentations have been shown to produce rather extensive reorganizations in the ventroposterior nucleus of the somatosensory thalamus of monkeys (Garraghty and Kaas, 1991; Jones and Pons, 1998; Florence et al., 2000) and humans (Davis et al., 1997). Limited recoveries have also been reported in the cuneate nucleus of the brainstem (Xu and Wall, 1997). However, only very limited recoveries have been reported in the visual thalamus after retinal lesions (Eysel, 1982; Darian-Smith and Gilbert, 1995). Reorganization in the auditory thalamus has not yet been adequately studied.

14.4 MOTOR CORTEX REORGANIZATIONS

Motor cortex includes primary motor cortex (M1) and several additional premotor fields. Each field represents body movements and muscles in a systematic pattern, as can be demonstrated by stimulating many cortical sites electrically with microelectrodes. Sites in medial M1 evoke leg, foot, and toe movements, followed laterally by cortex devoted to trunk and then forearm, hand, and digit movements. The most lateral parts of M1 represent face, tongue, and mouth movements. The large forelimb portion of M1 in monkeys and humans is devoted to digit, wrist, arm, and shoulder movements, but mainly digit movements.

Reorganizations of M1 have been studied recently in monkeys that suffered extensive accidental forelimb damage such that amputation was required for therapeutic reasons. There are few such monkeys, but some of these few have been studied years after the amputations for possible alterations in the organization of motor cortex. After a long-standing loss of a forelimb in monkeys, electrical stimulations with microelectrodes of the portion of M1 that normally produces movement of the hand and forelimb, evoked movements of the stump of the limb and the shoulder instead (Schieber and Deuel, 1997; Wu and Kaas, 1999; Qi et al., 2000). Similar alterations in the organization of M1 have been reported after motor nerve damage in rats (Sanes et al., 1988; Donoghue et al., 1990), and in M1 of humans after loss of an arm (e.g., Cohen et al., 1991). Thus, cortical sites normally devoted to the missing limb or body part come to evoke new movements.

Motor cortex also has a sensory map that at least roughly matches the motor map. Loss of a limb also deprives motor cortex of a major source of sensory activation, but the consequences of this differentiation on the sensory map in motor cortex have not yet been studied. Presumably, the sensory map, which depends in part on inputs from somatosensory areas that are altered in organization by amputations, is likewise reorganized in M1.

14.5 REORGANIZATIONS OF SOMATOSENSORY CORTEX FOLLOWING CHANGES IN SENSORY STIMULATION AND EXPERIENCE

Reorganizations of sensory representation also occur as a result of periods of changed sensory activity. Of course, sensory deafferentations are one way of changing activity in a sensory area, and these major changes in neural activity produce the large reorganization of sensory maps that are the easiest to detect. The sizes of alterations produced by other procedures should be proportional to the magnitude and duration of the sensory change, and this appears to be at least roughly the case. In practice, one of the easiest ways to produce a marked change in the activity of a sensory nerve or sensory area of cortex is by electrical stimulation, which is a powerful mode of stimulation. Electrical stimulation over a period of hours of a forelimb digit in rats increases the size of the cortical representation of that digit in S1 (Li et al., 1996), and electrical stimulation of neurons in area 3b of monkeys (Recanzone et al., 1992) or S1 of rats (Dinse et al., 1993) increases the cortical territory where neurons have the same receptive fields as the stimulated neurons. Increases in natural stimulation of sensory receptors and afferents result in a similar, although sometimes less dramatic, outcome. As a mother rat nurses her young, the cortical representation of the chest and belly skin increases in S1 (Xerri et al., 1994). An advantage of using nursing to stimulate the ventral fur of rats is that no training was necessary. However, Jenkins et al. (1990) trained monkeys to maintain contact with the fingertips on a rotating disk for moderate periods of time over many sessions, and found this sensory experience enlarged the representations of the fingertips in area 3b of somatosensory cortex. Changes in the detailed somatotopic organization and the proportion of cortex devoted to digits have also been reported after monkeys were trained to discriminate between various types of vibratory stimuli on the hand, and given many hours of this experience (Recanzone et al., 1992; Wang et al., 1995). Even the motor map in primary motor cortex was altered by extensive overtraining on a hand-grasp task in monkeys (Byl et al., 1996). Such alterations in the detailed organizations of sensory representations may follow most changes in sensory experience, especially those that involve training and improved performance on a task. However, the alterations often may be so limited in extent that they are difficult to detect by microelectrode mapping procedures.

14.6 MECHANISMS OF NEURAL PLASTICITY

Changes in neural circuits and cortical maps probably occur in two different ways. Most commonly, previously existing connections change in effectiveness by increasing the strengths of some synaptic connections and decreasing the effectiveness of others. In addition, new connections grow, even in the mature nervous system. Probably most reorganizations of sensory and motor maps involve both types of mechanisms. Synapses are always in the process of being lost and created, and induced changes in neural activity can cause axons to sprout or retract locally and produce more or less synaptic contacts, more or less synaptic spines on dendrites, and larger or smaller dendritic arbors. While the evidence for such changes in the microanatomy of brain circuits is limited (see Greenough et al., 1985; Fischer et al., 1998; Jones et al., 1996; Shepherd, 1996), local growth of axons and dendrites, and an overall increase or loss of synapses, is difficult to demonstrate without difficult, detailed, quantitative studies. More extensive new growth is easier to document; however, its occurrence is less common and largely under conditions of extreme deafferentation.

In many studies of neural plasticity, immediate changes of local somatotopy and receptive fields are often reported after nerve injury. These immediate consequences of injury have been termed the "unmasking of silent synapses" (Wall, 1977) or "disinhibition" (Calford and Tweedale, 1988). As the new responses represent a new balance of excitation and inhibition in a dynamic system as a reflection of the functions of existing circuits, one might question calling such rebalancing as plasticity. The "new" receptive fields are more like the part of an iceberg that had been under the surface and not seen. The loss of some afferents allows the "underwater" part of the receptive field to be seen.

Rapid changes in receptive fields also occur with increases and decreases in attention and motivation. States of attention and motivation are correlated with the activities of widely projecting neuromodulatory systems. When these modulatory inputs are added to the activity evoked by sensory inputs, neurons become more or less responsive to the sensory inputs. These neuromodulatory systems are important in neural plasticity (see below), but their immediate modulatory effects on ongoing neural responses are also part of the normal range of responsiveness of a system, and probably should not be considered as neural plasticity.

However, experience and sensory stimulation do lead to persisting changes in synaptic strengths that clearly qualify as a basis for neural circuit plasticity. Periods of strong stimulation of afferents can result in a persisting increase in synaptic strength, called long-term potentiation (Brown et al., 1988), so that subsequent stimulations are more effective. The induction of LTP depends on the depolarization of the postsynaptic neuron, and the release of a voltage-dependent block of the NMDA glutamate receptors so that calcium ions can enter the postsynaptic cell, mediating further changes and possibly the release of nitric oxide as a retrograde messenger that modifies the presynaptic cell (Brenman and Bvedt, 1997). Thus, structural modifications take place that potentiate and maintain the synaptic effectiveness between neurons. Such activity-induced alterations in synaptic strength are sometimes known as Hebbian plasticity (see Rauschecker, 1991) after the neuropsychologist, Donald Hebb (1949) who proposed learning is based on changes in synaptic strengths. The negative counterpart of LTP is long-term depression (LTD), a long-lasting decrease in the strengths of synapses after weak or discorrelated modes of stimulation (Kerr and Abraham, 1996). Thus, neural circuits are altered by activity-related mechanisms that either increase or decrease synaptic strengths. Such alterations are thought to be important, not only in learning, but also in most types of plasticity in cortical representations (Bronomano and Merzenich, 1998).

Neuromodulation is important in creating plasticity because the added excitement induced by the modulatory system can raise activity levels over the threshold needed for Hebbian plasticity. Because neuromodulatory systems increase the opportunity for modifying neural circuits, they have been considered permissive systems (Dykes, 1997).

Altering activity levels in neural systems also induces plasticity by up-regulating or down-regulating gene expression in neurons. For example, maintained low levels of neural activity cause neurons that make GABA as an inhibitory neurotransmitter to make less GABA, and neurons that respond to GABA via GABA receptors to make fewer GABA receptors (see Arckens et al., 1998 for review). Conversely, neurons, which maintain high levels of activity, make more GABA and more GABA receptors. The result is a tendency to increase or decrease neural activity toward a middle level, and thereby render some previously weak and ineffective synapses as fully effective, or some previously effective synapses as below threshold. This mechanism of altering neural circuits is one of probably a number of rather unselective activity-based ways of balancing systems.

Under conditions of extreme sensory deprivation, it may be only possible to restore activity levels by inducing the formation of new, active connections to replace the inactive connections. New growth in the central nervous system is normally restricted by growth-inhibiting factors released by glial cells (Schwab and Bartholdi, 1996). This potentially protective factor suggests that extensive new growth in the mature brain could often be detrimental. However, neural tissue that is deprived of major sources of activation may respond by releasing factors that promote growth in nearby neurons or reduce the growth-inhibiting factors. Whatever the molecular signals, there is much evidence of the sprouting of uninjured axons into inactive or deprived territories in the central nervous system. Most notably, there is evidence for the growth of horizontal connections in regions of somatosensory (Florence et al., 1998) and visual cortex (Darian-Smith and Gilbert, 1994) that have been deactivated after the loss of inputs from an arm or part of the retina. In addition, afferents from the skin may grow from terminations in their normal brainstem targets to nearby additional groups of neurons that have been deprived of afferents by injury (Florence and Kaas, 1995; Jain et al., 1999). Such new growth might be sparse, but it can serve as an activating framework that is reinforced by other cellular mechanisms for promoting neural activity levels. The major brain reactivations that occur after limb amputations and spinal cord damage appear to depend, in part, on the growth of new connections. Quite possibly, deactivations of cortex that occur as a result of local lesions and restricted strokes are compensated by new growth from active areas into deactivated areas of cortex. It even seems likely that some types of experience promote the addition of a few new neurons in the mature brain (Gould et al., 1999).

14.7 THE PERCEPTUAL AND MOTOR CONSEQUENCES OF PLASTICITY

There are three likely consequences of brain plasticity for perception and motor performance. First, the plasticity may result in modifications in neuronal circuits that allow them to perform or perform better on some desired or new task. In general, as one practices a motor or perceptual skill, performance increases. In conjunction with this improvement in performance, a number of studies have demonstrated that sensory or motor representations, in cortex, and the response properties of neurons in these representations, are altered in ways that could help mediate the improvement in performance. Typically, more neurons become devoted to processing the relevant sensory information and to controlling the appropriate muscle movement. In addition, neurons may change their response characteristics so that they respond more selectively to relevant stimuli. Neurons related to a task may also become more correlated in their discharge activity. All of these changes are what one would expect in a neural system as it is altered to better perform a task. Such changes are also those that emerge in self-organizing models of neural networks that learn. However, it is important to remember that the evidence is based on the correlation of neural and behavioral changes, and it would be useful to get more direct and compelling evidence. Yet, it is reasonable and logical to propose that practice changes sensory and motor representations, and that these changes mediate alterations in performance. The great tennis player got that way by reorganizing the brain through practice.

It is also reasonable to propose that neurons are involved in a "zero sum game." That is, as neurons rewire into new circuits they lose connections and influence in old circuits. Each neuron probably participates in a number of local circuits, but training in one task probably alters circuits so they function less well in other tasks and circuits. This is rarely bad, since the practiced ability is the one desired, and deteriorations in the unpracticed, unused abilities are unnoticed. While we can learn a number of closely related skills, top performers generally do not attempt to do so. An outstanding tennis player, for example, would not practice racquetball. It seems probable that such a misguided effort would devote neurons to racquetball that would have been useful in tennis.

It follows from this reasoning that some rewiring of sensory and motor circuits occurs as one goes about daily tasks. This is why even the most skilled performers need continuous practice. They repair the damage done to the circuits produced by other non-relevant experience. Presumably, the damage is greatest when neurons in the trained circuits are used in other incompatible ways. After learning one list of words by rote, it is best to avoid experience with another list. Sleeping or performing quite different tasks would do the least harm. Thus, practice and experience allow us to modify brain circuits in functionally adaptive ways, and bring brain circuits back to acceptable levels of performance after they deteriorate.

A second important consequence of brain plasticity is to compensate for brain injury and sensory loss. Neurons that have been deactivated by injury to brain pathways cannot possibly be of any use unless they are incorporated into new neural circuits. We have seen that deactivated neurons typically acquire responsiveness to sensory stimuli, indicating that they have been included in new circuits. Many of these new circuits are likely to be useful, and help compensate for the loss. Of course, training and experiences might be necessary to reform the new circuits in ways that they can be most useful. After restricted lesions of motor cortex, for example, so that circuits involved with moving the fingers are lost, training on finger movement skills helps reform motor cortex so that more neurons are devoted to controlling finger movements (Nudo et al., 1996). In certain neurological disorders where sensory or motor neurons are gradually and progressively lost over time, training and modifying existing circuits can greatly postpone the time where an acceptable level of performance is no longer possible. Impairments that are produced by stable brain lesions often can be followed by remarkable recoveries.

A third possible consequence of brain plasticity is apparent in many instances of major reorganizations of sensory and motor systems. In humans with an amputated arm, for example, inputs from the face or stump of the arm may activate cortex formerly devoted to the hand. Yet, these new neuronal circuits for the face or upper arm are not incorporated into circuits mediating face or upper arm perceptions. Instead, they can continue to function as parts of circuits devoted to the hand that are now inappropriately activated by face and arm inputs. The new circuits are a strange and maladaptive blend, with inputs and outputs that are functionally so misaligned that no useful compromise can occur. Activating the reorganized cortex leads to misperceptions and error, rather than compensation. Of course, some of the neurons in the reorganized system may be capable of usefully participating in new functions, but clearly many are not. Massive reorganizations seem to exceed the capacity of the system to usefully adjust.

Some misperceptions may be only distracting or even unnoticed. The sensation that a missing limb is still there is not usually a problem, and the blind spot produced by a restricted lesion of the retina is filled in by information from neurons around the lesion so that a hole in the visual scene is not seen. Yet, it might be more useful to be aware of what is missing, and a phantom limb can be painful. This pain, since it does not have a peripheral cause, can be quite difficult to treat.

What does all this mean in a practical sense? First, we should recognize that there are many effective training procedures for improving perceptual and motor skills, as well as behavioral therapies for assisting recoveries from brain injury and developmental disorders. However, these effective procedures have largely been empirically derived. Procedures are modified, and those that produce the best results are retained. Now, as we develop a theoretical understanding of brain plasticity and how it relates to desired and useful behavioral changes, we can design theoretically powerful approaches that would maximize the wanted changes in brain circuits, possibly using approaches that combine behavioral methods with more direct interventions in brain function. In addition, we can use the same information to reduce or eliminate unwanted and undesirable plasticity. Thus, we may be able to more effectively treat or prevent the local motor coordination problems or focal dystonias, such as "writer's cramp" that affect writers and skilled musicians as they over-perform and over-practice a task, possibly leading to maladaptive rewiring of sensory and motor systems. Likewise, we may be able to prevent brain changes that lead to phantom pain and tinnitus, a ringing in the ears and the hearing of sounds that are not there. Research on the plasticity of the mature, as well as the developing, brain should have some very practical consequences.

ACKNOWLEDGMENTS

This paper was prepared while the author was a Fellow at the Center for Advanced Study in the Behavioral Sciences. I am grateful for the financial support provided by The John D. and Catherine T. MacArthur Foundation, and research support by NIH Grant NS16446.

REFERENCES

Arckens, L., Eysel, U. T., Vanderhaeghen, J. J., Orban, G. A., and Vandesande, F., Effect of sensory deafferentation on the GABAergic circuitry of the adult cat visual system, *Neuroscience*, 83, 381, 1998.

Brenman, J. E. and Bredt, D. S., Synaptic signaling by nitric oxide, *Current Opinion in Neurobiology*, 7, 374, 1997.

Brodmann, K., *Vergleichende Lokalisationslehre der Grosshirnrinde*, Barth: Leipzig, 1909.

Brown, T. H., Chapman, P. F., Kairiss, E. W., and Keenan, C. L., Long-term synaptic potentiation, *Science*, 242, 724, 1988.

Buonomano, D. V. and Merzenich, M. M. Cortical plasticity: From synapses to maps, *Annual Review of Neuroscience*, 21, 149, 1998.

Byl, N. N., Merzenich, M. M., and Jenkins, W. M., A primate genesis model of focal dystonia and repetitive strain injury: I. Learning-induced dedifferentiation of the representation of the hand in the primary somatosensory cortex in adult monkeys, *Neurology*, 47, 508, 1996.

Calford, M. B. and Tweedale, R., Immediate and chronic changes in responses of somatosensory cortex in adult flying-fox after digit amputation, *Nature*, 332, 446, 1988.

Chino, Y. M., Receptive-field plasticity in the adult visual cortex: Dynamic signal rerouting or experience-dependent plasticity, *Seminars in the Neurosciences*, 9, 34, 1997.

Chino, Y. M., Kaas, J. H., Smith, III, E. L., Langston, A. L., and Cheng, H., Rapid organization of cortical maps in adult cats following restricted deafferentation in retina, *Vision Research*, 32, 789, 1992.

Cohen, L. G., Bandinelli, S., Findley, T. W., and Hallett, M., Motor reorganization after upper limb amputation in man: A study with focal magnetic stimulation, *Brain*, 114, 615, 1991.

Cohen, N. J. and Eichenbaum, H., *Memory, Amnesia, and the Hippocampal System*, M. I. T. Press: Cambridge, MA, 1993.

Darian-Smith, C. and Gilbert, C. D., Axonal sprouting accompanies functional reorganization in adult cat striate cortex, *Nature*, 368, 737, 1994.

Darian-Smith, C. and Gilbert, C. D., Topographic reorganization in the striate cortex of the adult cat and monkey is cortically mediated, *Journal of Neuroscience*, 15, 1631, 1995.

Dinse, H. R., Recanzone, G. H., and Merzenich, M. M., Alterations in correlated activity parallel ICMS-induced representational plasticity, *Neuroreport*, 5, 173, 1993.

Donoghue, J. P., Suner, S., and Sanes, J. N., Dynamic organization of primary motor cortex output to target muscles in adult rats: II. Rapid reorganization following motor nerve lesions, *Experimental Brain Research*, 79, 492, 1990.

Dykes, R. W., Mechanisms controlling neuronal plasticity in somatosensory cortex, *Canadian Journal of Physiology and Pharmacology*, 75, 535, 1997.

Ebner, F. F., Rema, V., Sachdev, R., and Symons, F. J., Activity-dependent plasticity in adult somatic sensory cortex, *Seminars in the Neurosciences*, 9, 47, 1997.

Eichenbaum, H., How does the brain organize memories?, *Science*, 277, 330, 1997.

Elbert, T., Pantev, C., Wienbruch, C., Rockstroh, B., and Taub, E., Increased cortical representation of the fingers of the left hand in string players, *Science*, 270, 305, 1995.

Fysel, U. T., Functional reconnections without new axonal growth in a partially denervated visual relay nucleus, *Nature*, 299, 442, 1982.

Fischer, M., Kaech, S., Knutti, D., and Matus, A., Rapid actin-based plasticity in dendritic spines, *Neuron*, 20, 847, 1998.

Flor, H., Elbert, T., Muhlnickel, W., Pantev, C., Wienbruch, C., and Taub, E., Cortical reorganization and phantom phenomena in congenital and traumatic upper-extremity amputees, *Experimental Brain Research*, 119, 205, 1998.

Florence, S. L. and Kaas, J. H., Large-scale reorganization at multiple levels of the somatosensory pathway follows therapeutic amputation of the hand in monkeys, *Journal of Neuroscience*, 15, 8083, 1995.

Florence, S. L., Massey, J., Hackett, T. and Strata, F., Thalamic contribution to cortical plasticity: the effects of forelimb amputation, *PNAS*, in press, 2000.

Florence, S. L., Taub, H. B., and Kaas, J. H., Large-scale sprouting of cortical connections after peripheral injury in adult macaque monkeys, *Science*, 282, 1117, 1998.

Garraghty, P. E. and Kaas, J. H., Functional reorganization in adult monkey thalamus after peripheral nerve injury, *Neuroreport*, 2, 747, 1991.

Garraghty, P. E., Hanes, D. P., Florence, S. L., and Kaas, J. H., Pattern of peripheral deafferentation predicts reorganizational limits in adult primate somatosensory cortex, *Somatosensory and Motor Research*, 11, 109, 1994.

Gilbert, C. D. and Wiesel, T. N., Receptive field dynamics in adult primary visual cortex, *Nature*, 356, 150, 1992.

Gould, E., Beylin, A., Tanapat, P., Reeves, A., and Shors, T. J., Learning enhances adult neurogenesis in the hippocampal formation, *Nature Neuroscience*, 2, 260, 1999.

Greenough, W. T., Hwang, H. M., and Gorman, C., Evidence for active synapse formation or altered postsynaptic metabolism in visual cortex of rats reared in complex environments, *Proceedings of the National Academy of Sciences U.S.A.*, 82, 4549, 1985.

Hebb, D. O., *The Organization of Behavior*, Wiley and Sons, New York, 1949.

Heinen, S. J. and Skavenski, A. A., Recovery of visual responses in foveal V1 neurons following bilateral foveal lesions in adult monkey, *Experimental Brain Research*, 83, 670, 1991.

Jain, N., Catania, K. C., and Kaas, J. H., Deactivation and reactivation of somatosensory cortex after dorsal spinal cord injury, *Nature*, 386, 495, 1997.

Jain, N., Catania, K. C., and Kaas, J. H., A histologically visible representation of the fingers and palm in primate area 3b and its immutability following long-term deafferentations, *Cerebral Cortex*, 8, 227, 1998.

Jain, N., Florence, S. L., Qi, H.-X., and Kaas, J.H. Growth of new brainstem connections in adult monkeys with massive sensory loss, *Proc. Natl. Acad. Sci. U.S.A.*, 97, 5546-5550, 2000.

Jenkins, W. M., Merzenich, M. M., Ochs, M. T., Allard, T., and Guic-Robles, E., Functional reorganization of primary somatosensory cortex in adult owl monkeys after behaviorally controlled tactile stimulation, *Journal of Neurophysiology*, 63, 82, 1990.

Jones, E. G. and Pons, T. P., Thalamic and brainstem contributions to large-scale plasticity of primate somatosensory cortex, *Science*, 282, 1121, 1998.

Jones, T. A., Kleim, J. A., and Greenough, W. T., Synaptogenesis and dendritic growth in the cortex opposite unilateral sensorimotor cortex damage in adult rats: A quantitative electron microscopic examination, *Brain Research*, 733, 142, 1996.

Kaas, J. H., What, if anything, is SI? Organization of first somatosensory area of cortex, *Physiological Reviews*, 63, 206, 1983.

Kaas, J. H., The functional organization of somatosensory cortex in primates, *Annals of Anatomy*, 175, 509, 1993.

Kaas, J. H., Plasticity of sensory representations in the auditory and other systems of adult mammals, In R. J. Salvi, D. Henderson, F. Fiorino, and J. Colletti, Eds., *Auditory System Plasticity and Regeneration*, Thieme Medical Publishers, New York, pp. 213, 1996.

Kaas, J. H., Florence, S. L., and Jain, N., Reorganization of sensory systems of primates after injury, *Neuroscientist*, 3, 123, 1997.

Kaas, J. H., Krubitzer, L. A., Chino, Y. M., Langston, A.L., Polley, E.H., and Blair, N., Reorganization of retinotopic cortical maps in adult mammals after lesions of the retina, *Science*, 248, 229, 1990.

Kerr, D. S. and Abraham, W. C., LTD: Many means to how many ends? *Hippocampus*, 6, 30, 1996.

Li, C. X., Waters, R. S., McCandlish, C. A., and Johnson, E. F., Electrical stimulation of a forepaw digit increases the physiological representation of that digit in layer IV of SI cortex in rat, *Neuroreport*, 7, 2395, 1996.

Merzenich, M. M., Kaas, J. H., Wall, J. T., Nelson, R. J., Sur, M., and Felleman, D., Topographic reorganization of somatosensory cortical areas 3b and 1 in adult monkeys following restricted deafferentation, *Neuroscience*, 8, 33, 1983.

Merzenich, M. M., Nelson, R. J., Stryker, M. P., Cynader, M. S., Schoppmann, A., and Zook, J.M., Somatosensory cortical map changes following digit amputation in adult monkeys, *Journal of Comparative Neurology*, 224, 591, 1984.

Nicolelis, M. A., Dynamic and distributed somatosensory representations as the substrate for cortical and subcortical plasticity, *Seminars in the Neurosciences*, 9, 24, 1997.

Nudo, R. J., Plantz, E. J., and Milliken, G. W., Adaptive plasticity in primate motor cortex as a consequence of behavioral experience and neuronal injury, *Seminars in the Neurosciences*, 9, 13, 1997.

Nudo, R. J., Wise, B. M., SiFuentes, F., and Milliken, G. W., Neural substrates for the effects of rehabilitative training on motor recovery after ischemic infarct, *Science*, 272, 1791, 1996.

Pons, T. P., Garraghty, P. E., Ommaya, A. K., Kaas, J. H., Taub, E., and Mishkin, M., Massive cortical reorganization after sensory deafferentation in adult macaques, *Science*, 252, 1857, 1991.

Qi, H-X., Stepniewska, I., and Kaas, J. H., Reorganization of primary motor cortex in macaque amputees, *Soc. Neuro. Abstr.* 25, 1999.

Rajan, R., Irvine, D. R., Wise, L. Z., and Heil, P., Effect of unilateral partial cochlear lesions in adult cats on the representation of lesioned and unlesioned cochleas in primary auditory cortex, *Journal of Comparative Neurology*, 338, 17, 1993.

Rauschecker, J. P., Mechanisms of visual plasticity: Hebb synapses, NMDA receptors, and beyond, *Physiological Reviews*, 71, 587, 1991.

Recanzone, G. H., Merzenich, M. M., and Dinse, H. R., Expansion of the cortical representation of a specific skin field in primary somatosensory cortex by intracortical microstimulation, *Cerebral Cortex*, 2, 181, 1992.

Robertson, D. and Irvine, D. R., Plasticity of frequency organization in auditory cortex of guinea pigs with partial unilateral deafness, *Journal of Comparative Neurology*, 282, 456, 1989.

Salmon, D. P. and Butters, N., Neurobiology of skill and habit learning, *Current Opinion in Neurobiology*, 5, 184, 1995.

Sanes, J. N., Suner, S., Lando, J. F., and Donoghue, J. P., Rapid reorganization of adult rat motor cortex somatic representation patterns after motor nerve injury, *Proceedings of the National Academy of Sciences U.S.A.*, 85, 2003, 1988.

Schieber, M. H. and Deuel, R. K., Primary motor cortex reorganization in a long-term monkey amputee, *Somatosensory and Motor Research*, 14, 157, 1997.

Schmid, L. M., Rosa, M. G., Calford, M. B., and Ambler, J. S., Visuotopic reorganization in the primary visual cortex of adult cats following monocular and binocular retinal lesions, *Cerebral Cortex*, 6, 388, 1996.

Schwab, M. E. and Bartholdi, D., Degeneration and regeneration of axons in the lesioned spinal cord, *Physiological Reviews*, 76, 319, 1996.

Schwaber, M. K., Garraghty, P. E., and Kaas, J. H., Neuroplasticity of the adult primate auditory cortex following cochlear hearing loss, *American Journal of Otology*, 14, 252, 1993.

Shepherd, G. M., The dendritic spine: A multifunctional integrative unit, *Journal of Neurophysiology*, 75, 2197, 1996.

Snow, P. J. and Wilson, P., Plasticity in the somatosensory system of developing and mature mammals: the effects of injury to the central and peripheral nervous system, *Progress in Sensory Physiology*, Autrum, H., Ottoson, D., Perl, E. R., Schmidt, R. F., Shimazu, H., and Willis, W. D., Eds., Springer-Verlag, Berlin, 1991.

Squire, L. R. and Zola, S. M., Episodic memory, semantic memory and amnesia, *Hippocampus*, 8, 205, 1997.

Tulving, E. and Markowitsch, H. J., Episodic and declarative memory: Role of hippocampus, *Hippocampus*, 8, 198, 1997.

Wall, J. T., Nepomucceno, V., and Rasey, S. K., Nerve innervation of the hand and associated nerve dominance aggregates in the somatosensory cortex of a primate squirrel monkey, *Journal of Comparative Neurology*, 337, 191, 1993.

Wall, P. D., The presence of ineffective synapses and the circumstances which unmask them, *Philosophical Transactions of the Royal Society of London Series B Biological Sciences*, 278, 361, 1977.

Wang, X., Merzenich, M. M., Sameshima, K., and Jenkins, W. M., Remodeling of hand representation in adult cortex determined by timing of tactile stimulation, *Nature*, 378, 71, 1995.

Wu, C. W. H. and Kaas, J. H., The organization of motor cortex of squirrel monkeys with longstanding therapeutic amputations, *Journal of Neuroscience*, 19, 7679, 1999.

Xerri, C., Stern, J. M., Merzenich, M. M., Alterations of the cortical representation of the rat ventrum induced by nursing behavior. *Journal of Neuroscience*, 14, 1710, 1994.

Xu, J. and Wall, J. T., Rapid changes in brainstem maps of adult primates after peripheral injury, *Brain Research*, 774, 211, 1997.

Zarzecki, P., Witte, S., Smits, E., Gordon, D. C., Kirchberger, P., and Rasmusson, D. D., Synaptic mechanisms of cortical representational plasticity: Somatosensory and corticocortical EPSPs in reorganized raccoon SI cortex, *Journal of Neurophysiology*, 69, 1422, 1993.

Index